Hydroides of the World

This book is dedicated to our late colleagues Nechama Ben-Eliahu, Phyllis Knight-Jones, Gottfried Pillai and Alexander Rzhavsky, who shared our fascination with serpulids.

Hydroides of the World

Elena Kupriyanova, Yanan Sun, Eunice Wong
and Harry A. ten Hove

CSIRO PUBLISHING

CRC Press
Taylor & Francis Group
Boca Raton London New York

CRC Press is an imprint of the
Taylor & Francis Group, an informa business

A catalogue record for this book is available from the National Library of Australia.

ISBN: 9781486311583 (pbk)
ISBN: 9781486311590 (epdf)
ISBN: 9781486311606 (epub)

Published in print in Australia and New Zealand, and in all other formats throughout the world, by CSIRO Publishing.

CSIRO Publishing
Private Bag 10
Clayton South VIC 3169
Australia

Telephone: +61 3 9545 8400
Email: publishing.sales@csiro.au
Website: www.publish.csiro.au
Sign up to our email alerts: publish.csiro.au/earlyalert

Published in print only, throughout the world (except in Australia and New Zealand), by CRC Press, with ISBN 9781032436692.

CRC Press
6000 Broken Sound Parkway NW, Suite 300, Boca Raton, FL 33487-2742
and
2 Park Square, Milton Park, Abingdon, Oxon, OX14 4RN
Website: www.routledge.com

CRC Press is an imprint of Taylor & Francis Group, LLC

Front cover: *Hydroides dirampha* (photo by Leslie Harris)

Edited by Joy Window (livinglanguage.wordpress.com)
Cover design by Cath Pirret
Typeset by Envisage Information Technology
Printed in Malaysia by Papercraft

CSIRO acknowledges the Traditional Owners of the lands that we live and work on and pays its respect to Elders past and present. CSIRO recognises that Aboriginal and Torres Strait Islander peoples in Australia and other Indigenous peoples around the world have made and will continue to make extraordinary contributions to all aspects of life including culture, economy and science. CSIRO is committed to reconciliation and demonstrating respect for Indigenous knowledge and science. The use of Western science in this publication should not be interpreted as diminishing the knowledge of plants, animals and environment from Indigenous ecological knowledge systems.

The paper this book is printed on is in accordance with the standards of the Forest Stewardship Council® and other controlled material. The FSC® promotes environmentally responsible, socially beneficial and economically viable management of the world's forests.

FSC www.fsc.org MIX Paper | Supporting responsible forestry FSC™ C007207

Mar23_01

Contents

Acknowledgements

This book is a culmination of several decades of research and would not have been possible without generous support of many colleagues and organisations. As early as in 1986, Harry A. ten Hove (HAtH) was supported by the Australian Museum visiting fellowship that allowed him to sample the shores of tropical Australia for serpulids and, along with Pat Hutchings (AMRI), to collect polychaetes at Lizard Island Research Station (LIRS). Collecting at LIRS in 2005 was funded by Australian Research Council (ARC) Discovery grant DP0558736 to Greg Rouse (Scripps Institution of Oceanography, California, USA (SIO)). More recently *Hydroides* material was collected during the Polychaete Workshop held in Lizard Island in 2013, supported by a grant from Lizard Island Reef Research Foundation (LIRRF). Special thanks are due to Anne Hoggett and Lyle Vail for hosting the workshop. Most recently our *Hydroides* research was supported by Australian Biological Resource Study (ABRS) National Taxonomy Research Grant Programme Grants RF213-19 and RG18-21 to Elena Kupriyanova (EK), as well as ABRS-2015/2016 Student Travel Grant STG15-01 to Yanan Sun (YS). The Department of Biological Sciences at Macquarie University awarded an International Macquarie University Research Excellence Scholarship (iMQRES) to YS.

HAtH would never have been able to process the large collections of serpulids collected by and/or sent to him for identification without the help of many students supervised by him and diverse technical assistants during his university career, some of them named in the text. Four of them worked specifically on aspects of the genus *Hydroides* – Eleonoor Creyghton, Stanley Criens, DirkJan Sluimers and Jossy Tromp – and are thanked for their unpublished Dutch MSc reports (1978–80).

We extend our sincere thanks to curators and collections managers of museums worldwide, especially Stephen Keable (AMRI, Australia) and Joke Bleeker (Naturalis, the Netherlands) for organising and handling loans, donations, and registrations of specimens crucial for this book. We thank Hironori Komatsu (NSMT, Japan), Luis F. Carrera Parra (ECOSUR, Mexico), Tarik Meziane (MNHN, France), Didier Van den Spiegel (MRAC, Belgium) and Dieter Fiege (Senkenberg, Germany) for loans of the *Hydroides* material not available in our museums.

We are grateful to the late Nechama Ben-Eliahu (Hebrew University of Jerusalem, Israel), Maria Ana Tovar-Hernández (Universidad Autónoma de Nuevo León, Mexico), Wagner Magalhães (Universidade Federal de Bahia, Brazil), Maria Capa (University of Balearic Islands, Spain), Vasily Radashevsky (Zhirmunsky Marine Biology Institute, Russia), Anja Schulze (Texas A&M, USA), Aylin Ulman (University of British Columbia, Canada), Eijiroh Nishi (Yokohama National University, Japan), Elena Lisitskaya (Kovalevsky Institute of Marine Biological Research, Russia), James Carlton (Williams College, USA), Erica Keppel (Smithsonian Environmental Research Centre, USA), Giuseppe Suaria (Italian National Research Council), Jixing Sui (MBMCAS, Qingdao, China), Joachim Langeneck (Università di Pisa, Italy), Maite Aguado (University of Göttingen, Germany), Melih Ertan Çinar (Ege University, Turkey), Priti Kubal (Eco-Echo NGO, Maharashtra, India), Mathan Magesh (formerly University of Kerala, India), Xuwen Wu (MBMCAS, Qingdao, China), Francesco Criscione (AMRI), Tulio Villalobos-Guerrero (ECOSUR, Mexico) and Torkild Bakken (University of Bergen, Norway) for assistance in various ways, from help with literature to mail discussions and loaned or donated material.

We thank our colleagues Greg Rouse (SIO), Alexander Semenov (White Sea Biological Station, Russia), Fredrik Pleijel (University of Gothenburg, Sweden), Leslie Harris (Natural History Museum of LA County, USA), Maria Ana Tovar-Hernández (Universidad Autónoma de Nuevo León), Maria Capa (University of Balearic Islands, Spain), Gustav Paulay (University of Florida, USA) and Eijiroh Nishi (Yokohama National University, Japan) for providing photographs of live animals and their tubes in the field.

We are thankful to colleagues who photographed specimens of *Hydroides* in museums worldwide when sending loans was impossible or impractical. Sergio Salazar-Vallejo (El Colegio de la Frontera Sur, Chetumal, Mexico) was not only material to HAtH's Mexican involvement and cooperation with Rolando Bastida-Zavala, but also photographed the type of *Hydroides ibarzalae* nov. spec. and the holotype of *H. dafnii* in MNHN during his most recent visit to Paris. Humberto Bahena Basave (El Colegio de la Frontera Sur, Chetumal, Mexico) photographed *H. mucronata* in Mexico; Xuwen Wu (IOCAS, China) – type specimens of *H. nanhaiensis*, *H. trilobula* and *H. xishaensis*; Dieter Fiege (Senckenberg) – a specimen of *H. monroi*; Jonathan Brecko (MRAC, Belgium) – the types of *H. monroi*; and Lenka Nealova (NHML) – types of *H. rostrata*. Laetitia Gunton (AMRI) photographed *H. uniformis* during COVID-19 lockdown in Sydney. Ton de Winter (Naturalis) helped HAtH to photograph specimens of *H. rzhavskyi* nov. spec., *H. sceptrifera* nov. spec. and *H. zibrowii* nov. spec., at Naturalis.

Bruno Pernet (California State University, California, USA) kindly raised larvae of *H. gracilis* and photographed their development. Michael Hadfield and Brian Nedved (University of Hawaii at Manoa, USA) provided photographs of larval development in *H. elegans* and so did Cesar Arenas-Mena (City University of New York, USA).

Thomas Perkins (formerly Florida Marine Research Institute, St Petersburg, USA) is thanked for his help with early unpublished descriptions of *Hydroides* spp. of Florida, USA, in part used in this book, and for his support to HAtH during his visit to Florida. We thank Alexei Ippolitov (Institute of Geology, Russia) and Olev Vinn (University of Tartu, Estonia) for their comments on fossil records and tube formation of *Hydroides*, the latter also for sharing images of *Hydroides* tube ultrastructures.

Geoff Read drew our attention to the feminine nature of the *Hydroides* name and Normand David (Québec, Canada) provided valuable International Commission on Zoological Nomenclature (ICZN) advice on some of our name evaluations in *Hydroides*.

Barry A. Vittor and Paul G. Johnson (Barry A. Vittor & Associates, Inc., Mobile, Alabama, USA), Guillermo San Martín (Universidad Autónoma de Madrid, Spain), Jeremy B.C. Jackson (Smithsonian Tropical Research Institute, Panama), John R. de Palma (U.S. Naval Oceanographic Office, Washington D.C., USA) are all thanked for sending material; as well as staff at Carmabi (Curaçao) for support during several visits (1970–89) of HAtH to the Netherlands Antilles.

We are grateful to Linda Ward (Smithsonian Institution, Washington D.C., USA) for making her extensive digital literature available. We are grateful to the staff of Naturalis library in general, but certainly in times of COVID-19, for providing necessary literature items. Adria Castellucci (AMRI) helped with obtaining a high-resolution scanned copy of the *H. norvegica* illustration by Gunnerus. Sameer Kumar Pati (Zoological Survey of India, Kolkata) provided a copy of the description of *H. vizagensis* by Lakshmana Rao, otherwise unavailable to us. We are grateful to Sue Lindsay (formerly AMRI and currently Macquarie University, Sydney), who helped with scanning electron microscopy for this project.

Leslie Harris enabled HAtH's work in LACM-AHF; Chris Glasby hosted EK's and YS's visit to MAGNT and helped in the field. Robin Wilson and Julian Finn (MV) hosted EK's visit to Melbourne Museum, Xinzheng Li hosted YS's visit to MBMCAS, João Miguel de Matos Nogueira hosted YS in Brazil, and Thierry Laperousaz facilitated EK's visit to SAM. Adrian Glover is thanked for hosting EK at NHM in London and for his help with photographing the type of *H. inornata*. We are grateful to the staff of the Kuwait Institute for Scientific Research (KISR) who made EK's visit to Kuwait possible and very enjoyable.

We thank Jennie Mallela (Australian National University) who organised the field trip to Mission Beach, Queensland, Australia, and invited EK to participate, and also thank both her and Laura Stoltenberg (Southern Cross University, New South Wales, Australia) for their help with collecting specimens during this trip. João Miguel de Matos Nogueira, Orlemir Carrerette, Felipe Ramon De Chiara and Carlos Diego (all from University of São Paulo, Brazil) helped YS with sampling in Brazil. Dmitry Mikhin helped with collecting of *Hydroides* in Adelaide and Port Lincoln, South Australia, Australia.

Nicholas Patocka and Zhijun Dong kindly shared their observation on *H. dianthus* communities in the field.

We thank Andre Minoche (Garvan Institute of Medical Research, Sydney) for his contribution to fixing problems during the mitochondrial genome assembly, Kenneth Halanych (Auburn University, Alabama, USA) for his suggestions on genome assembly and contribution of reference mitochondrial sequences for local Basic Local Alignment Search Tool (BLAST) searches, as well as Don Colgan and Andrew King (AMRI) for their suggestions on assembly software and annotation.

Most recently, AM volunteers Eva Garzisi, William Zhang, and Nika Mikhin helped with image editing and plate preparation for the book. Last but certainly not least, we thank Helmut Zibrowius for his collegiality, his sharing of material and his unending support over the decades.

1 Introduction

The annelid family Serpulidae Rafinesque, 1815 is a highly specialised group of marine segmented worms that has adapted to inhabiting self-secreted calcareous tubes (hence the common name 'calcareous tubeworms') attached to a wide range of hard substrates. Although these animals are found in all oceans at all depths and habitats and some form associations with live corals, one serpulid genus is of special concern and importance. The genus *Hydroides* Gunnerus, 1768 with 99 currently recognised valid species (Read & Fauchald 2020) is not only the largest serpulid genus, but is also one of the most ecologically and economically important groups of marine invertebrates. The genus includes a number of notorious biofoulers and common bioinvaders that travel around the world hitchhiking on ships' hulls and on floating plastic debris.

GENERAL BODY MORPHOLOGY

A glossary of serpulid terminology is given at the end of this book. The illustrated version by Wong *et al.* (2014) may be useful, too. Like all annelids, the segmented body of serpulids has dorsal and ventral sides, the latter marked by the location of the mouth externally and a nerve cord internally. On the sides (laterally), each segment bears parapodia enforced with two bundles of chitinous bristles (chaetae). In the abdomen, the chaetae in the bundle located closer to the dorsal side are called notopodial (dorsal) chaetae, and the chaetae in the bundle located closer to the ventral side are called neuropodial (ventral) chaetae. Like in all other serpulids, the body of all *Hydroides* species is divided into three regions: radiolar crown, thorax and abdomen (Fig. 1). The Sabellida (including the families Fabriciidae, Sabellidae and Serpulidae) have a unique feature, the chaetal inversion at the border of the thorax and the abdomen – the thorax bears notopodial (dorsal) chaetae and neuropodial (ventral) uncini, whereas in the abdomen their positions are inversed.

Radiolar crown

The radiolar crown – made of radioles each bearing rows of paired ciliated pinnules (Fig. 1) – can be very colourful in live animals (Figs 2, 3). The radioles are arranged in semicircles when in feeding position and attached to paired radiolar lobes located laterally on both sides of the mouth. The bases of the radioles may be joined with an inter-radiolar membrane, but this is very rare in *Hydroides* (Bastida-Zavala & ten Hove 2003a) unlike in closely related genera (e.g. *Serpula*). The second most dorsal radiole on one side is transformed into a smooth, thickened opercular peduncle (Fig. 1) that terminates into a distal operculum (Fig. 1) serving as a plug to the tube. The peduncle is inserted below and between the first and second normal radiole, outside the line of radioles. The most dorsal radiole on the opposite lobe is reduced and carries a club-shaped underdeveloped operculum called a

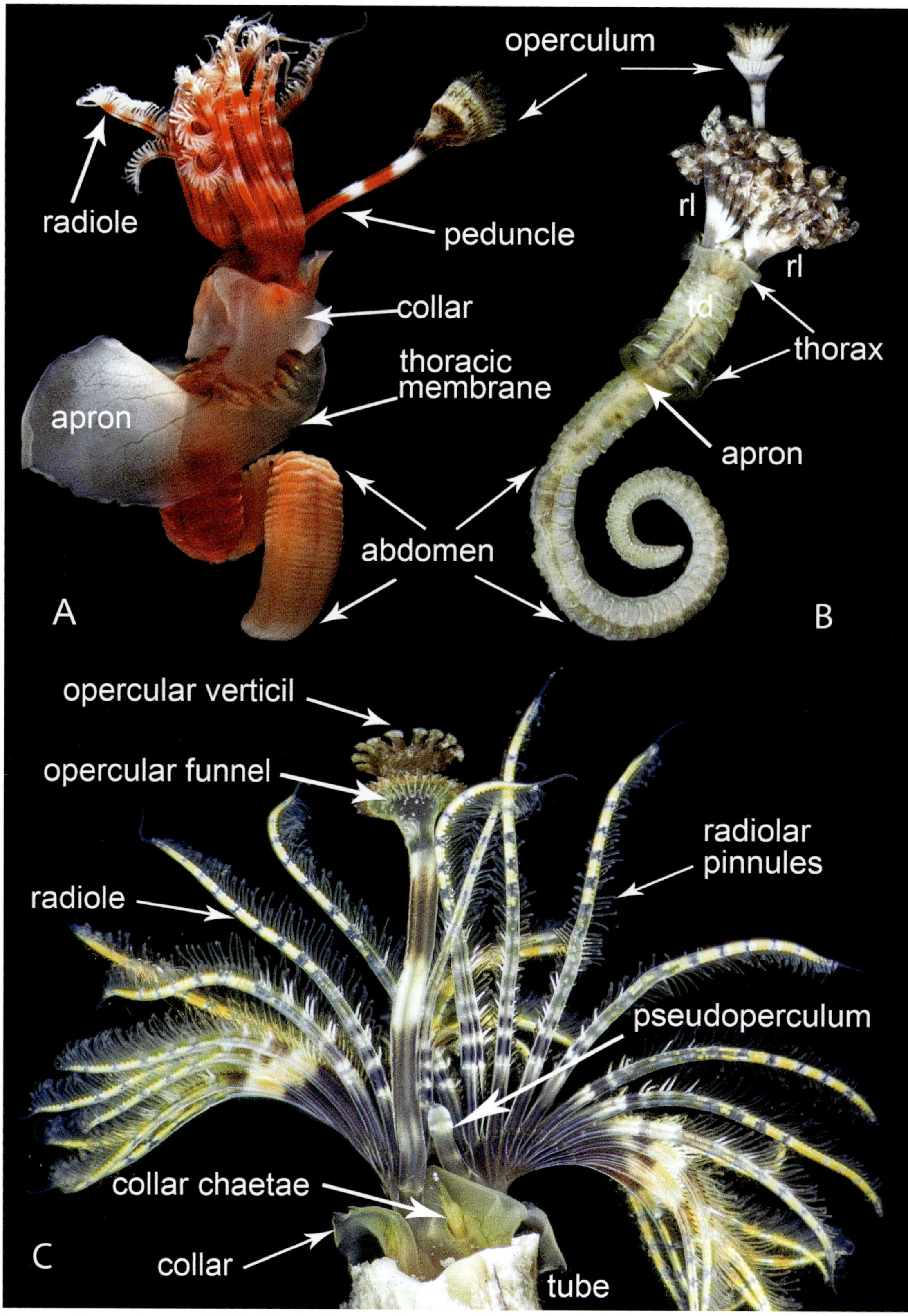

Fig. 1. Live complete specimens of *Hydroides*. A – *H. ezoensis* removed from tube, from the Sea of Japan; B – *H. elegans* removed from tube, from San Diego, California; C – *H. dirampha* in tube, from San Diego, California (photo L. Harris).

Fig. 2. Live specimens of *Hydroides*. A – *H. norvegica* in tube, Sweden (photo F. Pleijel); B – Anterior end of a specimen of *H. tuberculata*-complex, Lizard Island, Queensland, Australia (photo G. Rouse); C – *H. minax* in tube, Lizard Island, Queensland (photo A. Semenov); D – *H. trivesiculosa* in tube, Lizard Island, Queensland (photo G. Rouse); E – *H. lirs* in tube embedded into a live coral, Lizard Island, Queensland (photo A. Semenov).

Fig. 3. Live specimens of *Hydroides*. A – *H. exaltata* removed from tube, anterior end, Hong Kong (photo G. Paulay); B – *H. homoceros*, removed from tube, anterior end, Kuwait (photo E. Kupriyanova); C – *H. crucigera*, operculum needs cleaning to reveal verticil structure, Mazatlán, Mexico (photo M. Tovar-Hernández); D – *H. tambalagamensis*, Lizard Island, Queensland, Australia (photo A. Semenov); E – *H. longistylaris*, Hong Kong (photo G. Paulay); F – *H. albiceps*-complex, Hong Kong (photo G. Paulay).

Fig. 4. Opercular morphology of *Hydroides*. Top row: symmetrical verticil spines without lateral or external spinules (A – *H. alatalateralis*, B – *H. augeneri*), with modified spine tips (C – *H. dirampha*, D – *H. kimberleyensis*). Second row: symmetrical verticil spines with lateral (E – *H. norvegica*, F – *H. longispinosa*, G – *H. crucigera*) or external (H – *H. tambalagamensis*) spinules. Third row: asymmetrical verticil spines with distinct modified single dorsal spine (I – *H. brachyacantha*, J – *H. trivesiculosa*, K – *H. tuberculata*–complex, L – *H. minax*). Bottom row: asymmetrical verticil spines without spinules (M – *H. dianthus*), with a distinct modified dorsal spine and lateral spinules (N – *H. heterocera*), with double distinct modified dorsal spines bearing modified spine tips (O – *H. glasbyi*), with external spinules (P – *H. externispina*) (all photos E. Wong). Abbreviations: w – wing of opercular spines, bs – basal spinules, ts – tips of spines, ls – lateral spinules, ct – central tooth, es – external spinules, ds – dorsal spine.

pseudoperculum (Fig. 1), borne on a shortened peduncle. The pseudoperculum can develop into a functional operculum if the latter is lost (Okada 1933; Schochet 1973a, b). Two functional opercula can be found simultaneously; such bioperculate forms are common in *Hydroides* (Ichikawa & Takagaki 1942; Schochet 1973a, b; ten Hove & Ben-Eliahu 2005), or some taxa may have two rudimentary opercula (ten Hove & Ben-Eliahu 2005: table 1).

The operculum has traditionally been one of the most important features used in morphological classification of serpulids (ten Hove & Kupriyanova 2009). Its terminology for *Hydroides* was first proposed by ten Hove (1990) and updated by Bastida-Zavala & ten Hove (2003a). The distinctive operculum of *Hydroides* consists of a basal funnel (Fig. 1) composed of numerous fused radii and a distal crown of chitinous spines called a verticil (Fig. 1).

The funnel is covered with a thickened cuticle (similar to that found in closely related genera *Serpula* and *Crucigera*) and may be separated from the peduncle by a chitinised constriction. The funnel morphology varies in terms of number of radii and shape of distal tips of radii, which are most commonly pointed (Fig. 4), but can be also blunt to rounded (Fig. 4E, F), or T-shaped to anchor-shaped (Fig. 4N). The most distinct funnel radii have long, slender, deeply bifurcated tips as found in *H. furcifera* (Fig. 50). The verticil is sitting in the funnel or positioned on a stalk in the centre of the funnel.

The most variable feature of a *Hydroides* operculum is the verticil. Many verticils have a central disk, sometimes bearing a central spine (tooth) (Fig. 4F, cs) surrounded by main spines. The structure of the verticil varies in terms of symmetrical spine arrangement when all spines are identical in shape or size (Fig. 4A–H) or asymmetrical arrangement, when one (Fig. 4J–L, N) or two (Fig. 4O, P) main spines are distinct from the rest, or the size of the spines changes gradually from one verticil side to another (Fig. 4M). The main spines vary in shape, from thin and elongated (Fig. 4A) to thick, nearly conical (Fig. 4D, I, K) or valvular. The spines can also bear flattened wings (Fig. 4A, w) as well as lateral (Fig. 4E–G, ls) and external (Fig. 4H, P, es) spinules. The verticil spines of many taxa have a basal spinule each internally (Fig. 4B, C), rarely in the form of a wart-like knob.

The distal spine modifications are exemplified by variability of the spine tips that can be simple pointed (Fig. 4A, B, E–H, M, N), T-shaped (Fig. 4C), rounded to globular (Fig. 4D), or pickaxe-shaped. In cases where distinct dorsal spines are present, their shape varies from incurved hooks (Fig. 4I, K, L, N–P) to exceptionally large vesicular structures (Fig. 4J).

Opercular development in *Hydroides* is indirect (**sensu** ten Hove 1984, 1989), which means that juveniles develop a *Serpula*-type of operculum (single funnel) on a pinnulated radiole first and then the opposite radiole loses its pinnules and becomes a smooth peduncle with two superimposed funnels (e.g. ten Hove & Ben-Eliahu 2005).

Thorax

The thorax normally has seven chaetigerous segments (collar chaetae and six uncinigerous segments), but occasionally species show a variable number of thoracic segments: for example, three *Hydroides* species (*H. bannerorum*, *H. bisecta* and *H. sceptrifera* nov. spec.) have 7–9 chaetigers. The base of the radiolar crown is surrounded by a membranous collar that is tri-lobed, divided into one larger ventral and two smaller lateral lobes. The latter lobes continue into thoracic membranes that completely surround the thorax, reaching the end of it and fusing ventrally to form the ventral apron.

The first thoracic chaetiger is the collar segment bearing only a bundle of notopodial (dorsal) collar chaetae. The collar bundle of *Hydroides* consists of unmodified hair-like chaetae and modified (special) bayonet-type chaetae. Examined under a compound microscope, hair-like chaetae consist of a basal smooth shaft and a distal striated blade that resembles a limbus. Although such chaetae are traditionally termed 'capillary' and 'limbate', examination with a scanning electron microscope (SEM) shows that the 'limbus' is made of numerous microfibres (Fig. 6A, B). Moreover, all thoracic chaetae are limbate, but capillary chaetae are smaller than limbate ones. Special bayonet-type collar chaetae additionally bear one or two large bosses and/or a proximal rasp at the base of the distal blade (Fig. 5).

Starting from segment 2, limbate and capillary notopodial chaetae are accompanied by modified neuropodial (ventral) comb-shaped chaetae called uncini (Fig. 6C, D). In *Hydroides* thoracic uncini have a small number (5–6) of curved teeth in one row (saw-shaped). The shaft of an uncinal comb is deeply embedded into tissue but the teeth protrude from the body wall. Uncini are arranged side by side in a single transverse row on an elevated torus. These thoracic uncinigerous tori are positioned along the side of the thorax, but are widely separated anteriorly and

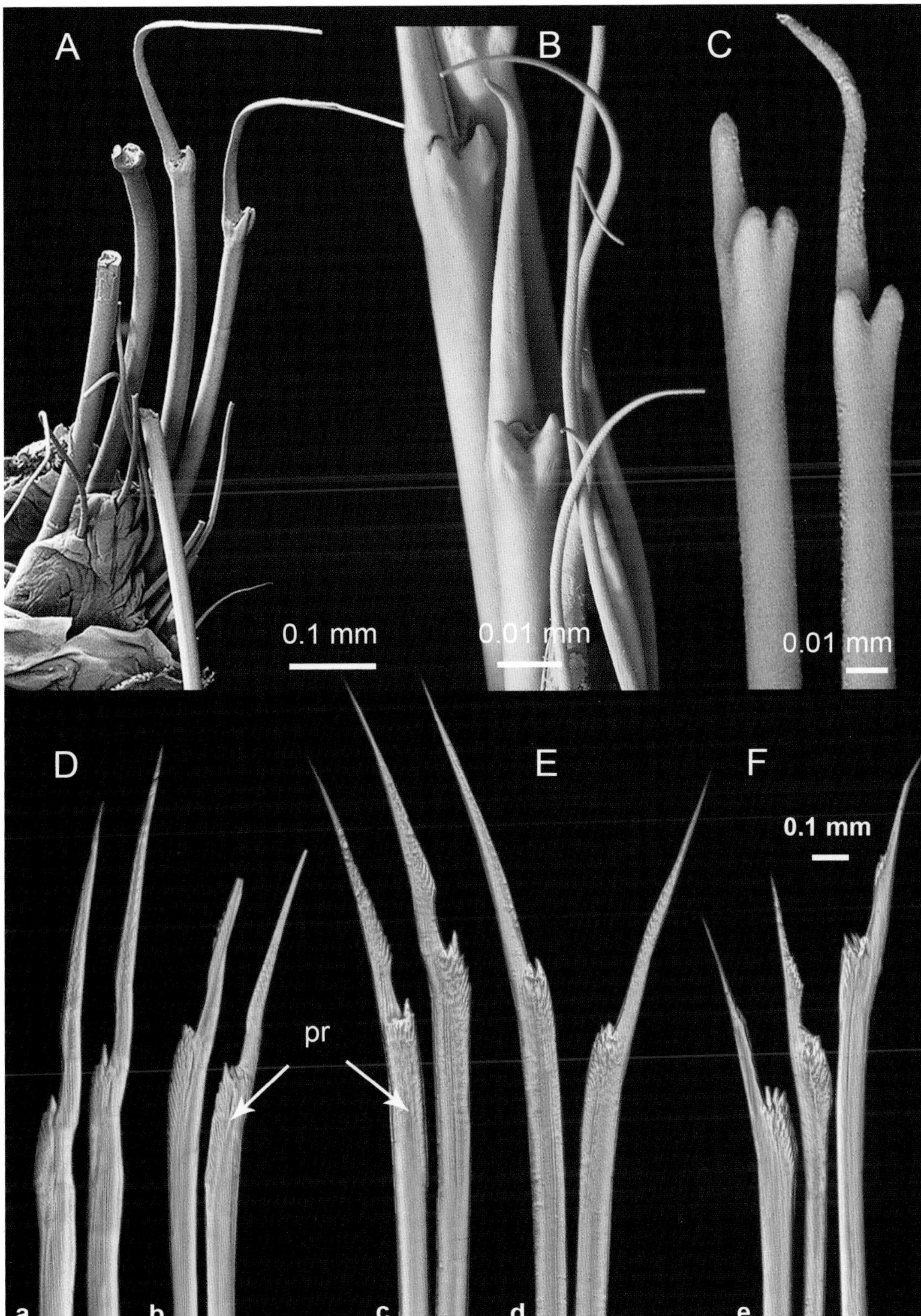

Fig. 5. Collar chaetae in *Hydroides*. A–C: Scanning electron microscopy images of bayonet special collar chaetae in *Hydroides*. A – *H. lirs* AM W.42366; B – *H. amri* AM W.25496; C – *H. amri* AM W.46928. D–F: Light microscopy images of bayonet collar chaetae of *H. elegans* (a – AM W.45550, b – AM W.46406): *H. longispinosa* (c – AM W.41995, d – AM W.41977); *H. multispinosa* (e – AM W.4236) (all photos E. Wong). Abbreviations: pr – proximal rasp.

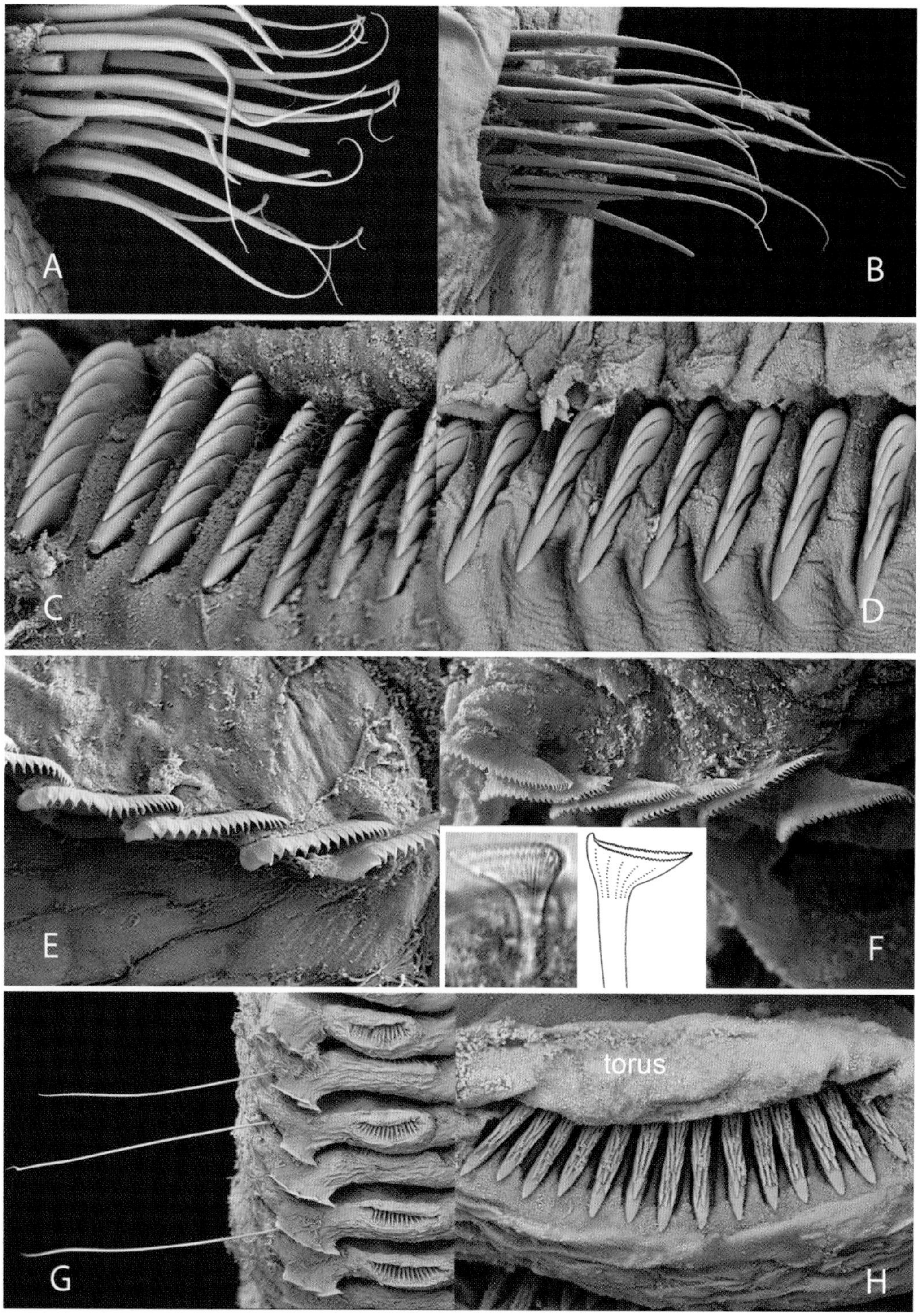

Fig. 6. Scanning electron microscopic images of chaetal characters in *Hydroides*. A, B – Thoracic chaetae in *H. glasbyi* W.46916 (A) and in *H. brachyacantha* AM W.46940 (B); C, D – Thoracic saw-shaped uncini in *H. nikae* AM W.46916 (C) and in *H. glasbyi* AM W.46892 (D); E, F – Abdominal flat trumpet-shaped chaetae in *H. nikae* AM W46916 (E) and *H. glasbyi* AM W.46892 (F). Inset: left – light microscopy image of abdominal chaeta of *H. dolabrus* giving an impression of two rows of teeth (photo A. Tovar-Hernandez); right – a typical line drawing interpretation (trumpet-shaped chaeta) of light microscopy image of abdominal neurochaeta of *H. alatalateralis* by Jones (1962); G, H – capillary chaetae and rasp-shaped uncini of posterior abdomen (G) and close-up view of posterior abdominal rasp-shaped uncini (H) in *H. amri* AM W.25496 (all photos S. Lindsay).

gradually converge posteriorly, forming a triangular depression on the ventral side of the thorax (Fig. 1, td).

Abdomen

The number of abdominal segments is very variable depending on size and age. The term 'trumpet-shaped chaetae', commonly used by various authors to describe the abdominal chaetae in the genera *Crucigera*, *Hydroides*, *Serpula* and *Spiraserpula*, is misleading. Although the distal parts of these chaetae under a compound microscope are widened into what in profile looks like a chalice or trumpet edged with two rows of thin teeth (Fig. 6, inset), examination with SEM shows that these chaetae are not hollow as the name might suggest, but rather flat, with a single row of marginal teeth (Fig. 6E, F). The second row of teeth observed under the compound microscope is an optical (refractory) illusion. Therefore, following ten Hove & Jansen-Jacobs (1984) and then ten Hove & Kupriyanova (2009), they are termed 'flat trumpet-shaped chaetae'. These abdominal chaetae usually become progressively longer towards the pygidium, and the posterior abdominal chaetae in *Hydroides* tend to be true capillaries (Fig. 6G). Uncini in the posterior abdomen have curved teeth in several rows (rasp-shaped, Fig. 6H).

2 Tubes

TUBE MORPHOLOGY

As opposed to some serpulid genera, morphology of *Hydroides* tubes is fairly simple, lacking complex sculpture, rows of denticles, transverse ridges and peristomes of varying complexity, regular pits or alveoli. Most species of the largest serpulid genus, irrespectively of their size, have uniform tubes, circular in cross-section (Fig. 7A) or more often with a flattened upper surface, usually bordered with two more or less distinct keels (Fig. 7B–G), making the tube cross-section trapezoid or nearly trapezoid. A third median keel can be present and, in very few cases (e.g. *H. pseudouncinata*), it is the only distinct keel on the upper surface. The anterior part of the tube, when growing away from the substrate, is always circular in cross-section. The tube surface is covered with faint transverse growth striations, slightly bending anteriorly on the upper side (Fig. 7A). Tubes completely attached to substrate have peripheral flanges (Fig. 7E, pf) increasing the area of tube attachment. The occurrence of internal tabulae, partitioning off damaged posterior tube parts, has rarely been mentioned in the literature, though Breton & Vincent (1999: 39, fig. 14) estimate the frequency of occurrence at 15–20% in tubes of *H. ezoensis*.

The colour of the tubes, especially of the newer anterior parts, is white (Fig. 7A–D, G), but in some cases tubes can look bluish because of a strongly coloured purple inner organic layer (e.g. in the *H. tuberculata*-complex; it has been mentioned for *H. perezi*, *H. rhombobula* and *H. uniformis* (Fig. 7E), but not in all tubes in the population). Some tubes have a distinct inner layer of spicules **sensu** Pillai (2009) or fish-net layer **sensu** Ippolitov *et al.* (2016): see Fig. 7F.

TUBE COMPOSITION AND MINERALOGY

The tubes are built of calcium carbonate intermixed with an organic matrix into the inorganic–organic composite material. The lumen is lined with an organic membrane. Thus, the organic tube component is represented by an organic matrix within the tube walls and an organic lining on the inner side of the tube wall.

In *Hydroides dianthus* the organic tube component is represented by both soluble and insoluble organic matrices (Tanur *et al.* 2010). Most of the soluble organic matrix (SOM) is composed of carboxylated and sulphated polysaccharides, which include acidic and non-polar neutral amino acids, the most abundant amino acids in SOM of *H. dianthus* being aspartic acid, glutamic acid, glycine and proline. A minor component, the insoluble organic matrix (IOM), is formed by collagen-containing fibres (Tanur *et al.* 2010). Bernhardt *et al.* (1985) showed that the decalcified insoluble organic matrix of *H. dianthus* was capable of becoming recalcified in an inorganic solution.

The inorganic component of serpulid tubes is represented by the crystals of calcite, aragonite, or a mixture of both crystal forms of calcium carbonate (Bornhold &

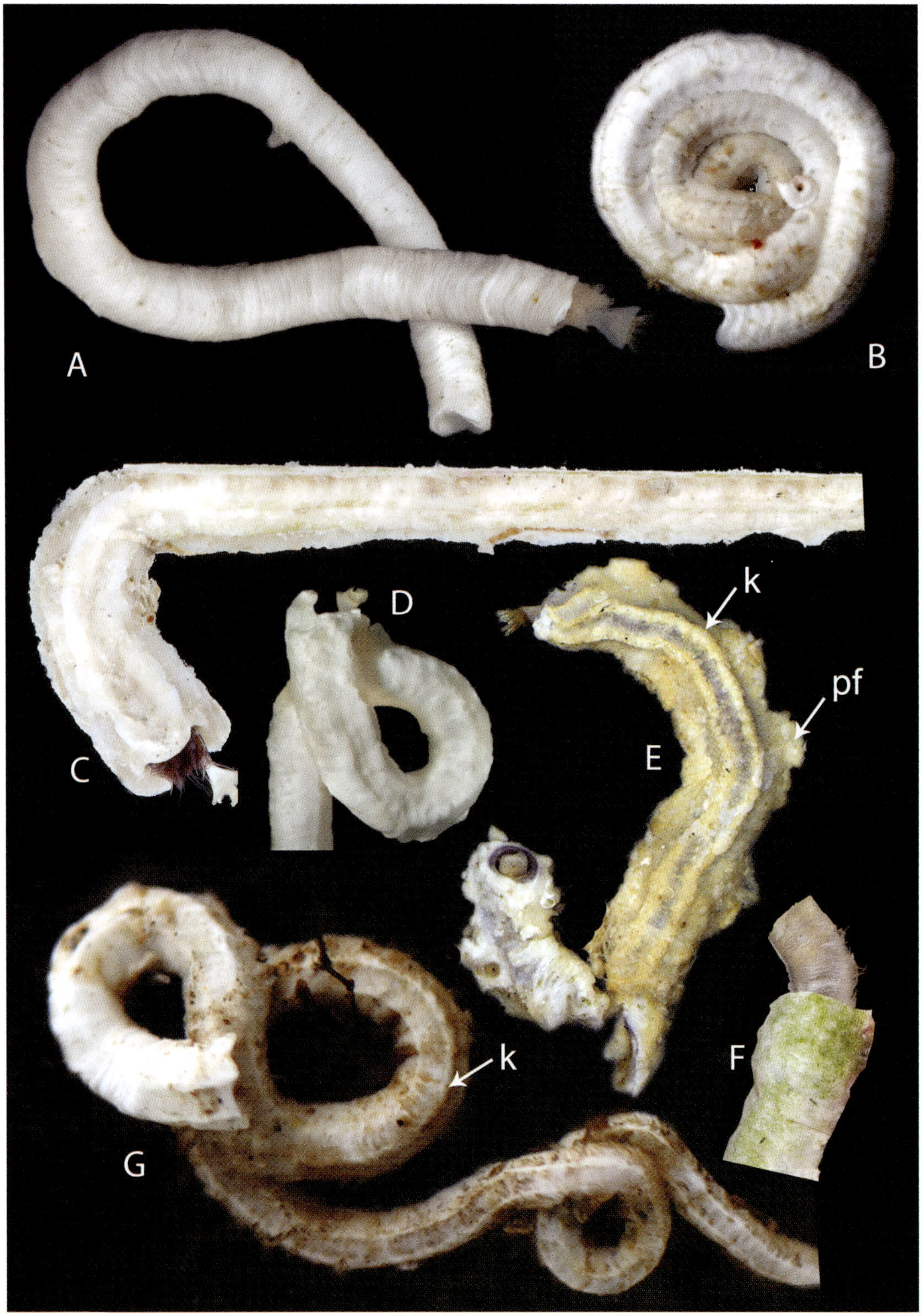

Fig. 7. General morphology of tubes in *Hydroides*. A – *H. norvegica*, Norway (photo E. Wong); B – *H. albiceps*, Hong Kong (photo G. Paulay); C – *H.* sp., probably *H. exaltata*, Hong Kong (photo G. Paulay); D – *H. albiceps*, Seychelles, AM W.28296 (photo E. Wong); E – *H. tambalagamensis*, Lizard Island, AM W.45050, tube fragment showing typical internal layer made of spicules (photo E. Wong); F – *Hydroides uniformis*, Vanuatu, AM W.47092 (photo L. Gunton); G – *Hydroides ezoensis*, Sea of Japan, Russia (photo L. Harris). Abbreviations: k – keel, pf – peripheral flange.

Milliman 1973; Smith *et al.* 2013; Ippolitov *et al.* 2014). Out of serpulids examined by Vinn *et al.* (2008), the mineral composition of *Hydroides* tubes varied from predominantly calcitic (with only a 1.6% aragonite component) in *H. norvegica* to predominantly aragonitic in *H. spongicola* (5% calcite). Mineralogical composition of tubes in most of the species of the genus remains unknown, although an unpublished study covering 36 species of *Hydroides* (Ippolitov *et al.* unpublished) found predominantly calcitic tubes with an aragonite content varying from 0 to 42% (normally 8–15%). The interspecific variations of crystal morph ratio in *Hydroides* may be partly due to habitat temperature variations (Smith *et al.* 2013).

It has been long understood that the calcite:aragonite ratio may significantly vary not only among species, but also within a species and even within a single tube during the ontogeny (Bornhold & Milliman 1973). Interestingly, the mineral composition of *H. elegans* tubes gradually shifted during ontogeny – juvenile tubes were predominantly aragonitic, whereas adult tubes were bimineralic with considerably higher amounts of calcite (Chan *et al.* 2015a). The mineralogical composition of the tube is spatially heterogeneous: the inner tube layer of the same species, *H. elegans*, was found to be calcitic, whereas the outer layer contained aragonite (Chan *et al.* 2015a).

TUBE ULTRASTRUCTURE

Regarding the ultrastructure, serpulid tubes can be either single- or multilayered; in the latter case adjacent layers differ in orientation, morphology and size of crystals, or in the quantity of homogeneous carbonate cement. Vinn *et al.* (2008), who studied 44 serpulids out of over 500 known species, distinguished four main groups of serpulid tube ultrastructures according to orientation of calcium carbonate crystals: isotropic, semi-oriented, oriented prismatic and oriented complex structures. In total, 13 ultrastructure types are currently recognised in serpulids (see Ippolitov *et al.* 2014 for review).

The three species of *Hydroides* studied by Vinn *et al.* (2008) show different numbers of layers with different types of structures. *Hydroides dianthus* (Fig. 8A–C) had a

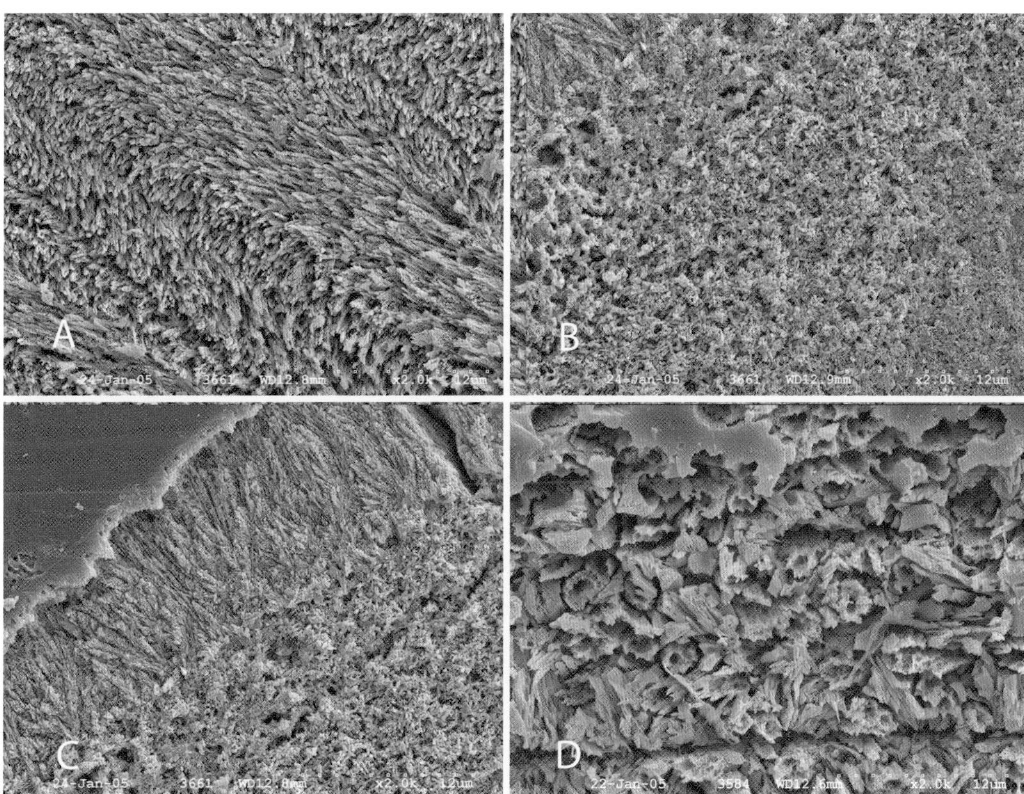

Fig. 8. Examples of ultrastructures in *Hydroides* tubes. *H. dianthus*: three-layered tube, A – lamello-fibrillar (LF); B – irregularly oriented platy (IOP); C – spherulitic prismatic (SPHP) and irregularly oriented platy (IOP) arrangement of crystals. *H. spongicola*: single-layered tube, D – semi-ordered irregularly oriented prismatic (SIOP) crystal structure (photos O. Vinn).

three-layered tube with inner lamello-fibrillar (LF) + middle irregularly oriented prismatic (IOP) + outer spherulitic prismatic (SPHP) structures; *H. norvegica* had a tube with crystal arrangement in two layers, irregularly oriented prismatic (IOP) + spherulitic prismatic (SPHP); while *H. spongicola* had a single layer with a semi-ordered irregularly oriented prismatic (SIOP) crystal structure (Fig. 8D).

In addition, all species of *Hydroides* examined in a yet unpublished study (Ippolitov *et al.* unpublished) are characterised by highly ordered structures, represented by a combination of an outer irregularly oriented prismatic (IOP) layer with a lamello-fibrillar (LF) inner layer. Many species also possess an outer SPHP or simple prismatic (SP) layer, and a SIOP layer in the inner half of the wall; some have a SP/SPHP inner layer (Ippolitov *et al.* unpublished). Thus, the true diversity of crystal structures in the genus still remains undescribed. It is also important to remember that the ultrastructural complexity does not remain constant during ontogenetic development, judging by the sequential appearance of unoriented ultrastructures followed by oriented ones in *H. elegans* (Chan *et al.* 2015a).

ORGANIC TUBE LINING

While calcareous serpulid tubes lack an external organic cover like the periostracum of molluscs or the cuticle of bryozoans, they possess an internal organic tube lining (Nishi 1993; Vinn *et al.* 2008). This organic membrane (insoluble in ethylenediaminetetraacetic acid, EDTA) is composed of collagen-containing fibres aligned in a criss-crossed structure with carboxylated and sulfated polysaccharides (Tanur *et al.* 2010).

The lining could protect the worm against roughness of the tube; this protection is important given that the animals quickly retract in the tube in response to shadow, touch or water movement – mechanical stimuli likely to be associated with predators (Dill & Fraser 1997). The role of the organic inner tube lining in tube formation is still poorly understood.

TUBE FORMATION

Although the process of serpulid tube deposition has been the subject of early studies, the true origin of crystals remained a mystery (Chan *et al.* 2015a). All serpulids, including *Hydroides*, have two calcium-secreting glands, one located on each side of the peristomium, under the fold of the collar. These glands produce both tubes and adhesive cement (Neff 1969, 1971; Nott & Parkes 1975). Ventral thoracic areas contain mucus-secreting epithelial cells that appear to be the major contributor of the organic part of the tubes (Hedley 1956a, b, 1958: *H. norvegica*). On the other hand, Neff (1969) hypothesised that the calcium-secreting glands of *H. dianthus* form the aragonitic outer part of the tube, while the thoracic ventral shield epithelium plays a role in the formation of the magnesium-calcitic inner layers.

Hedley (1956a, b) hypothesised that mucus-rich calcium carbonate is secreted from the glands in the form of a slurry and deposited on the tube aperture. This slurry solidifies sufficiently slowly to allow the undersurface of the collar, which is folded back over the aperture of the tube, to mould the calcite-saturated mucus, shaping the end of the tube. While this model describes the accretion of simple irregularly oriented tube ultrastructures, for more advanced and complex oriented ultrastructures, including those found in *Hydroides* spp., a matrix-controlled crystallisation model of biomineralisation was proposed by Vinn *et al.* (2009).

Tanur *et al.* (2010) hypothesised that inner organic tube lining aids tube formation, suggesting that *H. dianthus* uses the tube lining as a scaffold for the biomineralisation of calcium carbonate ($CaCO_3$). However, Vinn (2011) argues that this hypothesis does not agree with the formation of the tube under the fold of collar (Hedley 1958) and observed that growth towards the tube lumen also contradicts the proposed scaffold biomineralisation model.

Although space delineation is a well-accepted requirement for biologically controlled mineralisation (Simkiss & Wilbur 1989), it remains poorly understood. The mechanism of crystal formation was proposed by Chan *et al.* (2015a) based on *in vivo* studies of *H. elegans* post-metamorphic larvae. According to this model, appearance of aragonite is compartmented within the collar region, where the metamorphosed larvae maintain an elevated pH 9, at the *p*Ka for $CaCO_3^-$ formation, which meets physicochemical and thermodynamics constraints of calcium carbonate formation. Crystallisation takes place in the cells of calcium-secreting glands, combining Ca^{2+} ions pumped from the gut with HCO_3^{2-} derived from both metabolic and environmental carbon dioxide (CO_2) (Chan *et al.* 2015a). The initial stage of crystal formation may undergo an amorphous phase because, although aragonite was present in the collar region, newly deposited tubes also contained amorphous calcium carbonate (Chan *et al.* 2015b), which could serve as a supply for further growth of the preformed crystals previously deposited by the calcifying tubeworm. This model demonstrates how biomineralisation is a space delineation event in which calcium carbonate formation is an intracellular phenomenon.

FOSSIL RECORD

Although due to their calcareous tubes serpulids have the best fossil records among the mostly soft-bodied annelids, recognising members of *Hydroides* in fossil state is difficult due to the simplicity of their tube morphology (Ippolitov *et al.* 2014 and the section above). Multiple extant species of the closely related *Serpula* have tube morphology identical to that of *Hydroides*. Based solely on tube morphology, most ancient tubes attributable to *Hydroides* can be dated back questionably to the middle Palaeocene, ~61 Ma (Lommerzheim 1981) or, more reliably, to the Eocene ~56 Ma (Rovereto 1904) and became dominant in serpulid communities in the Neogene ~30 Ma (Schmidt 1950, 1951, 1955; Radwańska 1994). Recent examination of the fossil tube ultrastructures using scanning electron microscopy revealed the unequivocal presence of *Hydroides* in the Early Miocene (~21 Ma) of India (Ippolitov & Desai 2015). Thus, *Hydroides* is a geologically young serpulid genus, which arose and diversified in the Caenozoic, after the Great Cretaceous/Tertiary extinction. However, the *Serpula–Hydroides* clade is much more ancient as its member *Spiraserpula*, having highly specialised tube morphology, is dated back to the Middle–Late Jurassic epochs, indicating that true representatives of the *Serpula–Hydroides* stem must have already existed at that time (Ippolitov *et al.* 2014), despite not being clearly recognisable within the Mesozoic serpulid diversity.

3 History of studies

TAXONOMIC STUDIES

Hydroides was erected by Gunnerus (1768), with *H. norvegica* Gunnerus, 1768 as the only species (Fig. 9A). Philippi (1844) established *Eupomatus* for species that have an operculum like *Serpula*, but with a set of movable spines positioned distally in the middle of the *Serpula*-like funnel ('auf der obern Seite im Centrum mit beweglichen Spitzen versehen'). As the first species of this new genus he mentions *E. uncinatus*, with eight incurving smooth spines, as the second *E. pectinatus*; both his diagnoses and figures Q and R cannot be identified with certainty and we regard them to be *species inquirenda* (see there); Philippi does not mention the genus *Hydroides* at all in his attempted overview of serpulid genera.

Mörch (1863) was the first to comprehensively revise the serpulids, including three subgenera in the genus *Hydroides*, based on the shape of the verticil spines. He restricted the name *Eupomatus* to species with simple, smooth, incurving verticil spines; *Hydroides* for straight, multispined (apparently lateral spinules) verticil spines; and *Eucarphus* for the species with verticil spines like an awl, bearing a pair of terminal lateral spinules, sometimes with a median spike, hooked like a gallows. Altogether he described 10 new species, including two of Krøyer's unpublished species.

De Quatrefages (1866) erected two higher taxa, a genus (*Codonytes*) and a subgenus (*Polyphragma*) for specimens with duplicity (bioperculate), both overlooked by ten Hove & Ben-Eliahu (2005). The interpretation of his nominal genus *Codonytes* is reasonably straightforward: de Quatrefages's diagnosis (1866: 494, *Utrinque operculum unum, spinulatum*) in free translation would be 'one operculum at both sides, with spinules'. The genus was erected for (1) *E. dipoma* Schmarda, 1861 and (2) *C. infundibulum* Delle Chiaje, 1822; however, de Quatrefages clearly mixed up figures and names from the legend by delle Chiaje's plate 48 (published already in 1822: he used the name of a bryozoan, *C. infundibulum* (Figs 23 and 24) for figures 21–22, the serpulid *Sabella euplaeana* delle Chiaje, 1822, which is clearly recognisable as a *Hydroides*, probably *H. pseudouncinata pseudouncinata*. The generic name *Codonytes* was not used afterwards. Synonymies and descriptions given by de Quatrefages (1866: 507–508) under *Polyphragma* do not allow the attribution of his taxa with certainty to species level, but most of – if not all – the specimens he refers to we now would perceive to belong to *Hydroides*, such as his interpretations of *Serpula uncinata* Philippi, 1844 and *S. mülleri* Berkeley, 1834 (indeterminable *nomen inquirendum* but referred to *H. norvegica* by McIntosh 1923: 348); de Quatrefages' interpretation of *S. vermicularis* Linnaeus, 1767 seems to be a mix of real *Serpula* and maybe some *Hydroides*. The generic name got no following.

Grube (1878b) described two new species of serpulids with a two-tiered operculum under the genus *Serpula* – *S. minax* and *S. furcifera* – though he attributed them to

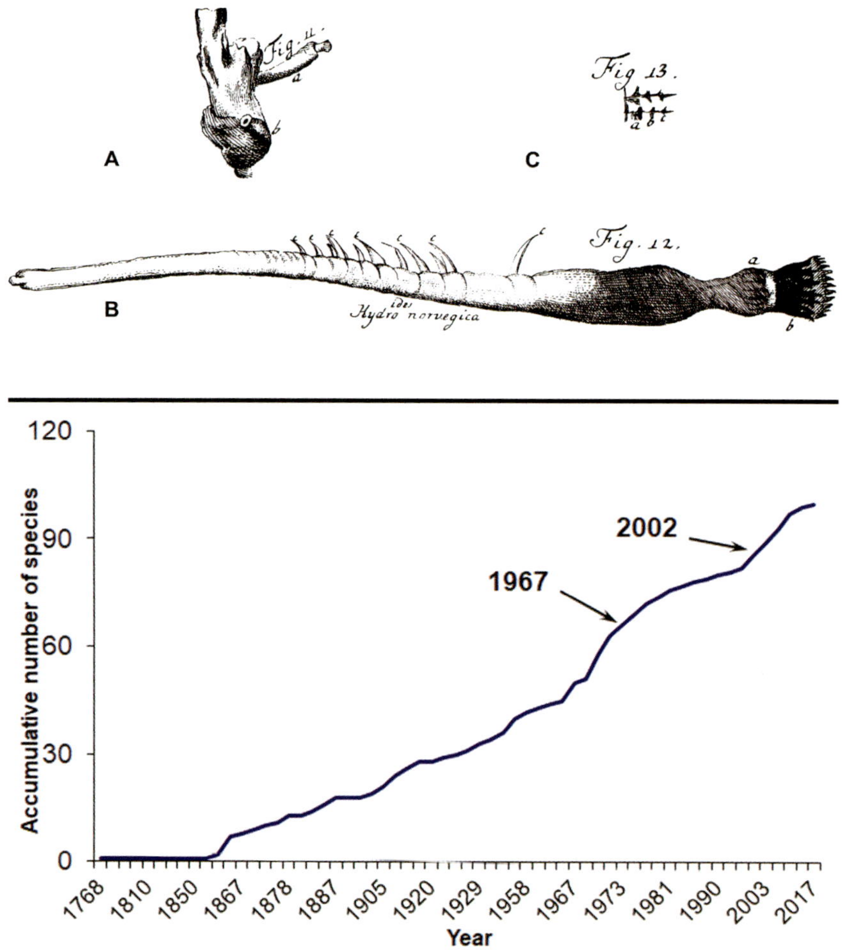

Fig. 9. History of studies of *Hydroides*. Top – Original drawings of *H. norvegica* by Gunnerus: A) tube with the anterior part of the worm, B) an entire animal, C) two verticil spines with the lateral spinules referred as a, b and c (from Gunnerus 1768); Bottom – Dynamics of species discovery in the genus before 2021.

Serpula (*Hydroides*) in his indexes. Bush (1905, 1910) described five new *Hydroides* species from the Western Atlantic and Eastern Pacific. She designated type species for the genera *Hydroides*, *Eupomatus* and *Eucarphus*. Moreover, she (Bush 1905: 225) erected the genus *Glossopsis* for *Serpula minax* Grube, 1878b, to recognise the asymmetrical arrangement of the verticil spines, as well as the genus *Schizocraspedon* for *Serpula furcifera* Grube, 1878b, to stress the unusual similarity of bifid shapes of both funnel radii and verticil spines of the operculum. Bush (1910: 498) renamed *Eupomatus uncinatus* Ehlers, 1887 [**non** (Philippi, 1844)] to *Eupomatus floridanus*. The use of the two new generic names got no following.

Pixell (1913) attempted the first major review of the genera *Hydroides* and *Eupomatus*, describing two new species. She reasoned that *Glossopsis* and *Eucarpus* should be included as synonyms in *Hydroides*, while retaining the genus *Eupomatus*. Thus, she mentioned 13 species for *Hydroides*, with lateral processes to the verticil spines, and seven for *Eupomatus*, without such lateral processes. Her subdivision within each nominal genus is based on whether the verticil spines are similar or not.

Chamberlin (1919: 476) still recognised five genera (*Hydroides*, *Eupomatus*, *Eucarphus*, *Schizoscraspedon* and *Glossopsis*) in his literature-based key to the serpulid genera. He considered *Polyphragma* to be a synonym of *Eupomatus*, and *Codonytes* as a synonym of *Eucarphus*.

Rioja (1941a, b, 1942, 1944, 1946, 1958) recorded numerous species (including six new ones) from both coasts of Mexico. He erected (1941b) the genus *Olgaharmania* for *H. glandifer* Rioja, 1941a, a peculiar species with a verticil supposedly consisting of a single bulbous spine. However, upon close inspection the 'single bulbous spine' is a composite of a huge dorsal hooked spine with four smaller verticil spines inserted halfway up, a verticil not essentially different from, for example, that of *H. albiceps*.

The generic names *Eucarphus*, *Glossopsis*, *Olgaharmania* and *Schizocraspedon* have not been used extensively and were subsequently – decades ago – considered redundant. By the early 1960s only two generic names, *Hydroides* and *Eupomatus* (including 23 species in Hartman 1959), were used to subdivide the group with a double funnel as operculum (Rioja 1958; Hartman 1959; Jones 1962). Some taxonomists already recognised *Hydroides* as the single generic name for these serpulids (Okuda 1934; Dew 1959; Straughan 1967a, b). Pillai (1960, 1961, 1971, 1972) formally synonymised *Eupomatus* with *Hydroides* – which was widely accepted – and described six species of *Hydroides*, after studying serpulids from Sri Lanka. In 1972, Pillai proposed a phylogeny of the genus (see 'Phylogeny', below). Based on a number of morphological trends (i.e. character transformation series), he discerned 12 monophyletic groups, but the phylogenetic relations between the groups remained unclear. He used 40 species for his analysis, but apparently many descriptions of Caribbean species were unavailable to him. Subsequently new methods of phylogenetic analysis have become available, and many more species have been described. Uchida (1978) proposed a new genus, *Protohydroides*, for *H. elegans*, based on the apparent absence of a verticil in some specimens and on the presence of collar chaetae with more than three teeth. However, the genus *Protohydroides* is considered to have no sound phylogenetic basis (ten Hove 1984), and this name has hardly ever been used. Thus, the consensus that *Hydroides* is a single genus has prevailed.

Between 1960 and 2000, a number of taxonomic studies of the genus *Hydroides* were conducted in various parts of the world. Zibrowius (1968, 1971a, 1972a, b, c, 1973a, b) exhaustively reviewed species from the Mediterranean and the Eastern Atlantic, including nine new species from this region, and one each from the Banda Sea (Indonesia) and the Yellow Sea (China). In Australia, Straughan (1967a, b, c) recorded 16 taxa of *Hydroides* from New South Wales and Queensland, including eight new species. Imajima (1976a, b, 1978, 1982) reviewed *Hydroides* species from Japan and Micronesia, describing three new species. Imajima & ten Hove (1984, 1986, 1989) added two more species, one from Japan and the other from the Solomon Islands. ten Hove (1990) described a new species from the Arabian (Persian) Gulf and proposed a standardised terminology for the *Hydroides* operculum. Bailey-Brock (1991) described a new species characterised by the absence of a funnel from Hawaii. Chen & Wu (1978, 1980), Wu & Chen (1981) and Sun & Yang (2000) described seven new species from China. In their study of the serpulids from the South China Sea, Fiege & Sun (1999) included eight species of *Hydroides* and were the first to illustrate some with SEM photographs.

From 2000, taxonomic studies of the genus *Hydroides* were mainly conducted in China, Mexico, and Australia. In Mexico, Bastida-Zavala & de León-González (2002) described a new species from Western Mexico, while Bastida-Zavala & ten Hove (2003a, b) provided the most detailed to date revision of *Hydroides* from the Western Atlantic Region, the Eastern Pacific Region and Hawaii, describing and naming in total six new species, as well as giving descriptions of six unnamed taxa, possibly new. Tovar-Hernández *et al.* (2016) added another new species for the region. In China, Sun & Yang (2014) in a book on Chinese polychaetes described and figured 35 taxa of *Hydroides*, of which seven were not assigned to a named species. In Australia, Pillai (2009) reviewed serpulids from Kimberley region, Western Australia, describing seven new species of *Hydroides*. The most recent morphological revision of *Hydroides* from Australia (Sun *et al.* 2015) reported 26 species, including three new, but Kupriyanova *et al.* (2015) and Sun *et al.* (2016) added two more species.

The trend of the last decade has been an increase of taxa number not only by discoveries of morphologically distinct species (accompanied or not by DNA sequence data), but also as a result of re-evaluation of the named species composition and resolution of species complexes. Application of molecular methods resulted in critical re-evaluation of wide distributions of nominal species revealing both previously unknown diversity (e.g. Sun *et al.* 2016, *H. brachyacantha*) and diversity hidden as result of previous synonimisations (e.g. Sun *et al.* 2017b, *H. operculata*). Sun *et al.* (2018) indicated in a molecular phylogeny study that a number of nominal species with reported wide distributions (e.g. *Hydroides albiceps*, *Hydroides minax*, *H. exaltata* and *H. tuberculata*) are in fact species complexes that need to be resolved.

Most recently Read *et al.* (2017) provided an update of all 107 at the time non-synonymised *Hydroides* names.

Such an update was needed because, as a result of an overlooked change in the wording of International Code of Zoological Nomenclature, the gender of genus *Hydroides* has from 1 January 2000 reverted to the original feminine. Read *et al.* (2017) also proposed *Hydroides gottfriedi* **nom. n.** to replace *H. rostrata* Pillai, 1971, the junior homonym of *H. rostrata* Iroso, 1921.

DYNAMICS OF SPECIES DISCOVERY

The dynamics of *Hydroides* species discovery is summarised in Fig. 9, bottom. For almost 200 years since *Hydroides* was erected by Gunnerus (1768), the number of species in *Hydroides* increased more or less gradually, while between 1967 and 2020 this number doubled. Until now, 99 species (including two subspecies of *H. pseudouncinata*) were considered valid (Read & Fauchald 2020).

The number of described valid species within the genus will continue to grow in the future. Currently, a number of *Hydroides* names in the literature are given in open nomenclature (see Sigovini *et al.* 2016 for the usage of open nomenclature qualifiers). Some of these are known species not identified beyond the genus by the authors of respective publications, while others are undescribed taxa. For example, Bastida-Zavala & ten Hove (2003a) included in their accounts four unnamed species (*Hydroides* spp. 1–4) and two *Hydroides* sp. cf. (confer = to compare, or to be compared with), *Hydroides* cf. *mucronata* and *H.* cf. *operculata*, to emphasise that the described specimens did not belong to the nominal taxa that they resemble. The unnamed species from Bastida-Zavala & ten Hove (2003a) were named and included in this book. Seven unnamed *Hydroides* spp. from Sun & Yang (2014) are accompanied by line drawings that leave space for plural interpretations; thus, we do not take responsibility to name those as full species. A checklist of intertidal polychaetes of Kuwait (Al-Kandari *et al.* 2019) mentions *Hydroides* nov. spec.; the record is based on a single specimen examined by EK.

Read *et al.* (2017) also included notes on seven species names that are not considered valid, while stressing (as explained by ICZN 1999, Article 23.9.6) that the inclusion of these names should not later be considered as new usages. These are *taxa inquirenda* (*Hydroides affinis*, *H. diplochone*, *H. euplaeana*, *H. exaltata vesiculosa*, *H. hexagona*, *H. pectinata* and *H. uncinata*). We also treat these names and the rejected name *Hydroides prisca* Pillai, 1971 (a juvenile *Hydroides* with a developing operculum) in the section '*Nomina inquirenda* and rejected names, occasionally still in use' later in this book.

4 Reproduction, development and life history

Species of *Hydroides* exhibit very similar life histories (Kupriyanova *et al.* 2001). Most information on life history comes from *H. elegans* (the most studied taxon) and *H. dianthus*. Only limited information can be found for *H. dirampha*, *H. ezoensis* and *H. fusicola* (see Kupriyanova *et al.* 2001; Qian *et al.* 2007; Cao *et al.* 2009, 2013; Leung *et al.* 2013; Shikuma *et al.* 2016), while little is known about the rest of *Hydroides* spp. beyond their basic taxonomic description.

REPRODUCTION

While sexes were traditionally considered separate in serpulids, studies revealed protandric hermaphroditism with a short intermediate stage in common species, including *Hydroides elegans* (see Ranzoli 1962, as *H. norvegicus*; Reish & Pernet 2009). Sequential hermaphroditism causes biased sex ratios and body sizes (females larger than males). The male to female ratio of juvenile *H. elegans* varied from 1:4 to 3:1 (Qiu & Qian 1998).

Oogenesis has been studied for *H. norvegica* (by Nordback 1956) and *H. dirampha* (by Mona 1992), and spermatogenesis for *H. norvegica* (by Franzén 1956) and *H. dirampha* (by Mona *et al.* 1994). Jamieson & Rouse (1989), in their review of polychaete spermatozoa, include the *nomen inquirendum H. hexagona*, and summarise one original paper by Colwin & Colwin 1961a (out of 10 papers on similar topics by these authors).

Mature egg sizes range from 45 µm in diameter in *Hydroides ezoensis* (see Miura & Kajihara 1981, 1984), *H. elegans* and *H. dianthus* to 67 µm in *H. fusicola*, the range typical for planktotrophic serpulids (Kupriyanova *et al.* 2001). Miles *et al.* (2007) showed the presence of significant heritability for egg size in *H. elegans*, which indicates potential for egg size to respond to varying selective pressures, at least in the direction of increase. Miles & Wayne (2009) further used artificial selection for increased egg size to demonstrate that such selection produced positive correlated responses in egg energy, fecundity and larval size at competence in *H. elegans*. The selection also resulted in earlier sex change as selected individuals were spending less time as males. Allen & Marshall (2014) tested in *H. dirampha* the relationship between egg size and performance across multiple life-history stages, including fertilisation, larval development, and post-metamorphosis survival and size. They found evidence of conflicting effects of egg size on performance: larger eggs had higher fertilisation under sperm-limited conditions, were slightly faster to develop pre-feeding, and were larger post-metamorphosis; however, smaller eggs had higher fertilisation when sperm was abundant, and faster planktonic development; and egg size did not affect post-metamorphic survival. The results indicate that egg size effects are conflicting depending on the environments within and among life-history stages. Apparently offspring size in this species may be a compromise between the overall costs and benefits of egg sizes in each stage and performance in any one stage is not maximised.

Sperms typical for externally fertilising serpulids with a spherical head, a midpiece containing spherical mitochondria and a flagellum were described for *H. norvegica* (see Claparède 1870, as *Eupomatus trypanon* (synonymised with *H. norvegica* by Zibrowius 1971a: 718); Franzén 1956), *H. dianthus* (see Colwin & Colwin 1961a, as *H. hexagona*), *H. ezoensis*, *H. fusicola* and *H. elegans* (see Matsuo & Yoshioshi 1983; Nishi 1992).

Female fecundity measured as the average number of mature eggs expelled by a female of *Hydroides dianthus* varied from 600 to 80 000 (Zuraw & Leone 1968; Leone 1970; Toonen & Pawlik 1994), whereas those of smaller *H. elegans* ranged from 1100 to 9050 (Qiu & Qian 1998). The estimated number of sperm released by *H. dianthus* was 62 million (Leone 1970). *Hydroides dianthus* responds to artificially elevated temperatures in the winter by developing gametes out of season (Turner & Hanks 1960); however, temperatures abnormally high for this species (26–30°C) decreased female fecundity (Leone 1970). Average fecundity of *H. elegans* was unaffected by temperature within the range of 15–30°C (Qiu & Qian 1998).

Gamete longevity was examined by Grave & Oliphant (1930), who showed that unfertilised eggs of *H. dianthus* (as *hexagonis*) retain their ability to be fertilised and develop into normal trochophores for 18–24 h, the extreme variability in longevity being 6–32 h. Spermatozoa, in dilute suspension or when expelled into seawater, live and retain the ability to fertilise eggs for 3–7 h.

Fertilisation is external: the gametes are released through nephridiopores and are delivered to the tube mouth with the help of ciliary beating in the faecal groove. Natural spawning events are rarely observed, but gamete release is stimulated by breaking the tubes and artificial fertilisation is easy to achieve in the laboratory. The rates of gamete transport and mixing, as well as fertilisation success have not been studied. Qian & Pechenik (1998) state that fertilisation is successful over a remarkably broad range of sperm concentration in *H. elegans* and over 95% of eggs were fertilised within 15 min after the gametes were mixed. Results of Pechenik *et al.* (2007) suggested that failure of early development at low salinity in *H. elegans* mainly reflects an inability of freshly fertilised eggs to complete metamorphosis and cleave, rather than a failure of fertilisation.

The ultrastructural details of fertilisation were examined in *H. dianthus* by Colwin & Colwin (1961a, b, c, as *H. hexagona*), who addressed the significance of the acrosome reaction and the events during the fusion of the gamete membrane. The fertilisation reaction has been studied in *Hydroides elegans* (Monroy 1954, as *H. norvegica*) and *H. norvegica* **sensu stricto** (Nordback 1956).

LARVAL DEVELOPMENT

Larval developmental events in *Hydroides* (see Fig. 10) and in other genera having planktotrophic larvae are very similar (reviewed by Kupriyanova *et al.* 2001). After fertilisation, the negatively buoyant eggs sink to the bottom, where they undergo equal spiral cleavage up to the blastula stage (Fig. 10A–C). All cleavages of the blastomeres up to the morula stage are synchronous, holoblastic and equal. The equal-sized spiral cleavage of *H. elegans* has been reconstructed from serial sections of nuclei-stained embryos (Arenas-Mena 2007). The order of cell divisions has been determined from the two-cell stage to the 80-cell stage, when gastrulation cell movements start to overlap with late spiral-cleavage divisions. The third cleavage in *H. elegans* is invariably sinistral (Arenas-Mena 2007).

The uniformly ciliated blastula (Fig. 10D) develops into a larva with a prototroch consisting of a single ring of cilia (Fig. 10E). The prototroch separates a rounded episphere from a conical hyposphere. The simple gut opens with the mouth below the prototroch and the anus exiting on the opposite side above the anal vesicle. The apical plate and the apical tuft of long rigid cilia become distinct. Cilia on the hyposphere are organised into the neurotroch, which runs from the ventral posterior surface to the mouth. Later, the prototroch develops three main ciliary bands: the upper and lower with shorter cilia and the middle with much longer cilia (Fig. 10F). A metatroch is developed at this stage. Between the prototroch and metatroch is a band of short feeding cilia. On the right side of the episphere, a cluster of red pigment cells forms an ocellus. The ventral longitudinal muscle and metatroch circular muscle form and protonephridia become visible. The trochophore continues to grow but does not undergo any significant changes. Next, the larva develops the left ocellus. After this the growth is confined to the hyposphere and the larva elongates and develops three chaetigerous segments (Fig. 10G). Before the settlement a small fourth trunk segment is delineated and paired radiolar rudiments appear

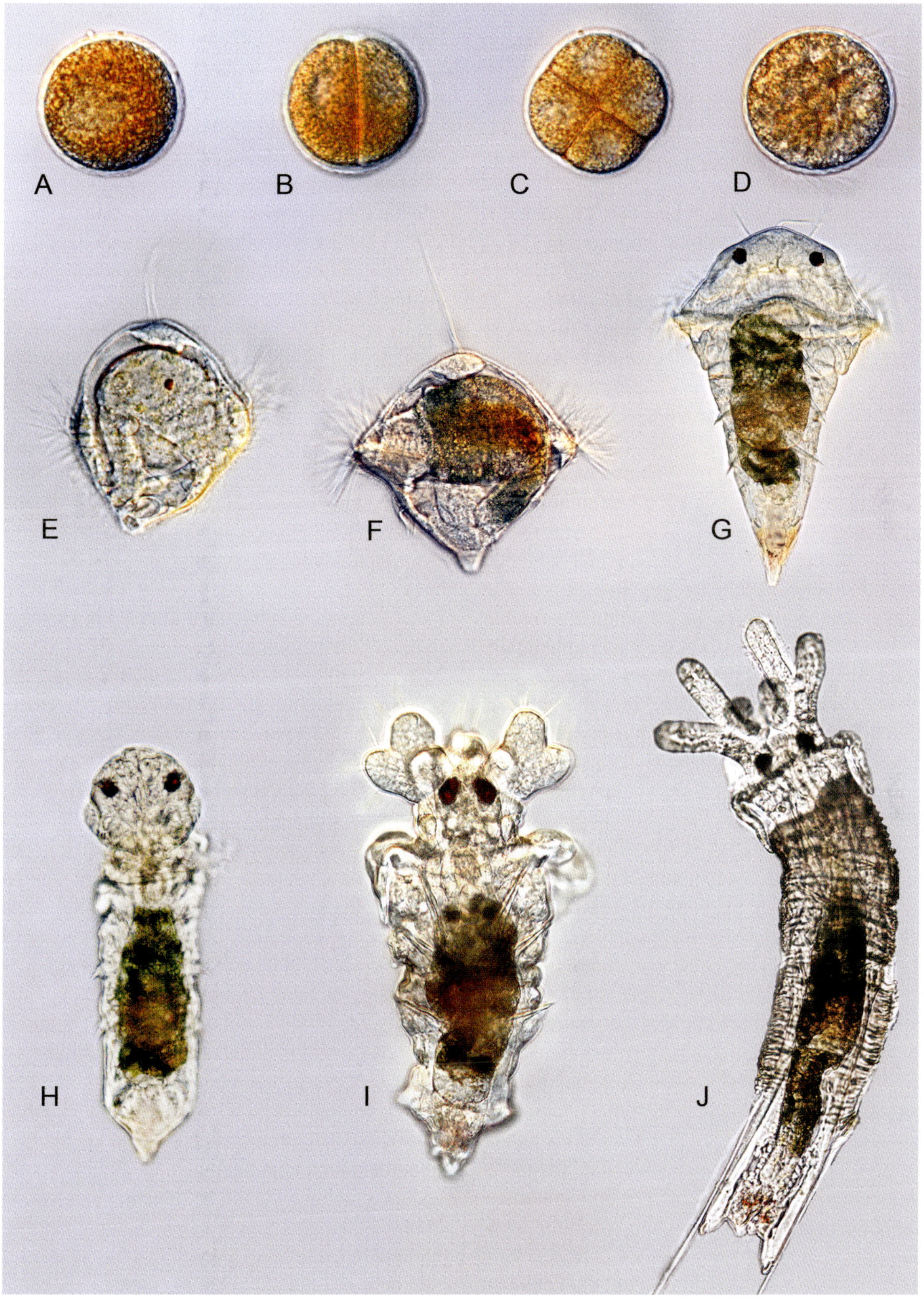

Fig. 10. Larval development and metamorphosis in *Hydroides gracilis*. A– unfertilised eggs; B – two-blastomer stage; C – four blastomere stage; D – uniformly ciliated blastula; E – trochophore 22 h after fertilisation; F – trochophore 3 days after fertilisation; G – metatrophophore 7 days after fertilisation; H, I – metamorphosing larvae; J – metaphorphosed settled juvenile starting calcareous tube (photo B. Pernet).

posterior to the metatroch. Most recently Arenas-Mena & Li (2014) examined expression of several transcription factors and developed methods for experimental exploration of the gene regulatory networks involved in trochophore development of *H. elegans*.

Planktotrophic larvae in *Hydroides* can last from 2 days to weeks, depending on temperature, salinity, and food type and availability (Qiu & Qian 1997; Toonen & Pawlik 2001b; Faimali *et al.* 2002). *Hydroides dianthus* and *H. elegans* develop to metamorphosis in 5 days at 24–35°C (Scheltema *et al.* 1981; Carpizo-Ituarte & Hadfield 1998). Although low temperature (15°C) led to longer duration of development of *H. elegans*, it did not affect survival from oocyte to trochophore stages and temperature was not a limiting factor for early development (Qiu & Qian 1997). However, for *H. ezoensis*, Cao *et al.* (2009) showed that high temperatures (30°C) lead to a significant increase in malformation and mortality in larvae, while 24°C was the optimal temperature for larval development in this species.

Lowered salinity lengthened the duration of development of *H. elegans* and reduced its survival (Qiu & Qian 1997). Metamorphic rates of *H. ezoensis* larvae were significantly lower at high (40 ppt) and at low (20 ppt) salinities (Cao *et al.* 2013). Both food concentration and diet composition affect larval planktotrophic development. Low food concentration lengthened the duration of development and reduced survival and settlement of *H. elegans* (Qiu & Qian 1997).

External metabolites released by some marine animals can make the environment either suitable or unsuitable for other organisms. Conditioning of water by adult *H. elegans* promotes normal development of larvae. Conditioning by *Mytilus* delays development and that by *Balanus* accelerates the development resulting in abnormalities (Srinivasagam 1966). The early development of *H. elegans* was affected if the eggs and sperms are treated with the extract of hemichordate *Ptychodera flava* Eschscholtz, 1825 (Whitin & Azariah 1982).

Metamorphic competence and metamorphosis

The three-chaetiger stage marks the onset of metamorphic competence in larvae that coincides with the transition to the demersal stage and with changes in larval behaviour. The easy repeatability of larval development up to a three-chaetiger (competent) stage indicates that the development to this stage is predetermined genetically. However, after the onset of competency, larvae require cues to proceed with further development. If such cues are unavailable, the competence period can be protracted and metamorphosis delayed. For example, competent larvae of *H. elegans* remain swimming for several weeks if they do not receive an appropriate cue (Bryan *et al.* 1997; Qian & Pechenik 1998). Exposure to metamorphic cues curtails the competency period, resulting in a shorter overall developmental period: larvae of this species can complete development in 5 days.

The metamorphosis – a set of morphogenetic events accompanying transition from a pelagic to an obligate sessile lifestyle – begins with the disappearance of the prototroch and is characterised by differentiation of the radiolar crown in the head region; the collar and thoracic membranes in the thoracic region; and the pygidium at the tip of the abdomen. The stages of metamorphosis have been described for *H. elegans* (Wisely 1958; Sentz-Braconnot 1964; both as *H. norvegica*). The onset of metamorphosis is preceded by increased contraction of the circular muscles below the eyes to constrict a definite head region. The ventral collar rudiment evaginates, forming a ventral fold, and at first two separate dorsoventral folds develop behind the metatroch. The collar folds fuse to form the adult collar. The chaetal sacs enlarge and the anterior pair begins to turn upwards. Uncini appear on the second chaetiger and later on the third. During settlement the head is reduced and the mouth and anus move to occupy their terminal positions. The prototroch, metatroch and feeding band disappear. The collar folds evaginate completely and the anterior pair of chaetae becomes collar chaetae. Stiff ciliary tufts are formed on the radiolar rudiments and all subsequently develop pinnules. Successful attachment and tube construction mark the completion of normal metamorphosis (Fig. 10J).

The onset of metamorphosis coincides with the shift into demersal life, but it is not completed until after settlement. Some initial events of metamorphosis, such as collapse of prototroch and development of radiolar buds, pygidium and thoracic membrane rudiments, can be observed even in the absence of settlement. However, tube formation depends on successful settlement, which itself depends on a number of cues. Lack of such cues may result in failure to complete normal metamorphosis. Bryan *et al.* (1997) observed several types of abnormal metamorphosis in *H. elegans*: (1) attached, production of radioles but no tube, (2) not attached, production of radioles but no tube, and (3) deformed development involving elongation of larvae and crawling behaviour but no tube or radiole production. Some metamorphosed *H. elegans* larvae (i.e. with radiole buds) were unable to secrete tubes but survived up to 2 months when fed single-celled algae (Carpizo-Ituarte & Hadfield 1998).

The response of larvae *H. elegans* to inhibitors of mRNA and protein synthesis was examined to determine if metamorphosis requires these molecular processes. When larvae were induced to metamorphose in the presence of a gene-transcription inhibitor or a translation inhibitor, *H. elegans* progressed through metamorphosis to the point of radiole formation. Thus, the induction of metamorphosis in *H. elegans* does not require *de novo* transcription or translation, and the juvenile worm develops in two phases. First, larvae respond to the inducer by attaching to the substratum, secreting a primary tube, resorbing the prototroch cilia, undergoing caudal elongation, differentiating the collar, and beginning secretion of the calcified tube. During the second phase, the juvenile develops radioles and begins to grow, requiring new mRNA and protein synthesis (Carpizo-Ituarte & Hadfield 2003).

Survival of *H. elegans* trochophores to metamorphosis increased with increasing algal concentrations from 0 to 10^6 cells mL^{-1}. No settlement occurred at concentrations $<10^3$ cells mL^{-1}. Larvae that were allowed to resume feeding after 10-day starvation did not settle. Such a period of food limitation exceeds the point at which *H. elegans* retain the ability to develop to competence (Qiu & Qian 1997). The timing of starvation (early or late in development) and the total number of days of feeding of *H. elegans* larvae had significant effects on the percentage of larvae that successfully completed metamorphosis. The larvae of this extremely successful fouling species are remarkably tolerant of starvation and seem to require approximately 5 days of continuous feeding to achieve high levels of metamorphic success (McEdward & Qian 2001).

Low salinity increased mortality during both the larval and early juvenile periods of *H. elegans* (see Qiu & Qian 1997, 1998). Delaying metamorphosis had significant adverse effects on juvenile survival of *H. elegans* (see Qiu & Qian 1998).

SETTLEMENT

Settlement behaviour has been described for larvae of *H. dianthus* and *H. elegans* (by Zeleny 1905a, respectively Wisely 1958). As metamorphosis starts, the behaviour of larvae changes from swimming to a slow exploration of the bottom of the dish. Hadfield *et al.* (1994) observed that pre-settlement larvae of *H. elegans* spent much time swimming across submerged surfaces, repeatedly contacting these surfaces with their apical tufts. After the demersal stage, the larvae settle by secreting an abdominal posterior mucous bag. Juveniles originally secrete a primary mucous tube, covering it later with calcareous matter secreted by the collar glands.

Settlement and recruitment are complex processes, determined by the interaction of biotic and abiotic factors that can serve as positive or negative cues, physical or chemical in nature. All factors can be divided into generic precondition factors and specific settlement cues. The former affect a range of larvae at various stages of development and an optimal combination of such factors ensures that larvae are in condition to respond to specific cues. The settlement cues are species-specific; they act during the settlement and affect only the larvae preconditioned (competent) to respond to them (reviewed in Qian 1999; Kupriyanova *et al.* 2001).

Generic precondition factors

Settlement is seasonal in temperate climates and the length of the settlement period coincides with the length of the reproductive period. In tropical species, reproduction and settlement take place throughout the entire year. In many places (Australia, New Zealand, California, Japan and China), peaks of *H. elegans* settlement occur in summer and autumn, but in Hong Kong settlement of this species peaks in early spring to early summer (Qiu & Qian 1997). The breeding season of *H. ezoensis* occurred from late May to September in warm-temperate Japan (Miura & Kajihara 1984).

During the settlement season, abiotic factors that affect larval development, behaviour and survival also affect the intensity of settlement. Settlement of *H. elegans* occurred at water temperature above 20°C and more intensive settlement occurred at stations having a greater concentration of dissolved oxygen (Reish 1961). Dew & Wood (1954) present evidence of the settlement periodicity of *H. elegans*, which may be related to tides and Daniel (1958) showed that larvae of this species could not settle in water currents exceeding 0.9–1.2 knots. *Hydroides dianthus* settled only when the high tide salinity was above 30 ppt and the low tide salinity was 20 ppt or over (Hill 1967).

Physical properties associated with the abiotic substratum, such as its colour and its roughness, also affect settlement preferences (e.g. Wisely 1959; Sentz-Braconnot 1968; Straughan 1969). Decreased pH neither affects embryo and larval development nor alters metamorphosis, but adversely affects larval calcification at the time of metamorphosis and early juvenile growth in *H. elegans* (Lane *et al.* 2013). Hypoxia induces abnormal larval development, probably due to the reduction in clearance rate, and affects biofilm–larva interaction in *H. elegans*. Only about 5% of the larvae settled on the biofilms developed under hypoxia, regardless of dissolved oxygen levels of the seawater. The delayed larval development and potential alteration of biofilms due to hypoxia may explain why the recruitment of *H. elegans* declines during summer (Shin *et al.* 2013).

Specific settlement factors

Although some *Hydroides* species are notorious opportunistic foulers that colonise any available substrate, others may have very specific habitat requirements. For example, *H. spongicola* preferentially if not exclusively inhabits the sponge *Neofibularia nolitangere* (Duchassaing & Michelotti 1864; see also Bastida-Zavala & ten Hove 2003a: 122). Such substratum-specific settlement usually results from active substratum selection by settling larvae and the settlement cues are chemical in nature. Pawlik (1992) distinguishes three categories of substratum-specific settlement: settlement on microbial films, associative settlement and gregarious settlement.

Settlement on biofilms

The presence of a bioorganic film has been long recognised as a prerequisite for larval settlement in fouling marine invertebrates (e.g. Scheltema 1974). It is well documented that larval settlement of *H. elegans* is mediated by marine biofilms: complex agglomerates of bacteria, diatoms, fungi and protozoa enmeshed in a matrix of extracellular polymers (e.g. Hadfield 2011). Which particular components of the biofilm affect settlement and the underlying mechanism have been addressed in the studies of the last two decades. The induction of *H. elegans* larval settlement by marine biofilms has been mainly attributed to bacteria and diatoms.

The bacterial component of marine biofilms plays an important role in the induction of larval settlement in *H. elegans* and bacterial metabolites comprise the chemical signal for larval settlement (Harder *et al.* 2002). Age and composition of bacterial films are important for larval settlement (Hadfield *et al.* 1994; Harder & Qian 1999).

Clean surfaces exposed to seawater go through a succession of changes advancing to a complex microbial community (Mitchell & Kirchman 1984) and thus it is not surprising that recruitment is affected by the immersion time and film age (Keough & Raimondi 1995). Larvae of *H. elegans* distinguish among the films ranging in age from 1 to 3 days. Larval settlement curves closely parallel population growth curves for bacteria, suggesting that settlement is quantitatively dictated by bacterial density (Hadfield *et al.* 1994; Huang & Hadfield 2003). Most recently, bacterial density in biofilms was shown to be positively correlated with biofilm age (Lema *et al.* 2019). Moreover, Shikuma & Hadfield (2006) showed that increasing densities of a stable community of biofilm bacteria, rather than a shift in dominant species composition, are responsible for the primary colonisation of submerged surfaces by this species in the succession of macrofouling invertebrates.

However, initial surface characteristics appear to be less important for biofilm community composition. Regardless of initial surface properties, after 10 days all surfaces were equally attractive to *H. elegans* larvae (Huggett *et al.* 2009). In the study by Chung *et al.* (2010) both age and substratum altered the bacterial community structures of the biofilms, but age had a greater effect in shaping the bacterial community structure than did the substratum.

Composition of bacterial film communities affects larval settlement in a number of ways. Different film-forming species and strains vary in their ability to promote larval settlement in *H. elegans* (Lau & Qian 1997). Of a series of bacterial strains isolated from marine biofilms, 13 induced settlement, 11 gave mixed results and 10 others did not stimulate the settlement of *H. elegans*. Moreover, the amount of settlement induced by monospecific strains is less than that induced with natural multispecies films (Unabia & Hadfield 1999).

Bacteria vary significantly in their ability to induce metamorphosis, but larval settlement of *H. elegans* could be induced by bacteria in a wide range of taxa, and phylogenetic relationships among marine-biofilm bacterial species are not predictive of their inductive capacity. Two bacterial species separated by a genetic distance of 30% shared the greatest capacity for inducing metamorphosis in *H. elegans*, while two strains with only a 3% divergence were very different: one strain induced metamorphosis strongly while another one did not (Huang & Hadfield 2003). The most inductive bacterial species was effective in much lower biofilm densities than weakly inductive bacteria. Similarly, 38 bacterial isolates from a marine biofilm investigated for their effects on larval settlement of *H. elegans* by Lau *et al.* (2002) belonged to three phylogenetic branches: *Proteobacteria* (26 isolates), Gram-positive (eight isolates) and *Cytophaga–Flexibacter–Bacteroides* (four isolates). Most of the isolates were affiliated to the genera *Vibrio* (seven isolates), *Alteromonas* (eight isolates) or *Pseudoalteromonas* (eight isolates). About 42% of the isolates were non-inductive and the rest contained highly and moderately inductive strains. Isolates affiliating to the same genus might have tremendously different activities for the induction of larval settlement.

Analysis of bacterial diversity of the natural biofilm community by Vijayan & Hadfield 2020 revealed dominance of *Bacteroidetes*, *Verrucomicrobiae*, *Alpha*- and *Gammaproteobacteria*, and members of *Sphingomonadaceae* and *Rhodobacteraceae*. These natural biofilms induced settlement in nearly 100% of larvae of *H. elegans*. But single-species biofilms of only six of the 40 bacterial isolates, all in the genus *Pseudoalteromonas*, induced settlement. However, *Pseudoalteromonas* was present at very low relative abundance (< 0.01%) in the natural biofilm. These findings reinforce the conclusion that larvae of *H. elegans* respond to various cues from different bacteria.

Huang *et al.* (2007b) reported that the conditions of bacterial biofilm formation may affect its induction of larval settlement. Cells of the marine bacterium *Pseudoalteromonas spongiae* grown in the medium containing both yeast extract and peptone (YP-grown) were highly inductive to larval settlement of *H. elegans*, whereas *P. spongiae* cells grown in the medium containing only peptone (P-grown) were not. Only the induction-capable YP-grown *P. spongiae* formed a well-developed biofilm, while the P-grown did not.

Compositions of the bacterial biofilms may vary greatly between near-shore and offshore oceanic habitats (Lema *et al.* 2019). Biofilms were highly diverse across inshore sites, while relatively low diversity was found at offshore sites. The localised distribution of *H. elegans* is not determined by 'selection' to locations by specific bacteria, but is rather linked to the local ecology and oceanographic features that affect the development of dense biofilms and the occurrence of larvae.

Chung *et al.* (2010) demonstrated that the larval settlement response cannot be predicted by the overall chemical composition of the biofilm alone as both successional changes in the bacterial community structure and the chemical profiles of subtidal biofilms affect larval settlement response of *H. elegans*. While the type of substratum more strongly affected the chemical profile, the extracts of biofilms induced larval settlement in a biofilm-age-dependent manner: extracts originating from different substrata of the same age showed no differences in larval settlement (Chung *et al.* 2010).

Analysis of associated microbiomes at multiple life-history stages (adults, their tubes, spawned eggs, 6-day-old nectochaete larvae) of *H. elegans* by Vijayan *et al.* (2019) showed that the eggs were almost devoid of bacteria, and bacteria on larvae were confined to the feeding apparatus. The absence of a consistent microbiome between stages and lack of bacteria on eggs demonstrates that *H. elegans* lacks a vertically transmitted microbiome, and bacteria found at each stage came from a varying environment.

Biofilms killed by heat, ultraviolet radiation or chemical fixatives were no longer inductive, suggesting that the bacterial metabolic activity in biofilms is essential for the biofilms to be inductive for larval settlement (e.g. Hung *et al.* 2005). Soluble, dialysable, heat-stable bacterial products induced settlement and metamorphosis more slowly (Unabia & Hadfield 1999).

Little is known about the molecular changes that occur in the animal upon stimulation by bacteria. Recent studies examined endogenous molecular mechanisms that regulate metamorphosis in *Hydroides* (e.g. Huang *et al.* 2007a; Xu *et al.* 2009; Wang & Qian 2010; Shikuma *et al.* 2016).

In contrast to bacteria-derived settlement cues, the nature of diatom-derived settlement cues is less understood. According to Lam *et al.* (2003) the putative diatom-derived settlement cue was water-insoluble and associated with the film surface. The induction of larval settlement by diatoms is not related to their viability, as the inductive effect prevailed even after heat treatments killing the diatoms. Contrary to marine bacteria, diatom-induced larval settlement cues are composed of heat-stable surface components, such as capsular extracellular polysaccharides. Diatom-derived larval settlement cues were identified as extracellular polymers mainly consisted of polysaccharides with a small proportion of proteinaceous components (large proteins or glycoproteins), the latter not important as settlement cues (Lam *et al.* 2005).

Associative settlement

Associative settlement (**sensu** Crisp 1974) is the specific or enhanced settlement of one species on another. Algae serve as negative or positive settlement cues for *H. elegans* larvae when waters were conditioned by 12 species of macroalgae (Walters *et al.* 1996). Compounds released by four of 12 tested macroalgae killed larvae of *H. elegans* or inhibited their settlement; the remaining eight algal species prevented settlement, had no effect or even stimulated settlement (Walters *et al.* 1996). Larval settlement of *H. elegans* was also inhibited by brown algal phlorotannins (Lau & Qian 1997). Similarly, larval settlement of *H. elegans* on otherwise suitable substrata decreased in the presence of the harpacticoid copepod *Tisbe japonica* Ho, 1982 (Dahms *et al.* 2004).

Sponge compounds have the antifouling effect on fouling communities; for example, *Mycale adhaerens* (Lambe, 1893) and *Callyspongia (Euplacella) pulvinata* (Lindgren, 1897) inhibited larval settlement of *H. elegans* by excreting toxic waterborne compounds (respectively Lee & Qian 2003 and Dobretsov *et al.* 2004). Densities of macrofoulers, including of *Hydroides elegans*, were suppressed on the surfaces of the sponges *Haliclona cymaeformis* (Esper, 1806), *Haliclona* sp. and *Callyspongia* sp., while densities of bacteria on the surfaces of *H. cymaeformis* and *Callyspongia* sp. were similar to those on the reference surfaces and were more than double the densities of bacteria on the surfaces of *Haliclona* sp. compared with a reference surface (Dobretsov *et al.* 2005).

The only example of settlement stimulation of serpulid larvae by an invertebrate host was demonstrated for *H. elegans*. This species in Hong Kong is also found living on the bryozoan *Bugula neritina* (Linnaeus, 1758) and the samples derived from this bryozoan induce metamorphosis (Bryan *et al.* 1998). Apparently *H. elegans* may gain a refuge from predation by associating with *B. neritina* colonies both from its structural and chemical attributes.

Other organisms can affect larval settlement of *H. elegans* indirectly by modifying biofilm composition. Dahms & Qian (2005) reported differences in larval settlement on biofilms treated by either the predatory copepod *Tisbe japonica* or the detritivorous copepod *Nitokra spinipes* Boeck, 1865. The density of settled larvae decreased on biofilms modified by increasing numbers of copepods, which reduced bacterial densities and caused changes in the bacterial community. Bacteria associated with the red sponge *Mycale adhaerens* had either an inhibitory or a neutral effect on larval settlement of *H. elegans* in contrast to natural biofilms, which harbour 65% of bacteria with inductive effect (Lee & Qian 2003). Bacterial strains isolated from seaweeds and invertebrates negatively affected settlement of *H. ezoensis* larvae. Of the 192 bacterial strains isolated from seaweeds 62 isolates were inhibitory against larval settlement, while 37% of the 127 bacterial strains isolated from invertebrates were inhibitory against larval settlement. The high proportions of host-associated bacteria producing antifouling compounds suggest that these bacteria may help the host organism in the defence against fouling (Ma *et al.* 2009).

Gregarious settlement

Larvae of *H. dianthus*, *H. elegans*, *H. ezoensis* and *H. dirampha* settle in response to the members of their own species to form conspecific aggregations (Scheltema *et al.* 1981; Okamoto *et al.* 1998; Hadfield & Paul 2001; Toonen & Pawlik 1994, 1996, 2001a, b). Such gregarious settlement relies on chemical cues associated with adults and was not demonstrated between members of a single cohort of settling larvae of *H. dianthus* (Mullineaux & Garland 1993). The settlement cue was soluble in water and was associated with the body of live adults in *H. dianthus* and *H. elegans* (Toonen & Pawlik 1996; Bryan *et al.* 1997). Live worms removed from their tubes and amputated radiolar crowns of live worms resulted in greater settlement response than dead worms, empty tubes or biofilmed slides. Newly settled juveniles began to elicit gregarious settlement after approximately 96 h (Toonen & Pawlik 1994, 1996).

Toonen & Pawlik (1994) found that females of *H. dianthus* produced larvae that settled in two different ways: one type colonised uninhabited substrata (founders), whereas the other settled only in response to cues associated with conspecifics (aggregators). There was a distinct subordination of larvae that responded to a biofilmed substratum; once these individuals were removed, the remaining aggregators delayed settlement in the absence of conspecific-associated clues.

Previous studies have suggested that the colonisation of new substrata occurs because larvae become 'desperate' to settle after searching unsuccessfully for conspecifics. This hypothesis is based on an energetic model in which non-feeding larvae search for specific substrata as long as their energetic reserves allow, but begin to accept sub-optimal habitat rather than exhaust their reserves and die without metamorphosing. Toonen & Pawlik (2001b) have shown that competent feeding larvae of the gregarious tube worm *H. dianthus* do not exhibit decreased substratum-specificity throughout a prolonged planktonic period. Neither altered feeding regimes nor larval starvation leads to decreased substratum-specificity and, furthermore, starving competent larvae results in a loss of competency rather than larval desperation. Thus, larval desperation is unlikely to be a general explanation for the initiation of monospecific aggregations of fouling marine invertebrates.

Some authors argue that aggregations may also arise as a result of passive deposition of larvae. Walters *et al.* (1997) demonstrated that larvae of *H. elegans* did not settle faster or in greater numbers on surfaces already occupied by adults or their tubes. However, in the field in moving water there was significantly more settlement in tube crevices than expected by chance; this probably results from hydrodynamics. The authors concluded that dense aggregations of *H. elegans* found on hard surfaces in bays and estuaries probably result from passive deposition of larvae in crevices beside tubes of conspecific individuals, followed by selective attachment in these locations if the bioorganic film is acceptable (Walters *et al.* 1997).

Chemical nature of settlement inducers

The chemical nature of the cues from multi-species natural microbial biofilms that induce larval attachment of *H. elegans* has not been fully characterised. The cue that triggers settlement is not unique to a single bacterial taxon and is likely to be an insoluble, surface-bound material (Huang & Hadfield 2003). Contact with biofilm is necessary for larval settlement in *H. elegans* (Hadfield *et al.* 2014) as larvae did not respond to water-borne metabolites when prevented from contacting the bacterial film surface, showing that soluble cues emanating from a biofilm do not affect settlement behaviour. Lau *et al.* (2003) concluded that bacterial extracellular polymers serve as indicators of substratum suitability on larval settlement of *H. elegans*. Hung *et al.* (2009) isolated compounds from natural biofilms and characterised the first chemical cues, a hydrocarbon (6,9-heptadecadiene) and a fatty acid (12-octadecenoic acid), inductive to larval attachment.

Larval metamorphic inducers isolated from adults of *H. elegans* were composed of free amino acids such as aspartic acid, glutamic acid, serine, histidine, glycine, arginine, alanine, asparagine, glutamine and taurine. The metamorphic response to an artificially prepared sample in identical concentrations of these amino acids was very similar to the response to the natural isolates (Harder & Qian 1999). A new monoacyl glycerol isolated from the methyl alcohol extract of adults acts as a metamorphosis-inducing substance in *H. ezoensis* larvae. However, since acyl glycerols are common primary metabolites, they have been hypothesised to act as the second messengers, not as a primary cue of metamorphosis (Watanabe *et al.* 1998).

Beckmann *et al.* (1999) tested whether the metamorphic response was caused by direct dissolved free amino acids perception or by induction of amino acid-utilising bacteria. The results suggested that the larval response had been exclusively triggered by an inductive bacterial film rather than by direct larval perception of amino acids. If putative signalling compounds are a nutrition source for settlement-inducing bacteria, an investigation of the efficacy of chemical metamorphic cues is unreliable (Beckmann *et al.* 1999). Similarly, Jin & Qian (2005) demonstrated that amino acid exposure modulates the bioactivity of biofilms for larval settlement of *H. elegans* by altering the bacterial community. Aspartic and glutamic acids increased bacterial abundance, modified the bacterial community structures, and elevated the inductive effect of the biofilms. Alanine and asparagine increased, while isoleucine decreased the bioactivities of the biofilms by changing the bacterial species composition. Leucine, threonine and valine did not have any effect.

The metamorphic response in *H. elegans* larvae can be induced by caesium ions (Cs^+) or potassium ions (K^+), but such a response is much slower than the response to biofilms (Carpizo-Ituarte & Hadfield 1998). Larval settlement was blocked if the biofilms were first exposed to certain lectines, indicating that larvae settled in response to particular surface polysaccharides or glycoproteins in the bacterial films.

The cyclic nucleotide phosphodiesterase inhibitor isobutyl methylxanthine (IBMX) and the non-specific phosphodiesterase inhibitors theophylline and papaverine induced normal metamorphosis in *H. elegans* (Bryan *et al.* 1997; Holm *et al.* 1998; Pechenik & Qian 1998; Qian & Pechenik 1998). Apparently IBMX triggers larval settlement by elevating the cellular adenosine 3′,5′-cyclic monophosphate (cAMP) level in this species. The results of Li *et al.* (2008) indicated that the IBMX-induced cAMP changes and the cAMP-dependent protein kinase gene mediate larval development and settlement of *H. elegans*.

Gamma-aminobutyric acid (GABA), choline chloride, dihydroxyphenyl L-alanine (L-DOPA) and potassium chloride evoked a low percentage of settlement but abnormal metamorphosis (Bryan *et al.* 1997). L-DOPA, D-DOPA and neuroactive molecules such as epinephrine and norepinephrine induced larval metamorphosis in *H. ezoensis* (Okamoto *et al.* 1995, 1998).

Polyaromatic hydrocarbons (PAHs), the main components of crude oil pollution, have an adverse effect on the settlement of *H. elegans* larvae (Paul *et al.* 1998). Larval settlement may be inhibited directly, or through compounds that regulate the growth of bacteria. Settlement of *H. elegans* is inhibited by brown algal phlorotannins and two related compounds: tannic acid and phloroglucinol. Phlorotannins, tannic acid and phloroglucinol were inhibitory to *H. elegans* larval settlement by suppressing the growth of certain marine bacteria that induced the settlement. However, some of the bacteria that induced low levels of larval settlement were resistant to these compounds (Lau & Qian 1997). A branched-chain fatty acid (12-methyl tetradecanoic acid known as 12-MTA) produced by a marine bacterium from deep-sea sediments inhibited the larval settlement of *H. elegans* (Xu *et al.* 2009). The adenylate cyclase activator forskolin inhibited responses of larvae of *H. elegans* to inductive bacterial biofilms (Holm *et al.* 1998). In conclusion, the larval settlement can be inhibited by compounds that regulate the growth of bacteria.

JUVENILE GROWTH AND MATURATION

Annelids are thought to generate new body segments sequentially from a posterior growth zone. Seaver *et al.* (2005) demonstrated the distinct nature of larval versus juvenile growth zones in *H. elegans*. Larval segments arise from dividing cells located in lateral regions of the body, rather than from a localised posterior growth zone. Post-metamorphic segments, in contrast, originate from a classical posterior growth zone, with the exception of four posterior thoracic segments of *H. elegans*, which appear to arise from an area in the middle of the body.

The rate of post-settlement juvenile growth is well documented for fouling species (e.g. *H. dianthus*: Grave 1933; *H. elegans* (often as *H. norvegica*): Paul 1937, 1942; Grave 1933; Dew 1958, Behrens 1968; Sentz-Braconnot 1968; and *H. ezoensis*: Miura & Kajihara 1981). Tubes of juveniles grow rapidly but the growth slows down in later life. For *H. elegans* Paul (1937) reported a growth rate of 14 mm in 9 days. During the first 3 months *H. dianthus* increases in length by 54 mm but in the next 9 months only 12 mm are added (Grave 1933).

Under natural conditions the growth is much slower in winter than in summer and the growth rate varies according to temperature and/or population density (*H. dianthus*: Leone 1970; *H. elegans*: Qiu & Qian 1998), the pollution (*H. elegans*: Moran & Grant 1984), and availability of food (*H. dianthus*: Leone 1970).

Juvenile growth may be determined not only by the factors acting during the post-settlement period but also by those experienced during the larval stage. In particular, in *H. elegans* juvenile growth rates are affected as a result of food limitation during development or delayed metamorphosis (Qian & Pechenik 1998).

Because serpulids reach sexual maturity after they achieve a certain body size, maturation correlates with growth rate and depends on the factors controlling it. Animals living under conditions sub-optimal for growth reach maturity slowly and some never reach it. For instance, at salinity ≥ 25 ppt and temperature ≥ 20°C, the first spawning of *H. elegans* occurred on day 16 after settlement. Both low temperature and low salinity led to slower growth and subsequently to a longer time to maturation (Qiu & Qian 1998). Spawning of *H. elegans* reared in the laboratory was observed on average 40 days after fertilisation at 23°C (Matsuo & Ko 1981). However, according to Paul (1937, 1942) *H. elegans* reaches maturity within 9 days after settlement in Madras, India, at 25.5–29.5°C, while Vuillemin (1965) mentions maturation within 4 weeks (Lake of Tunis, Tunis). *Hydroides dianthus* matured within 7–8 weeks at Woods Hole, USA (Grave 1933), and within 6 weeks in Lagos, Nigeria (Hill 1967).

The smallest mature worm of *H. dianthus* was 13 mm long (Hill 1967). Tubes of reproductive individuals of *H. elegans* were usually longer than 1.2 cm (Qiu & Qian 1998). The life span of the aquarium-reared specimens of *H. elegans* was 1 year (Grave 1933; Smith & Haderlie 1969).

Mortality rates vary for different life history stages and two peaks of increased associated mortality – the early embryonic and early juvenile stages – have been reported for serpulids with planktonic larvae. Qiu & Qian (1997), who studied salinity tolerances among various developmental stages of *H. elegans*, demonstrated that the trochophore stage was slightly more tolerant to salinity stress than the embryonic and early juvenile stages. Juveniles of *H. elegans* were more vulnerable to low salinity (20 ppt) within 1 day of settlement than when older. The early developmental stages are more sensitive to environmental stress than late juveniles and adults and the juveniles are most vulnerable at the onset of benthic life (Qiu & Qian 1998).

5 Ontogeny: reversible asymmetry, compensatory regeneration and duplicity

REVERSIBLE ASYMMETRY AND COMPENSATORY REGENERATION

The term 'reversible asymmetry' comes from experimental studies and stands for the alternating development of functional opercula and regeneration of pseudopercula, leading to an alternation of position (left or right) between them. The process has been studied extensively, often by amputating opercula and studying their regeneration – hence the term 'compensatory regeneration' or 'compensatory regulation' (e.g. Zeleny 1902, 1905a, b, 1911, 1927; Okada 1933; Ludwig 1941; Abeloos 1949, 1952a, b, c, d, 1954a, b; La Greca 1950; Ludwig & Ludwig 1954; Ludwig 1957a, b; Cresp 1964; Green 1972; Puccia & Durante 1973; Schochet 1973a, b).

DUPLICITY

The repeated reversal of symmetry as described above leads to a stage with two-tier opercula (Fig. 11), one mature functional operculum not yet shed and the other, a younger developing operculum, nearly always smaller and on a shorter peduncle. Occurrence of bioperculate animals has been repeatedly found confusing by taxonomists. For example, Schmarda (1861) considered the bioperculate condition as a character of specific level (he named *Eupomatus dipoma* after its two opercula), while de Quatrefages (1866) considered it to be a character of generic character as he erected a genus *Codonytes* to include species with two opercula. However, already Claparède (1870: 164) grasped the ontogenetic process involved in duplicity. Ludwig (1941) and Schochet (1973a, b) described the process leading to bioperculate specimens in detail. Nevertheless, Mohammad (1981: 130) reported presence of two opercula in *H. homoceros* as malformation.

Apparently the bioperculate stage is transient and fairly brief (compare Ludwig 1941). Such specimens are likely to be a stage in the regular replacement of an 'old' operculum, often overgrown by algae or damaged, by a new one on the other side of the radiolar crown. *Hydroides elegans* appears to have particular propensity for developing bioperculate specimens (Rioja 1919; Sentz 1962; Alcázar & San Martín 2016).

Schochet (1973a) reported 3–12.65% (up to 13.6% in juveniles) bioperculate specimens in *H. dianthus*. Bioperculate specimens were documented at rates of 4.7% for *H. dianthus* and 4.1% for *H. elegans* by Sandonnini *et al.* (2021). However, Ichikawa & Takagaki (1942) reported from 0.9 to 29.9% of bioperculate specimens in the population of *H. ezoensis*. Kubal *et al.* (2012) reported two bioperculate specimens out of six collected specimens of *H. inornata* (as *H. operculata*).

MALFORMATIONS

Malformations, either teratological or ontogenetic, are not uncommon. For instance, Hartman (1952, figs 1, 2)

Fig. 11. Opercular duplicity in *Hydroides*. A – *H. elegans* with two developed opercula (photo C. Arenas-Mena); B – *H. ezoensis* with two equally developed opercula; C – *H. ezoensis* with partially developed pseudoperculum (photo L. Harris).

show an operculum of *H. elegans* (as *H. norvegica*) with two superimposed funnels carrying a single verticil, for the same species Ludwig & Ludwig (1954, fig. 4) show two opercula on a single peduncle. An operculum of *Hydroides sanctaecrucis* from the Pacific coast of Panama with a double verticil on a single funnel was mentioned in Bastida-Zavala & ten Hove (2003b). Most recently, two specimens of *H. elegans* were reported as trioperculate, bearing one *Serpula*–type operculum, a typical *H. elegans* operculum, and a pseudoperculum (Sandonnini *et al.* 2021).

BIOPERCULATE CHIMAERA

Ten Hove & Ben-Eliahu (2005) coined the term 'bioperculate chimaera' in the genus *Hydroides* to indicate the ontogenetic state of having two different types of two-tier opercula: a small simple operculum with undifferentiated spines (so called *H.* '*prisca*'-type, see *H. prisca* Pillai, 1971 rejected name) located on the opposite radiolar lobe from an operculum typical for *Hydroides* cf. *albiceps*, *H. brachyacantha* and *H.* cf. *brachyacantha*, *H. dirampha*, *H. homoceros*, *H. minax*, and *H. trivesiculosa*.

6 Economic and ecological impacts

The genus *Hydroides* is one of the most economically and ecologically important groups of marine invertebrates as it includes notorious members of fouling communities and common bio-invaders (Thorp *et al.* 1987; Zibrowius 1992, 2002; Lewis *et al.* 2006; Link *et al.* 2009; Otani & Yamanishi 2010; Tovar-Hernández *et al.* 2009).

NUISANCE FOULING SPECIES

A number of *Hydroides* species are capable of colonising a wide range of natural and artificial substrates and of forming dense aggregations of calcareous tubes on underwater structures, making them important fouling nuisance species (Fig. 12). Serpulid foulers constitute a significant financial burden to shipping industries, navigation, power plants and marine aquaculture due to costs associated with the removal of tubes from artificial structures. Millions of dollars are spent annually to prevent the fouling of marine organisms, especially larval settlement of *Hydroides*, on artifical structures (Dürr & Watson 2010).

Fouling interferes with navigation and shipping industries by decreasing ship speed, while increasing the weight and drag of buoys, and fouled docks require regular cleaning maintenance in harbours around the world (Ben-Eliahu & ten Hove 1992; Wang & Huang 1993; Ulman *et al.* 2019). Tube aggregates attach to underwater seawater intake pipes of power plants reducing water flow and causing blockages (Qiu & Qian 1997; Zibrowius 2002). *Hydroides elegans* is the major fouler of submarine seawater piping systems on boats based in Sydney, and is problematic for the new submarines built in Adelaide (Lewis, pers. comm.).

In marine aquaculture the key impact is the direct fouling of stock causing physical damage, biological competition and environmental modification, while infrastructures are also damaged. The conservative estimates of economic loss to the aquaculture industry are 5–10% of production costs attributed to biofouling (Fitridge *et al.* 2012). Economic impact of *Hydroides* foulers manifests itself as revenue lost from fouled commercial molluscs, such as abalones and pearl oysters. For example, outbreaks of introduced *H. elegans* caused serious damage to cultured oyster crops in Japan (Arakawa 1971). *Hydroides* tubes damage aquaculture infrastructure by clogging nets and cages, forming high biomass on submerged surfaces. In Hong Kong, during the spring and summer months, fish farmers need to clean their nets and cages that are heavily fouled by *H. elegans* every other week (Qian & Xu 2012).

Foulers can modify ecosystem dynamics and species assemblages through competition for space and food. They can contribute to jellyfish blooms as demonstrated by Dong *et al.* (2018), who found that dense colonies of *Aurelia coerulea* von Lendenfeld, 1884 (Cnidaria, Scyphozoa) polyps occurred mainly on reefs formed by

Fig. 12. *Hydroides* aggregations and fouling. A – tubes of *H. norvegica* on a settlement plate in Bergen, Norway (photo M. Capa); B *H. sanctaecrucis* aggregation from Playa Vicente Superior Lagoon, Oaxaca, Mexico (photo R. Bastida-Zavala); C – close-up view of *H. ezoensis* aggregations from coasts of Yokohama, Japan (photo E. Nishi); D – *H. elegans* covering a hull of a ship, Spain (photo J. Murcia); E – A *Pinna nobilis* shell covered with *H. dianthus* tubes, Spain (photo J. Murcia); F – *H. elegans* aggregations on a fender from Eel Pond, Woods Hole, Massachusetts, USA (photo D. A. Carlton).

aggregations of *H. dianthus*, providing a modified, complex habitat suitable for the settlement of *A. coerulea* polyps.

INVASIVE SPECIES

The same features that make a number of *Hydroides* species troublesome biofoulers also make them successful invasives. Species of the genus common in biofouling communities and quickly forming large conspecific aggregations on anthropogenic surfaces are also easily translocated to new localities on ship hulls and floating marine debris (Tovar-Hernández *et al.* 2009). Out of 18 polychaetes on the list of 100 'worst invasives' of the Mediterranean Sea (Streftaris & Zenetos 2006), eight species belonged to the genus *Hydroides*. Hull fouling is the most likely mode of translocation for *Hydroides* species as they are able to tolerate the common copper-based antifouling paints used on ship hulls (Lewis & Smith 1991). Despite suggestions (e.g. Thorp *et al.* 1987; Zibrowius 1992, 2002; Lewis *et al.* 2006) that transport in ballast water is another possible vector, Ben-Eliahu & ten Hove (2011) doubt that ballast water is the source of the alien serpulid populations, given the low survival rate in the ballast tanks over long distances. Moreover, there is no direct evidence that serpulid trochophores were ever found in ballast waters.

If translocated taxa become established in a remote locality, they can significantly expand their ranges and become invasive non-indigenous species. The criteria for distinguishing introduced non-indigenous and native species include a new record for a well-studied area, a new localised occurrence spreading or showing a population explosion, disjoint species distributions, occurrence predominantly in ports and records from ship-hull biofouling (e.g. Chapman & Carlton 1991; Wolff 2005; Ben-Eliahu & ten Hove 2011). However, a wide, even supposedly 'cosmopolitan' distribution of species is often automatically taken as evidence of their invasive status. Thus, a critical assessment of non-indigenous species records is needed.

While true cosmopolitan shallow-water species ('eucosmopolitan' **sensu** Darling & Carlton 2018) are rare or non-existent (see Hutchings & Kupriyanova 2018), Darling & Carlton (2018) proposed the term 'neocosmopolitan' for species that achieve extensive geographical ranges only through anthropogenic dispersal. In contrast, for pseudocosmopolitan taxa (again **sensu** Darling & Carlton 2018) a broad distribution is a reflection of uncertain taxonomic status. Such species remain 'cosmopolitan' only as long as taxonomic uncertainty persists, and a molecular investigation may splinter them into geographically restricted species. We suggest that all reported wide distributions of species of *Hydroides* should be re-evaluated to establish the either neocosmopolitan or pseudocosmopolitan status of such widely distributed species instead of them being automatically attributed to human-mediated translocations.

Hydroides species commonly reported as invasive

In total, 10 nominal *Hydroides* species have been reported as either introduced or invasive outside of their natural ranges. Of these, seven species (*H. elegans*, *H. brachyacantha*, *H. dianthus*, *H. dirampha*, *H. ezoensis*, *H. operculata* and *H. sanctaecrucis*) are commonly reported as successful invasives established in numerous remote coastal environments outside of their natural ranges. In addition, three species (*H. homoceros*, *H. heterocera* and *H. minax*) have been reported as established Lessepsian migrants to the Levant basin of the Mediterranean (Çinar 2006; Ben-Eliahu & ten Hove 2011). Some other nominal *Hydroides* (e.g. *H. albiceps*: see Zibrowius 1979; *H. crucigera*: see Bastida-Zavala 2008; *H. dipoma*: see Zibrowius 1973b: 36 and Ben-Eliahu & ten Hove 2011; and *H. steinitzi*: see Ben-Eliahu & ten Hove 2011: 17) have been incidentally reported from ship-hull fouling, but apparently never established and spread.

Confirmed invasive species of *Hydroides*

Out of seven commonly reported worldwide invaders, only five are confirmed. *Hydroides elegans* is by far the best-known cryptogenic neocosmopolitan species. It has been reported from most subtropical locations around the world: Mediterranean–Atlantic, Indo–West Pacific including all around Australia, tropical Pacific America, East Atlantic and South Africa (ten Hove & Kupriyanova 2009). Biofouling has been shown as a major mode of dispersal for this nominal species (Pettengill *et al.* 2007), but recent data suggest that this taxon consists of at least two genetically distinct cryptic species (Grosse *et al.* 2021). The most recent data of Pazoki *et al.* (2021) provided a possible indication of heteroplasmy in *H. elegans* where two distinct mitochondrial lineages with about 11% of genetic divergence seem to coexist in the same individuals.

Hydroides dianthus, originally described from New England, USA, has been subsequently reported in eastern and southern USA, Brazil, China, Japan and the Mediterranean (Sun *et al.* 2017a). It is a confirmed invader, although the nominal taxon is a complex of at least two species translocated worldwide: a second genetic lineage has been recently reported from Texas and the Black Sea off the Crimea Peninsula (Sun *et al.* 2017a). Also, the haplotype variability analysis suggests that its native range may be the Mediterranean (Sun *et al.* 2017a).

Hydroides dirampha was originally described from and is common in the Caribbean (Bastida-Zavala & ten Hove 2003a). According to Zibrowius (1971a, 1994) the circum(sub)tropical distribution of *H. dirampha* is due to a long history of being ship-transported. Almost all non-Caribbean records are from harbours and ships' hulls (e.g. Day 1967 (as *H. lunulifera*); Zibrowius 1971a; Imajima 1978; Hayes *et al.* 2005; Bastida-Zavala & ten Hove 2003b; Sun *et al.* 2012b; 2015; Bensari *et al.* 2020).

Hydroides ezoensis has originally been described from Northern Japan and is well known throughout the Japanese islands, the Russian coast of the Sea of Japan and the east coast of China (see Zibrowius 1978; Thorp *et al.* 1987). It was first recognised associated with oysters imported from Japan to France in 1973 (Gruet *et al.* 1976; Zibrowius 1978) and then was introduced to the United Kingdom and Australia (Zibrowius & Thorp 1989; Hayes *et al.* 2005; Glasby *et al.* 2007); most recently it arrived in Dutch harbours (Faasse *et al.* 2020).

Hydroides sanctaecrucis was originally described from St Croix, Caribbean, is known on both sides of Mexico and Panama, and was recently ship-transported to Singapore, tropical Australia (Lewis *et al.* 2006; Sun *et al.* 2015), Hong Kong (Sun *et al.* 2012a) and Taiwan (Ferrario & Minchin 2017). Reports of this species in the Mediterranean (Rech *et al.* 2018) are not substantiated (Langeneck *et al.* 2020).

Questionable invasive species

The other commonly reported invasive taxa, *H. brachyacantha* and *H. operculata*, are apparently examples of pseudocosmopolitan rather than invasive taxa as there is no evidence that they have been translocated by human-mediated means. Instead, records of these taxa from remote geographical localities belong to distinct species (Sun *et al.* 2015, 2017b).

Hydroides brachyacantha was originally described from the Pacific coasts of Mexico and although it was reported from (sub)tropical localities around the world, this taxon is a complex of an unknown number of species that in Australia alone includes two species (*H. amri* Sun *et al.* 2015 and *H. nikae* Sun *et al.* 2016), both morphologically and genetically distinct from *H. brachyacantha*, which also comprises two cryptic lineages (Sun *et al.* 2016).

Hydroides operculata, originally described from the Gulf of Aden, Somalia, was subsequently reported from South and East Africa, India, Pakistan, Sri Lanka, Hong Kong and tropical Australia. as well as being a Lessepsian migrant to the eastern Mediterranean. However, Sun *et al.* (2017b) revealed three distinct clades corresponding to regionally distributed species from Arabian (Persian) Gulf, India and Hong Kong, and Australia. These results suggest a natural phylogeographic pattern and argue against the invasive status of *H. operculata*.

Three nominal Indo-Pacific species – *H. minax* described from Philippines, *H. heterocera* from the Red Sea and *H. homoceros* with types coming from as far afield as the Maldives, Sudanese Red Sea and Zanzibar – have been reported as Lessepsian migrants to the Eastern Mediterranean (Ben-Eliahu & ten Hove 1989, 1992; Zibrowius & Bitar 1981; Laubier 1966; Zibrowius 1971a; Çinar 2006). The latter two species have also been reported from the Persian (Arabian) Gulf (Wesenberg-Lund 1949; Al-Kandari *et al.* 2019). The invasive and taxonomic status of these species still needs to be confirmed with molecular data, especially for *H. minax*, which has been recently revealed as a species complex (Sun *et al.* 2018).

MODEL ORGANISMS FOR RESEARCH

Larval ecology

Hydroides elegans has almost all the required characteristics for a laboratory testing animal, such as short generation time, fast growth, high fecundity, repeat spawning ability, potential to reproduce all year round, and ease of culturing (for both larvae and adults). The adults in the laboratory release gametes when their tubes are broken; eggs are easily fertilised in lab conditions and develop into trochophores within 24 h. With sufficient food supply, trochophores can develop into competency to attach and metamorphose within 4–5 days. It takes settling larvae only 9–12 h to complete attachment and metamorphosis. The same adult can release mature gametes every 7–10 days with a life span of about 8–10 months; their metamorphosis-competent larvae develop in ~5 days at 25°C and it takes only about 1–2 weeks for a juvenile to develop into a sexually mature adult that can release several thousands of small eggs. Thus, *H. elegans* has been declared an important model organism (Hadfield 1998; Nedved & Hadfield 2008) and has been used routinely in the last two decades in hundreds of experimental embryological, larval ecology and biofouling studies. Other *Hydroides* species, such as *H. ezoensis* and *H. dianthus*, have also acted as model organisms for larval ecology research (e.g. studies of mechanisms of gregarious settlement by Toonen & Pawlik 1994, 1996, 2001a, b).

Mechanisms of biomineralisation

Hydroides species, especially *H. elegans*, also provide outstanding models for studies addressing mechanisms of biomineralisation (e.g. Chan *et al.* 2015a, b). Molecular mechanisms of biomineralisation were addressed by Batzel *et al.* (2016), who searched transcriptomes for carbonic anhydrase (CA), the metalloenzyme responsible for biomineralisation in animals. They recovered 12 CA genes, two of which were expressed within the collar corresponding to the location of calcium-secreting glands, supporting the role of CAs for biomineralisation processes.

Ocean acidification

The oceans are becoming unfavourable for calcification due to the stress associated with ocean acidification and associated declining saturation state of $CaCO_3$. *Hydroides elegans* has been used as a model in studies aimed at understanding biomineralisation organisms' susceptibility to a changing climate. As expected, CO_2-driven ocean acidification (OA) alters and weakens integrity of the tubes of *H. elegans* (Chan *et al.* 2012), affecting tube growth in both adults and reared offspring (Díaz-Castañeda *et al.* 2019). Decreased pH does not alter metamorphosis but compromises the juvenile calcification of *H. elegans* (Lane *et al.* 2013). The lowered pH value of 7.8 altered the tube ultrastructure, volume and density as well as decreased the tube hardness and elasticity, reducing the crushing force required for breaking the tube, thus leading to a lower ability to withstand predator attacks (Li *et al.* 2014). Despite this, animals can physiologically recover from the impacts of OA on tube calcification, composition, density, hardness and stiffness when returned to optimal conditions (Meng *et al.* 2019). Moreover, *H. elegans* might be robust to the environmental change because its juvenile development positively responded to temperature and the interaction between temperature and salinity. Tubeworms could be resilient to the projected near-future decreased pH or salinity as long as surface seawater temperature rise at least by 4°C (Chan *et al.* 2013). It implies that *H. elegans* is adaptable to the rising temperature and might further spread as a result of global warming, causing the biofouling problem in temperate coastal regions (Hou 2013).

Interestingly, transgenerational response to low pH depends on parental gender. Paternal and maternal low pH experiences had opposite and additive effects on *H. elegans* offspring. Growth rates of offspring of fathers or mothers raised in low pH were higher or lower respectively, but there was no difference when both parents were from low pH. Gender differences may result in different selection pressures for each gender (Lane *et al.* 2015).

Ecotoxicology

Hydroides elegans larvae have been widely used as a test organism for both acute and chronic ecotoxicity bioassays. Toxic effects of an antibiotic flucloxacillin on fertilisation and early development of *H. elegans* were demonstrated by Arumugam (2012). Results of Lau *et al.* (2007) indicated that the early development of *H. elegans* is highly sensitive to tributyltin. Gopalakrishnan *et al.* (2007) examined the toxicity of copper, aluminium, lead, nickel and zinc on fertilisation and embryogenesis of *H. elegans*, revealing that fertilisation rate drastically decreased at the highest metal concentration and embryos showed differential stage-specific responses to metals. Devakumar *et al.* (2017) revealed adverse effects of the silver nanoparticles (AgNPs) on the fertilisation and early development of *H. elegans*. Cao *et al.* (2008) examined acute toxicity of cadmium, copper, zinc and palladium ions to trochophores of *H. ezoensis*.

Due to the popularity of copper-based antifouling paints, copper contamination is a potential threat to the marine environment. Xie *et al.* (2005) showed that the embryogenesis and larval stages of *H. elegans* are more sensitive to copper stress than adults, and that algal sorption reduces soluble copper concentration and may affect the larval toxicity result. Copper toxicity is influenced by dissolved organic matter (DOM). Qiu *et al.* (2007) examined the complexation of copper by fulvic acid (FA), one of the major components of DOM. Fulvic acid was protective against copper toxicity to the larvae of *H. elegans*, and such an effect was caused by the reduction in labile copper due to Cu–FA complexation. Bao *et al.* (2010) showed that copper affected larval settlement mainly by acting on biofilm development, including changes in bacterial survival, the bacterial community composition and the chemical profiles of young biofilms in the environment.

Thilagam *et al.* (2008) described a bioassay using the gametes and embryos of *H. elegans* to test industrial and domestic effluent for potential toxicity and to monitor coastal water pollution. Since early life stages of *H. elegans* are highly sensitive to contaminants, they are suitable for the assessment of effects of low pollution levels. Vijayaragavan & Raja (2018) developed a toxicity test to examine the effects of the pesticide phorate on fertilisation and early developmental of *H. elegans* and found that the rate of successful embryonic development decreased with increased concentration of phorate in seawater. The results indicate that the early development of *H. elegans* is highly sensitive to phorate.

Leung & Cheung (2017) assessed the clearance rate larval and adult stages of *H. elegans* on coastal phytoplankton and concluded that both larvae and adults can consume and utilise different phytoplankton, except the diatom *Thalassiosira pseudonana*. Given the fast-colonising ability that allows easy manipulation of abundance, *H. elegans* is considered practically and biologically ideal for tackling algal blooms.

7 Barcoding, phylogeny and genome structure

BARCODING

Species within *Hydroides* are distinguished mainly by the morphology of their complex opercula and, to a lesser degree, by the structure of their tubes and details of collar chaetae morphology. However, the opercula show high intraspecific variability that might overlap with interspecific variability (Bastida-Zavala & ten Hove 2003a, b). Moreover, significant ontogenetic variability is accompanied by the variability observed in often regenerating opercula (see *H. 'prisca'* in ten Hove & Ben-Eliahu 2005). These factors can make morphology-based species delimitation problematic.

It is often claimed that DNA 'barcoding' (the idea that a species can be uniquely characterised by a short DNA fragment and then identified by comparing such a fragment to a reference DNA sequence; see Hebert *et al.* 2003) is a reliable, cost-effective alternative to routine taxonomic identification by experts even rendering traditional identification tools redundant (e.g. Mishra *et al.* 2016). A 650 base pair fragment of the mitochondrial cytochrome oxidase subunit I (COI) was proposed as a standard barcoding marker for animals (Hebert *et al.* 2003) because of its variability among closely related taxa and supposed ease of amplification. Indeed, using this COI fragment has been effective in distinguishing species and revealing cryptic species of numerous polychaetes (Barroso *et al.* 2010; Carr *et al.* 2011). However, development of molecular tools for species identification has been hampered by the fact that in some polychaetes, such as the families Cirratulidae, Nephtyidae, Spionidae and especially Serpulidae (Pleijel *et al.* 2009; Nygren & Pleijel 2011; Carr *et al.* 2011; Sun *et al.* 2012a), the barcoding fragments of COI are extremely difficult to amplify due to lack of appropriate taxon-specific primers. All initial attempts to amplify COI fragments in serpulids (including *Hydroides*) using the 'universal' barcoding primers HCO2198–LCO1490 ('Folmer primers' (Folmer *et al.* 1994) failed (see Sun *et al.* 2012a: 539–540). The first nine serpulid COI sequences reported by Carr *et al.* (2011) were a result of bacterial contamination as proven by Sun *et al.* (2012a). However, Sun *et al.* (2012a) successfully amplified COI for three species of *Hydroides* using the polychaete-specific primers polyLCO/polyHCO designed by Carr *et al.* (2011). Using these sequences as references, Sun *et al.* (2012a) designed the first *Hydroides*-specific primers (Hydro-COIF/Hydro-COIR, Table 1), and generated 460 bp COI barcoding fragments from 11 out of 14 tested species.

Fortunately, next-generation sequencing technologies have made generation of mitochondrial genomes cheaper and faster. New *Hydroides*-specific COI primers were designed by Sun *et al.* (2018), thus obtaining reference sequences for effective primer design. The new primer pair (Table 1) has been used to sequence 581 bp COI fragments of 48 species of *Hydroides* (Sun *et al.* 2018).

Table 1. Barcoding primers designed to amplify COI fragments in *Hydroides*

Primer name	Primer sequence	Reference
Hydro-COIF	TCWRTWRTKACDGTKCATGCTA	Sun *et al.* (2012a)
Hydro-COIR	CMRYAGGWTSAAARAACCTAGTA	
COI HyCOF190	TCNRTNTTKACDGTKACATGCTA	Sun *et al.* (2018)
HyCOR886	ACCCYATYATHCCRATAGARCACAT	

The 17-fold higher mean inter- as compared to intra-lineages genetic divergence in COI, and the rarity of intermediate divergences indicated that the COI barcodes had high discrimination power for *Hydroides* species delineation. COI barcoding on 48 morphospecies of *Hydroides* revealed that several morphospecies have more than one genetic lineages, such as *H. albiceps*, *H. exaltata*, *H. longispinosa*, *H. minax*, *H. tuberculata* and *H. operculata*-complex, indicating a cryptic species diversity in *Hydroides* that was overlooked by the current taxonomic system. The potential presence of cryptic species means that numerous species are still awaiting formal descriptions.

Because of the problem with obtaining COI sequences before the new primer had been designed, other variable markers, such as mitochondrial cytochrome *b* (*cytb*) and nuclear ITS, have been tested and successfully used for barcoding in *Hydroides* (Sun *et al.* 2016). The importance of proper molecular markers for successful molecular identification is exemplified by Rech *et al.* (2018), who identified *Hydroides* specimens associated with floating debris in the Lagoon of Venice as tropical *H. sanctaecrucis*. The authors suggested its presence in the Mediterranean might have been overlooked due to misidentification as common *H. dianthus*. Ironically, the authors did not even examine the specimens they sequenced although *H. sanctaecrucis* is easily distinguished from *H. dianthus* by opercular morphology. Moreover, to substantiate their claim, the authors used 18S sequences, accepting an identity ≥ 97% with sequences of *H. sanctaecrucis* in GenBank. It is unclear why COI barcoding was not used, given numerous COI sequences of both *Hydroides* species available on GenBank (Sun *et al.* 2017a). Langeneck *et al.* (2020) correctly pointed out that very conservative 18S rDNA sequences are ill-suited for molecular barcoding because in closely related species they have identity close to 100%. The low sequence identity shows that the specimens in Rech *et al.* (2018) study did not belong to *H. sanctaecrucis*, and the species name should be removed from checklists of species non-indigenous for the Mediterranean (Langeneck *et al.* 2020).

Most importantly, as an identification technique for previously characterised organisms, DNA barcoding is dependent on morphology-based traditional taxonomy and thus it invariably fails for undescribed species. But even for well-known valid species, identification using molecular barcodes crucially depends on the reliability of morphological identification of original reference specimens. The specimen selected for sequencing is first given a name based on morphology and the sequence is assumed to be associated with the name. The most reliable way to obtain a DNA barcode that accurately represents a species is to base it on the type specimen of that species (reviewed in Jinbo *et al.* 2011). In all other cases mistakes are possible, even when the voucher specimen is identified by an expert, making morphology-based identification tools crucial for development of molecular identification tools.

PHYLOGENY

Even the earliest morphological evidence indicated that the genera *Serpula*, *Crucigera* and *Hydroides* are closely related (e.g. Rioja 1919, 1934, 1947). These taxa form a monophyletic group united by the presence of synapomorphies such as the funnel-shaped operculum made of fused radii, presence of the pseudoperculum, bayonet-shaped special collar chaetae and flat trumpet-shaped abdominal chaetae (ten Hove 1984; ten Hove and Jansen-Jacobs 1984). The monophyly of the *Serpula–Crucigera–Hydroides* clade was further supported by the results of serpulid phylogenetic studies based on morphological (Kupriyanova 2003), molecular (Lehrke *et al.* 2007) and combined datasets (Kupriyanova *et al.* 2006, 2008). The monophyly of the genus *Hydroides* has also been strongly supported by all molecular-based phylogenetic analyses to date: *cytb*, 18S and 28S rRNA of eight species of *Hydroides* (Kupriyanova *et al.* 2008); COI of 10 species (Sun *et al.* 2012a); COI, *cytb* and 18S rRNA of 18 species

(Tovar-Hernández *et al.* 2016); COI, *cytb*, ITS, 18S and 28S rRNA for 284 individuals of 44 morphospecies (Sun *et al.* 2018), and mitochondrial genomes of 10 species (Sun *et al.* 2021).

The phylogenetic relationships within the genus *Hydroides* were, however, another matter. Early hypotheses of relationships were based exclusively on proposed transformation series in the opercular morphology, implying that the similarity in morphology of the complex verticil can serve as indication of close relationships among species. Jones (1962) proposed the first hypothetical scheme of evolutionary transformations within *Hydroides*, dividing the 21 species of *Eupomatus* into four groups according to the morphology of the verticil: (1) species with verticil spines of different sizes, (2) species with asymmetrically arranged verticil spines, (3) species with straight verticil spines, and (4) species with verticil spines ending in T-shaped tips. As *Eupomatus* was later synonymised with *Hydroides*, Pillai (1972: figs 2–4) proposed an alternative evolutionary scheme for 42 species of *Hydroides*, dividing them into two major categories according to the presence or absence of modified dorsal spines, and then further into 12 groups. Neither of the two proposed schemes, however, resulted from formal phylogenetic analyses. Moreover, with the synonymy of *Hydroides* and *Eupomatus* and exponentially increasing numbers of newly discovered species of *Hydroides* in subsequent years (as well as synonymisation of other species; see Read *et al.* 2017), phylogenetic relationships remained contentious.

Interestingly, none of the molecular phylogenetic analyses within *Hydroides* supported the morphology-based phylogenetic grouping proposed by either Jones (1962) or Pillai (1972). Instead, DNA sequence-based phylogenies recovered close relationship among species from Indo-Pacific region (Sun *et al.* 2012a, 2018, 2021), eastern Pacific, western Atlantic (Tovar-Hernández *et al.* 2016; Sun *et al.* 2018, 2021) and the Mediterranean (Sun *et al.* 2018, 2021). The strongly supported clade including *Hydroides elegans–H. longispinosa–H. norvegica* (the type genus) was recovered at the basal position of the genus phylogenetic tree (Sun *et al.* 2018, 2021). Thus, the conclusion is that phylogenetic relationships within *Hydroides* are more concordant with geographical distribution rather than morphological expectations (Sun *et al.* 2018) and a high degree of homoplasy exists in opercular morphology (Fig. 13).

GENOME STRUCTURE

Mitochondrial genomes (mitogenomes) have been frequently used in the phylogenetical and evolutionary studies of annelids (see Weigert *et al.* 2016 for review). Early studies on mitogenome structures lead to the notion that gene order is conserved across annelids (Jennings & Halanych 2005). While further published mitochondrial genomes of annelids supported this notion, few exceptions in rearrangements of genes or gene blocks were reported (reviewed in Sun *et al.* 2021). The mitogenome of the serpulid *Spirobranchus giganteus* showed extremely and fast substitution rates and the most pronounced gene rearrangements deviating from the proposed conserved gene order pattern of Annelida (Seixas *et al.* 2017); this called for studies to determine if this exception is an isolated case for the species or it represents a pattern for the entire family Serpulidae.

Investigation on mitogenomes of 10 species of *Hydroides* revealed several unusual features (Sun *et al.* 2021; Fig. 14). The mitogenome size of *Hydroides* is substantially larger than those reported for other annelids, except for *S. giganteus* (22 058 bp; Seixas *et al.* 2017). Numerous large intergenic non-coding regions (over 100 bp), which are rarely found in annelids, contribute to the large mitogenomes size in this genus. All mitogenomes of *Hydroides* exhibited nucleotide composition bias of a highly positive GC skew combined with a highly negative AT skew – a feature again consistent with that found only in the mitogenome of *S. giganteus*, but not in other available annelid mitogenomes. The order of protein-coding genes of *Hydroides* exhibited extensive rearrangements when compared to those of other annelids, including *S. giganteus*. Moreover, the gene orders varied significantly even among the 10 mitochondrial genomes of *Hydroides*, especially with respect to the location of tRNAs (Sun *et al.* 2021). The conclusion is that the unusual gene order found in *S. giganteus* (Seixas *et al.* 2017) is not restricted to the individual species, but is likely to be found family-wide.

Phylogenetic analyses based on protein-coding genes recovered *Hydroides* as a monophyletic group sister to *Spirobranchus* with a long branch, and sister to the fan worm Sabellidae. Yet the Serpulidae + Sabellidae clade was grouped with Sipuncula, not with Fabriciidae as expected based on results of all previous analyses. The results indicate that, while analyses of mitochondrial genes are useful for studies of phylogenetic relationships

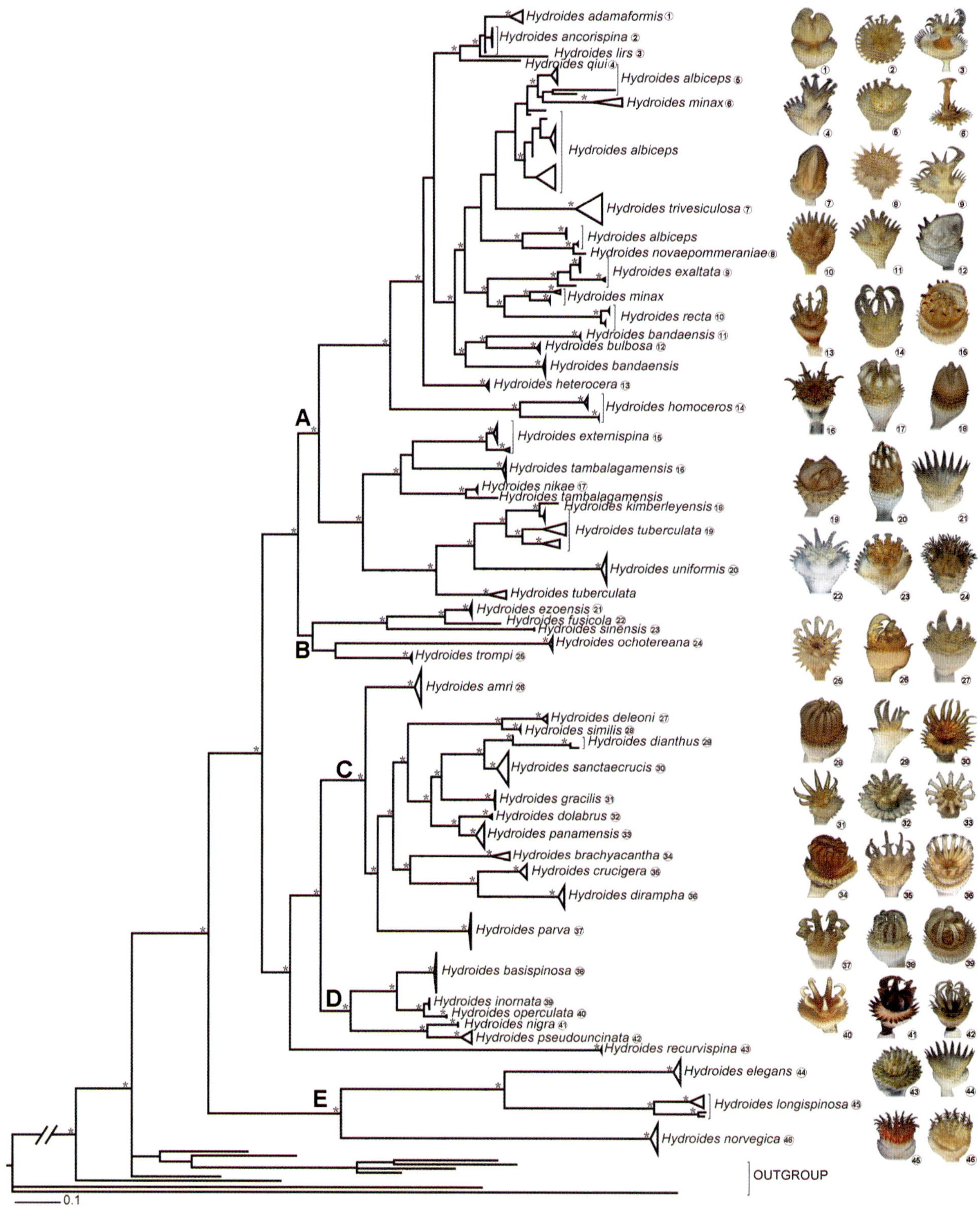

Fig. 13. Phylogeny of *Hydroides* based on Bayesian analysis of 45 morphospecies using a combined dataset of five molecular markers. The analysis yielded five main distinct clades (A, B, C, D, E) roughly corresponding to main geographical regions in Fig. 16 and 70 distinct molecular species-level lineages. Photos on the right show opercula of each species. Numbers in circles correspond to species names in the tree (modified from Sun *et al.* 2018).

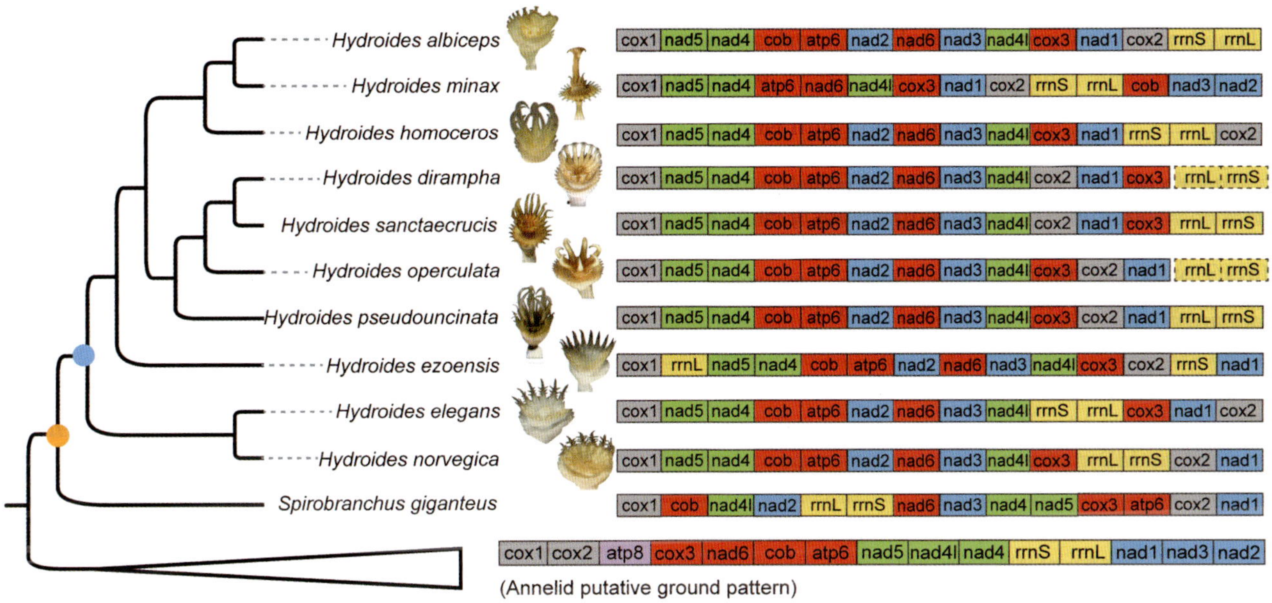

Fig. 14. Phylogeny of *Hydroides* based on 13 protein-coding genes of mitochondrial genomes (modified from Sun *et al.* 2021).

within *Hydroides* and Serpulidae in general, mitogenomes alone are insufficient to resolve the phylogenetic position of Serpulidae within Annelida due to its high base substitution rates (Sun *et al.* 2021).

Almost nothing is known about the structure of genomes of *Hydroides*. Although a raw genome of *H. elegans* has been published by Shikuma *et al.* (2016), the annotation of the genome is yet to be made available.

8 Biogeography

PATTERNS OF GEOGRAPHICAL DISTRIBUTION

All species of *Hydroides* are distributed in shallow-water coastal locations in temperate to tropical waters between 43.3°N and 35.3°S (Fig. 15). Most published records are from scuba diving depths, only 22 taxa have been collected deeper than 50 m, but this most certainly is a collecting bias. *Hydroides norvegica* is the only exception occurring from 3 m down to 650 m (Rzhavsky *et al.* 2014), and to the highest latitude at 63.4°N. It is the most cold-tolerant based on type locality, with a 20° latitudinal gap to all other species type localities, although its distribution extends south into the Mediterranean (Zibrowius 1971a), where it occurs only in depths of 30–380 m (temperature submergence). The coldest Arctic record of *H. norvegica* are also deepest. All recorded depth distributions are restricted to shelf depth, with the exception of *H. capensis* (350 m, McIntosh 1925 as *H. uncinata* (*norvegica*) var. *multispinosa*) and *H. caribensis* (250 m, herein). The record of 830 m for *H. heterocera* by Ilan *et al.* (1994) has been questioned by Ben-Eliahu & ten Hove (2011); that of *H. norvegica* of 4808 m (by Fauvel 1909) has been proven incorrect by Zibrowius (1977), yet still is mentioned in checklists, such as Budaeva *et al.* (2014).

Table 2 shows type localities of *Hydroides* species in biogeographical realms **sensu** Spalding *et al.* (2007). Seventy-three species have been described from tropical regions, including 49 from the Pacific and Indian Oceans (26 from Central Indo-Pacific, one from Eastern Indo-Pacific, seven from Tropical Eastern Pacific, 15 Western Indo-Pacific) and 24 from tropical Atlantic. In contrast, only 33 have been described from temperate regions (15 from temperate Northern Pacific, 10 from temperate Northern Atlantic, two from temperate South America, two from temperate South Africa, and three from temperate Australasia). The northern-most species is *H. norvegica* described from northern Norway, whereas none of the species has been described from the Antarctic regions and *H. plateni* from South America is southernmost (42°S) species of the genus.

Phylogeography

By linking molecular data of *Hydroides* to their native distributions, Sun *et al.* (2018) demonstrated a geographical pattern where *Hydroides* spp. from the eastern Pacific and western Atlantic were more closely related to each other than to those collected from other regions, reflecting the Indo-West Pacific/Atlanto-East Pacific faunal separation commonly observed in other warm-water coastal marine invertebrates (Ekman 1953). Molecular data also suggested that the peak of diversification occurred during the Miocene; this is concordant with the common appearance of tube fossils of the *Hydroides* morphotype (Ippolitov *et al.* 2016) at that time. The divergence estimates indicated that most opercular

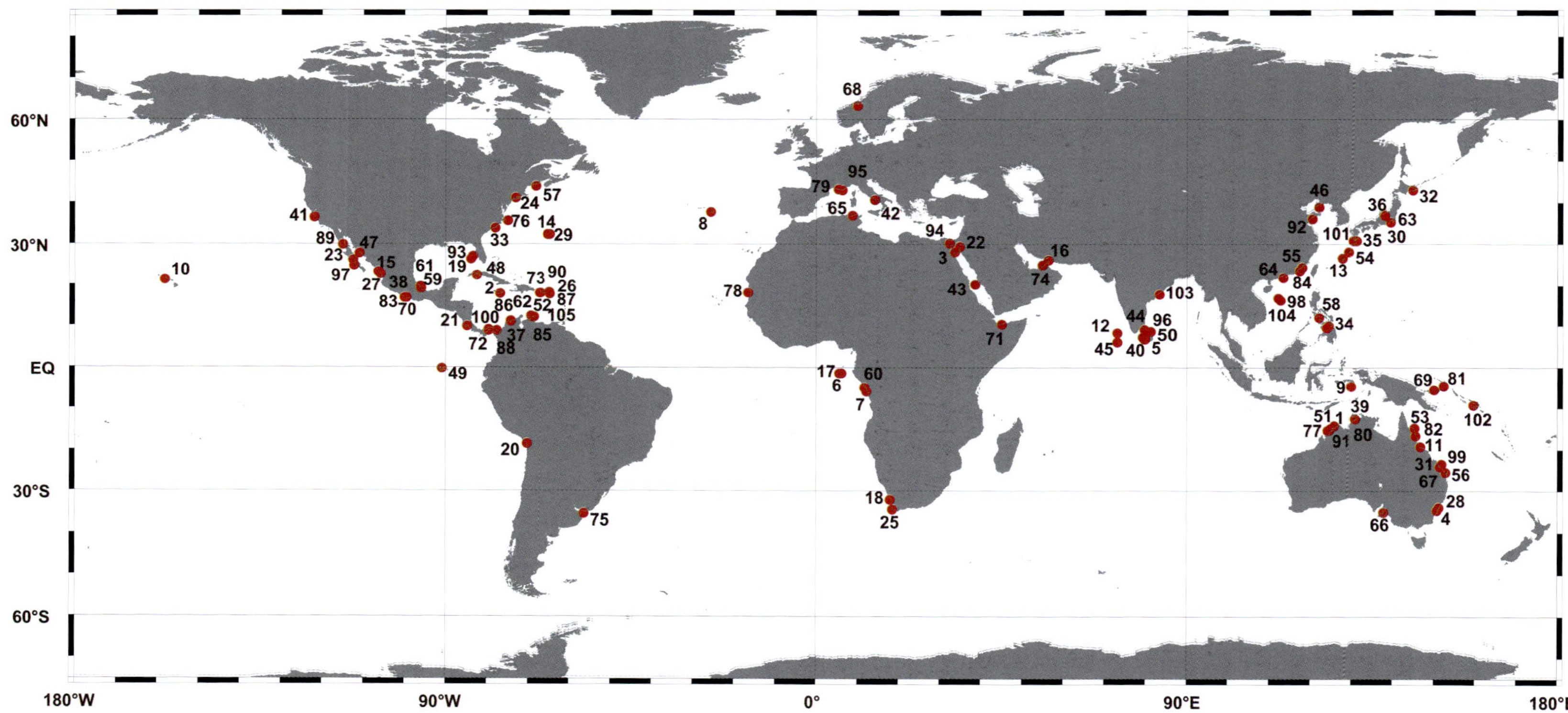

Fig. 15. Map of type localities of all *Hydroides* species currently considered valid. Numbers correspond to valid species names arranged in alphabetical order: 1. *H. adamaformis*, 2. *H. alatalateralis*, 3. *H. albiceps*, 4. *H. amri*, 5. *H. ancorispina*, 6. *H. arnoldi*, 7. *H. augeneri*, 8. *H. azorica*, 9. *H. bandaensis*, 10. *H. bannerorum*, 11. *H. basispinosa*, 12. *H. bifurcata*, 13. *H. bisecta*, 14. *H. bispinosa*, 15. *H. brachyacantha*, 16. *H. bulbosa*, 17. *H. calopoma*, 18. *H. capensis*, 19. *H. caribensis*, 20. *H. chilensis*, 21. *H. crucigera*, 22. *H. dafnii*, 23. *H. deleoni*, 24. *H. dianthus*, 25. *H. dipoma*, 26. *H. dirampha*, 27. *H. dolabrus*, 28. *H. elegans*, 29. *H. elegantula*, 30. *H. exaltata*, 31. *H. externispina*, 32. *H. ezoensis*, 33. *H. floridana*, 34. *H. furcifera*, 35. *H. fusca*, 36. *H. fusicola*, 37. *H. gairacensis*, 38. *H. glandifera*, 39. *H. glasbyi*, 40. *H. gottfriedi*, 41. *H. gracilis*, 42. *H. helmata*, 43. *H. heterocera*, 44. *H. heterofurcata*, 45. *H. homoceros*, 46. *H. huanghaiensis*, 47. *H. humilis*, 48. *H. ibarzabalae*, 49. *H. inermis*, 50. *H. inornata*, 51. *H. kimberleyensis*, 52. *H. lambecki*, 53. *H. lirs*, 54. *H. longispinosa*, 55. *H. longistylaris*, 56. *H. malleolaspina*, 57. *H. microtis*, 58. *H. minax*, 59. *H. mongeslopezi*, 60. *H. monroi*, 61. *H. mucronata*, 62. *H. mucronatoides*, 63. *H. multispinosa*, 64. *H. nanhaiensis*, 65. *H. nigra*, 66. *H. nikae*, 67. *H. nodosa*, 68. *H. norvegica*, 69. *H. novaepommeraniae*, 70. *H. ochotereana*, 71. *H. operculata*, 72. *H. panamensis*, 73. *H. parva*, 74. *H. perezi*, 75. *H. plateni*, 76. *H. protulicola*, 77. *H. pseudexaltata*, 78. *H. pseudouncinata africana*, 79. *H. pseudouncinata pseudouncinata*, 80. *H. qiui*, 81. *H. ralumiana*, 82. *H. recta*, 83. *H. recurvispina*, 84. *H. rhombobula*, 85. *H. rzhavskyi*, 86. *H. salazarvallejoi*, 87. *H. sanctaecrucis*, 88. *H. sceptrifera*, 89. *H. similis*, 90. *H. similoides*, 91. *H. simplidentata*, 92. *H. sinensis*, 93. *H. spongicola*, 94. *H. steinitzi*, 95. *H. stoichadon*, 96. *H. tambalagamensis*, 97. *H. tenhovei*, 98. *H. trilobula*, 99. *H. trivesiculosa*, 100. *H. trompi*, 101. *H. tuberculata*, 102. *H. uniformis*, 103. *H. vizagensis*, 104. *H. xishaensis*, 105. *H. zibrowii*.

Table 2. Distribution of *Hydroides* in geographical coastal regions (realms) proposed by Spalding *et al.* (2007). Type localities were used to assign each species to a biogeographical realm.

Species	Type locality	Biogeographical realms **sensu** Spalding *et al.* 2007
Hydroides adamaformis	Kimberley, WA, Australia	Central Indo-Pacific
Hydroides alatalateralis	Port Royal, Jamaica	Tropical Atlantic
Hydroides albiceps	Gulf of Suez	Western Indo-Pacific
Hydroides amri	NSW, Australia	Temperate Australasia
Hydroides ancorispina	Sri Lanka	Western Indo-Pacific
Hydroides arnoldi	Isla Annobón, Equatorial Guinea	Tropical Atlantic
Hydroides augeneri	Zaire (DR Congo)	Tropical Atlantic
Hydroides azorica	Ilha de São Miguel, Azores	Temperate Northern Atlantic
Hydroides bandaensis	Banda, Indonesia	Central Indo-Pacific
Hydroides bannerorum	Hawaii, USA	Eastern Indo-Pacific
Hydroides basispinosa	Townsville, Qld, Australia	Central Indo-Pacific
Hydroides bifurcata	North of the Maldive Islands	Western Indo-Pacific
Hydroides bisecta	Sesoko Island, Okinawa, Japan	Central Indo-Pacific
Hydroides bispinosa	Bermuda	Tropical Atlantic
Hydroides brachyacantha	Mazatlán, Mexico	Temperate Northern Pacific
Hydroides bulbosa	Strait of Hormuz, Oman	Western Indo-Pacific
Hydroides caribensis	Sanibel Island, FL, USA	Tropical Atlantic
Hydroides calopoma	Isla Annobón, Equatorial Guinea	Tropical Atlantic
Hydroides capensis	Cape Town, South Africa	Temperate Southern Africa
Hydroides chilensis	Arica, Chile	Temperate South America
Hydroides crucigera	Puntarenas, Costa Rica	Tropical Eastern Pacific
Hydroides dafnii	Elat, Gulf of Aqaba, Israel	Western Indo-Pacific
Hydroides deleoni	Baja California Sur, Mexico	Temperate Northern Pacific
Hydroides dianthus	New Jersey to Massachusetts, USA	Temperate Northern Atlantic
Hydroides dipoma	Cape of Good Hope, South Africa	Temperate Southern Africa
Hydroides dirampha	Saint Thomas, US Virgin Islands	Tropical Atlantic
Hydroides dolabrus	Mazatlán, Mexico	Temperate Northern Pacific
Hydroides elegans	Port Jackson, NSW, Australia	Temperate Australasia
Hydroides elegantula	Bermuda	Tropical Atlantic
Hydroides exaltata	Honshu, Japan	Temperate Northern Pacific
Hydroides externispina	Heron Island, Qld, Australia	Central Indo-Pacific
Hydroides ezoensis	Hokkaido, Japan	Temperate Northern Pacific
Hydroides floridana	Cape Fear, North Carolina, USA	Temperate Northern Atlantic
Hydroides furcifera	Ubay, Pandanon, Philippines	Central Indo-Pacific
Hydroides fusca	Tanegashima, Japan	Temperate Northern Pacific
Hydroides fusicola	Japan	Temperate Northern Pacific
Hydroides gairacensis	Gairaca, Santa Martha, Colombia	Tropical Atlantic

continued

Table 2. *continued*

Species	Type locality	Biogeographical realms sensu Spalding *et al.* 2007
Hydroides glandifera	Playa Caleta, Acapulco, Mexico	Tropical Eastern Pacific
Hydroides glasbyi	Darwin, NT, Australia	Central Indo-Pacific
Hydroides gottfriedi	Sri Lanka	Western Indo-Pacific
Hydroides gracilis	Pacific Grove, California, USA	Temperate Northern Pacific
Hydroides helmata	Gulf of Naples, Italy	Temperate Northern Atlantic
Hydroides heterocera	Red Sea	Western Indo-Pacific
Hydroides heterofurcata	Sri Lanka	Western Indo-Pacific
Hydroides homoceros	Multiple Indian Ocean localities	Western Indo-Pacific
Hydroides huanghaiensis	Yellow Sea, off Dalian, China	Temperate Northern Pacific
Hydroides humilis	Gulf of California, Mexico	Temperate Northern Pacific
Hydroides inermis	Isla Santiago, Galapagos	Tropical Eastern Pacific
Hydroides inornata	Sri Lanka	Western Indo-Pacific
Hydroides ibarzabalae	Cuba	Tropical Atlantic
Hydroides kimberleyensis	Kimberley, WA, Australia	Central Indo-Pacific
Hydroides lambecki	Curaçao, Netherlands Antilles	Tropical Atlantic
Hydroides lirs	Lizard Island, Qld, Australia	Central Indo-Pacific
Hydroides longispinosa	Amami, Japan	Central Indo-Pacific
Hydroides longistylaris	Zhangzhou, Fujian Province, China	Central Indo-Pacific
Hydroides malleolaspina	Hervey Bay, Qld, Australia	Central Indo-Pacific
Hydroides microtis	North America	Temperate Northern Atlantic
Hydroides minax	Philippines	Central Indo-Pacific
Hydroides mongeslopezi	Veracruz, Mexico	Tropical Atlantic
Hydroides monroi	Pointe Noire, Congo, West Africa	Tropical Atlantic
Hydroides mucronata	Veracruz, Mexico	Tropical Atlantic
Hydroides mucronatoides	Bonaire, Netherlands Antilles	Tropical Atlantic
Hydroides multispinosa	Sagami Bay, Honshu, Japan	Temperate Northern Pacific
Hydroides nanhaiensis	Macao, Guangdong, China	Central Indo-Pacific
Hydroides nigra	Pointe Meloula, Tunisia	Temperate Northern Atlantic
Hydroides nikae	Edithburgh, SA, Australia	Temperate Australasia
Hydroides nodosa	Gladstone, Qld, Australia	Central Indo-Pacific
Hydroides norvegica	Trøndelag region, Norway	Temperate Northern Atlantic
Hydroides novaepommeraniae	New Britain	Central Indo-Pacific
Hydroides ochotereana	Acapulco, Mexico	Tropical Eastern Pacific
Hydroides operculata	Gulf of Aden, Berbera, Somali	Western Indo-Pacific
Hydroides panamensis	Panama City, Western Panama	Tropical Eastern Pacific
Hydroides parva	Puerto Rico	Tropical Atlantic
Hydroides perezi	Pearling Banks near Dubai	Western Indo-Pacific
Hydroides plateni	La Plata, Argentina/Uruguay	Temperate South America
Hydroides protulicola	Off Cape Hatteras, NC, USA	Temperate Northern Atlantic

Species	Type locality	Biogeographical realms **sensu** Spalding *et al.* 2007
Hydroides pseudexaltata	Kimberley, WA, Australia	Central Indo-Pacific
Hydroides pseudouncinata africana	Rio de Oro, Mauritania	Tropical Atlantic
Hydroides pseudouncinata pseudouncinata	East off Île Gaby, Marseille, France	Temperate Northern Atlantic
Hydroides qiui	Darwin, NT, Australia	Central Indo-Pacific
Hydroides ralumiana	New Britain, Papua New Guinea	Central Indo-Pacific
Hydroides recta	Cairns, Qld, Australia	Central Indo-Pacific
Hydroides recurvispina	Acapulco, Mexico	Tropical Eastern Pacific
Hydroides rhombobula	Dongshan, Fujian Province, China	Central Indo-Pacific
Hydroides rzhavskyi	Netherlands Antilles	Tropical Atlantic
Hydroides salazarvallejoi	Cabo de la Aguja, Colombia	Tropical Atlantic
Hydroides sanctaecrucis	Saint Croix, US Virgin Islands	Tropical Atlantic
Hydroides sceptrifera	Bonaire, Netherlands Antilles	Tropical Atlantic
Hydroides similis	Baja California, Mexico	Temperate Northern Pacific
Hydroides similoides	Isla Magueyes, Puerto Rico	Tropical Atlantic
Hydroides simplidentata	Kimberley, WA, Australia	Central Indo-Pacific
Hydroides sinensis	Qingdao, China	Temperate Northern Pacific
Hydroides spongicola	Venice, Florida, USA	Tropical Atlantic
Hydroides steinitzi	Suez Canal, Egypt	Western Indo-Pacific
Hydroides stoichadon	Parc Nacional de Port Cros, France	Temperate Northern Atlantic
Hydroides tambalagamensis	Tambalagam Lake, Sri Lanka	Western Indo-Pacific
Hydroides tenhovei	Baja California Sur, Mexico	Temperate Northern Pacific
Hydroides trilobula	Xisha Islands, China	Central Indo-Pacific
Hydroides trivesiculosa	Heron Island, Qld, Australia	Central Indo-Pacific
Hydroides trompi	Panama Canal, Panama	Tropical Eastern Pacific
Hydroides tuberculata	Tanegashima Island, Japan	Temperate Northern Pacific
Hydroides uniformis	Solomon Islands	Central Indo-Pacific
Hydroides vizagensis	Visakhapatnam Harbour, India	Western Indo-Pacific
Hydroides xishaensis	Xisha Islands, South China Sea	Central Indo-Pacific
Hydroides zibrowii	Santa Marta, Colombia	Tropical Atlantic

diversification and distribution patterns occurred during the Miocene (5–23 Ma). The lineages formed from the Pliocene onwards showed subtle morphological differences or overlapped in the extent of variation with respect to morphology. Historical geographical reconstruction based on molecular data and calibrated by fossil suggests that the most probable distribution of the most recent common ancestor of *Hydroides* was in the western Atlantic and the Indian Ocean, which was a tropical Indo-Mediterranean region during the Mid-Palaeocene – the so-called west Tethyan region (Sun *et al.* 2018; Fig. 16). During the closure of the Tethys in the early Miocene, the central Indo-Pacific region was a centre of palaeodiversity of *Hydroides*, evidenced by the highly localised cladogenesis. Also, the central Indo-Pacific region appeared to act as a source for the nearby Indian Ocean and temperate Australia since the Miocene via dispersal. Further paleogeographical studies based on *Hydroides* fossil tubes would provide a test for this biogeographical hypothesis.

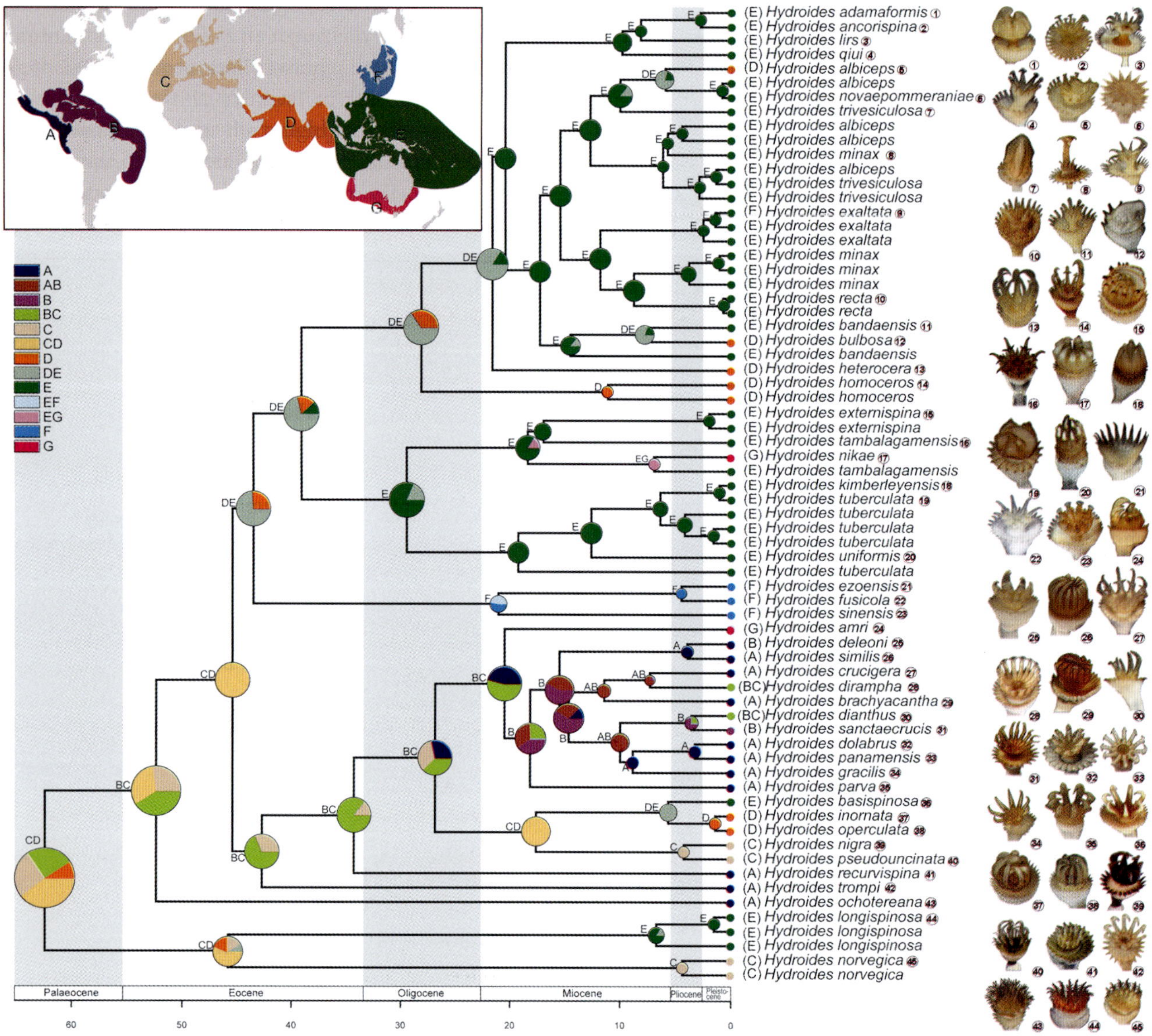

Fig. 16. Biogeographical reconstruction of ancestral ranges in *Hydroides*. Map on top-left corner illustrates the used biogeographical regions. Pie charts and letters at nodes represent the probabilities of the most likely ancestral ranges. Letters to the left of the species names indicate current biogeographical distributions. A – East Pacific; B – West Atlantic; C – East Atlantic and Mediterranean; D – West Indo–Pacific; E – Central Indo-Pacific; F – Temperate Northern Pacific; G – Temperate Australasia (modified from Sun *et al.* 2018).

9 Methods of collecting, examination, preservation and identification

COLLECTING

Because most *Hydroides* species are subtidal, they can be collected on the substrate in shallow water while snorkelling or examining tidal pools, and by scuba diving or by dredging. Substrate obtained in the field is examined for tubes. Tubes of *Hydroides* are fairly easily recognisable as such in the field, even underwater while scuba diving, but they may not be immediately obvious as they usually attach to substrates like the undersides of flat rocks, mollusc shells and human-made objects. Distinguishing empty and inhabited tubes is often impossible in the field. Tubes collected with the substrate are preferably returned to the lab alive in seawater. Animals in intact tubes on original substrate can easily survive in the lab for 24–48 h, especially if aerated. If substrate items are too large to take along from the field, the tubes can be taken off by applying pressure with a scalpel or even a coarse preparation needle to the attachment area of the tube, preferably on convex curving sections of the tube. If surfaces are smooth (e.g. glass or mussel shells), it is possible to remove the tube intact or nearly intact. Otherwise, large tube fragments with the worm inside come off the substrate, but animals in damaged tubes are viable for a short period of time only, so they need to be fixed sooner to avoid fast decomposition.

EXAMINATION AND PRESERVATION

To identify animals to species level, specimens need to be examined under a stereomicroscope in a Petri dish filled with seawater in case of live animals or in preservative liquid in case of preserved specimens. Under no circumstances should live animals or fixed/preserved specimens be examined without water/fixative or allowed to get dry.

If conditions in the field permit, or the animals have been transported to the lab alive, the tubes and the animals should be photographed first. Removing live individuals from their tubes without tube damage is almost impossible, so it is important to examine and photograph intact tubes before they are broken to extract animals. Live animals normally spontaneously come out of the tube to feeding position when left in a Petri dish under a stereomicroscope. Alternatively, posterior parts of the tube can be broken off carefully piece by piece, thus forcing the animal towards the tube mouth and into the feeding position.

A fine artist brush with partly shortened, cut-off hairs often needs to be used to clean the operculum from silt trapped on the verticil spines (e.g. Fig. 3C). It is often difficult to view the inside of the verticil if the spines are too close to each other and mask the central disk. Squeezing the insertion of the verticil carefully with a pair of watchmaker's tweezers will open up the verticil and bring the characters inside to view; the spines will return to their original position afterwards.

Fixation and preservation methods depend on further intended purposes of samples. Specimens aimed for a morphological study only are fixed in a 4% solution of formaldehyde in seawater for 24–48 h, ideally after relaxation in magnesium chloride or in seawater gradually diluted with ethanol to a 5–10% concentration. Animals are then rinsed in distilled water and preserved in 70–80% ethanol. Ideally, the posterior abdomen should be cut off and fixed in 90–100% ethanol for molecular study, while the rest of the body is fixed in formalin as above. Direct preservation in ethanol only is an option when proper sample processing is not possible but samples are destined for molecular work. Freezing of fresh samples is not recommended as they defrost rapidly and it is difficult to examine and properly preserve them after defrosting.

Careful labelling is critical and the filled label should include tentative field identification, date, locality/station number and fixative. It is important to use alcohol and ink proof waterproof paper. The labels should be inserted inside the vials with well-sealing lids to prevent evaporation of the alcohol.

For illustrations, traditional ink drawing (pencil sketching followed by India ink tracing) is currently being supplemented or gradually replaced by digital tracing of scanned pencil sketches, using a drawing pad. Advances in digital photography and universal availability of microscope-mounted digital cameras and Z-stacking software also result in photographs of live or preserved specimens, rather than line drawings being used in species descriptions. Use of scanning electron microscope (SEM) micrographs helps to illustrate chaetal, opercular and soft body characters with precision and objectivity.

IDENTIFICATION

A cautionary note on identification

The most important thing to be remembered when trying to identify a specimen is that names attributed to specimens in collections, certainly for members of difficult species complexes such as *Hydroides albiceps*, are hypotheses only. Based on the knowledge at the time of identification and the perceptions of characters which are changing with better equipment, names are given to the best of the knowledge at the time. For instance, the accessibility of SEM, digital photography techniques (especially focus stacking) and the use of DNA sequence data have changed our insights during the past decades. Previously widely distributed or 'cosmopolitan' species are now being (re)described as suites of regionally distributed (new) species, and species names previously synonymised with such 'cosmopolitan' species names are being resurrected; see for instance the *H. operculata* story (Sun *et al.* 2017b).

As a result of these advances, names initially given to specimens in older collections often turn out to be a wrong choice if scrutinised against the contemporary insights. This is evidenced by quite a few of our selected synonymies below, for example *H. basispinosa*, *H. bispinosa* and certainly in the enormous number of different species in earlier times classified under the *nomen inquirendum Hydroides uncinata* (Philippi 1844). A further complication are the misidentifications of juveniles as discussed under the rejected name *Hydroides prisca* Pillai, 1971. We attempted to highlight the known identification errors (partly changing insights, partly caused by high pressure to provide names for, e.g., ecological papers) in our selected synonymies, also as a warning that a mistake is quicker made than explained and corrected. It is important to stress that, although together we have seen a lot of museum material, we cannot be comprehensive in our listed errors, and there still will be more to be found (also in material identified by one of us in earlier days).

10 Diagnostic characters used for identification

In their taxonomic revision of *Hydroides* from the Western Atlantic region, Bastida-Zavala & ten Hove (2003a) suggested a standardised format for the full (re)description and provided characters (with character states) and measurements to be included in such descriptions.

The most useful diagnostic characters for species identification in the taxonomic key for the genus, below, are those of the operculum. They are found in the proximal funnel (number of radii and shape of radial tips) and in the distal verticil (mostly arrangement, number and shape of main spines, as well as presence, number and positions of spinules). Opercular colouration in both funnel and verticil can be diagnostic (e.g. *Hydroides nigra* has distinctly dark operculum). The funnel is reduced to a small number of triangular processes in *H. mucronatoides* nov. spec. (Fig. 77, previously recorded as *Hydroides* cf. *mucronata*) or even absent, as in *H. bannerorum* (Fig. 26). The presence of a constriction between peduncle and opercular funnel, and the degree of its chitinisation and colour, can serve as auxiliary characters.

The chaetation patterns are mostly too uniform in the genus to be useful for species discrimination. The exception is the shape of the bayonet collar chaetae (Fig. 5) – while they bear two large teeth in most species, one- and three-toothed bayonets are also found or only small teeth form a proximal rasp (e.g. *Hydroides elegans*). The thoracic uncini are saw-shaped in all species except for *H. bannerorum*, with rasp-shaped uncini. The number of thoracic chaetigers is seven in all species except for *H. bannerorum* (7–9), *H. bisecta* (9), and *H. sceptrifera* nov. spec. (6–9).

Tubes of *Hydroides* have been poorly documented and thus are not used here in the key and diagnoses. However, tubes merit more attention in species descriptions and an understanding of tube ultrastructure is important for fossil studies.

It is recommended to first identify a specimen using the dichotomous key and then confirm the identification with species information provided in the taxonomic part. This information includes an illustration (an original photograph unless a species is known based on a holotype that has not been designated or has been lost) and consistently ordered as type information (with museum deposition if known), etymology of name, selected synonymy, diagnosis, taxonomic remarks and geographical distribution.

For most species selected synonymies have been given; these include the original description, most frequently used other names, earliest and most recent revisions, well-illustrated descriptions, first and last DNA records, and other special interest records such as ecological remarks. No effort has been made to be complete – this would be impossible anyway for taxa like *H. elegans* and *H. dianthus* with each having 400 or more records under various names known to us. A 'complete' synonymy may accompany species described in the last two or three decades and new species designated herein.

Diagnoses given for all taxa have been limited to the characters of the operculum (exceptionally of special collar chaetae, if other than a bayonet with two teeth and for numbers of thoracic chaetigers other than seven). Numbers of both funnel radii and verticil spines are variable within limits, and sometimes can be used in species distinction (e.g. see morphometric comparisons as given by Bastida-Zavala & ten Hove 2003a figs 9, 19, 31), but the amount of data available is very unequally distributed among taxa. Here we therefore only give the estimated median followed by the known range in full sized specimens: for example, 13 (11–16) verticil spines and 34 (26–44) funnel radii in *H. alatalateralis*. Actual ranges may start with lower numbers, sometimes occurring in juvenile specimens; exceptional aberrations are given as, for example, 34 ((25)29–41) radii (*H. basispinosa*), where 34 is the median, 25 a juvenile, and 29–41 the range in presumably adult specimens.

In the distribution section, we made an effort to give geographical ranges from north to south, west to east. Usually a combination of the position in the taxonomic key, illustration, characters of the diagnosis and distribution data is sufficient to identify the specimen, certainly at least to the complex in which the taxon belongs. Nevertheless, if in doubt one should always consult a full description from the selected synonymies for a definite identification, the original description (if detailed enough) and/or recent redescriptions (if available). Sometimes there are quite obvious differences among taxa in, for example, size, numbers of radioles and (abdominal) segments, and more rarely in details of the shape of collar chaetae.

For newly described taxa, the diagnosis is followed by a full description. For some species rarely reported in the literature, but in reality common, lists of not yet reported localities have been given. These are mostly from the Caribbean, for decades a region of special interest for one of us (HAtH), and intended for a never published paper on *Hydroides* from Florida, USA, by ten Hove & Perkins (e.g. *H. elegantula*, *H. microtis*, *H. mongeslopezi*). If relevant, a more complete description has been added, basically from the same source.

11 Key to the species of *Hydroides* worldwide

The following key assists with species identification of *Hydroides* specimens, using diagnostic features observed under a dissection and/or compound microscope. Please refer to 'General body morphology' for graphical illustrations as required.

1a Opercular funnel absent. Verticil with 16–19 straight spines similar in size, with pointed tips..*Hydroides bannerorum* (Fig. 26)
1b Opercular funnel present 2
2a Funnel replaced by 4 (2–7) globular processes ....*Hydroides mucronatoides* nov. spec. (Fig. 77)
2b Funnel typical crown made of multiple radii 3
3a Spines of opercular verticil all similar in shape, spine size may change gradually.. 4
3b Spines of opercular verticil dissimilar, one or more dorsal spines distinctly different.................................. 55
4a Spines of opercular verticil all identical in size and shape 5
4b Spines of opercular verticil all similar in shape, but becoming gradually smaller from dorsal to ventral side 90
5a Both verticil spines and funnel radii thin elongated and bifurcated at mid-length..*Hydroides furcifera* (Fig. 50)
5b Spines of verticil and funnel radii not bifurcated at mid-length 6
6a Spines of verticil lacking external or lateral spinules, internal spinules may be present.................................... 7
6b Spines of verticil with external or lateral spinules, internal spinules may be present 35
7a. Verticil spines very short, similar in size to tips of funnel......................................*Hydroides plateni* (Fig. 90)
7b Verticil spines significantly larger than tips of funnel radii 8
8a Verticil spines bulbous, rounded or conical.................. 9
8b Verticil spines elongated... 12
9a Verticil with three rounded spines only ....*Hydroides trilobula* (Fig. 112)
9b Verticil with 5 or 6 spines pointed distally, with or without rounded knobs 10
9c Verticil with 7 or 8 spines with rounded tips..*Hydroides uniformis* (Fig. 116)
10a Verticil spines with distal knob each...*Hydroides kimberleyensis* (Fig. 66)
10b Verticil spines distally pointed, but without distal knobs 11
11a Five verticil spines rhomboid in cross-section, collar chaetae with two conical teeth only .... *Hydroides rhombobula* (Fig. 98)
11b Five verticil spines triangular in cross-section, collar chaetae with two big teeth and 2 to 3 small teeth.. *Hydroides xishaensis* (Fig. 117)
12a Verticil spines with distinctly modified (T-shaped, conical, pickaxe shaped) tips 13
12b Verticil spines with simple pointed or blunt tips......... 21
13a Verticil with dilated or globular tips.............................14
13b Verticil tips with laterally or radially positioned projections 17
14a Verticil composed of 12–16 spines with distinctly globular tips..........................*Hydroides microtis* (Fig. 72)
14b Verticil composed of 7–9 uniform spines with dilated or blunt tips 15

15a Internal spinules absent, funnel tips rounded .... *Hydroides novaepommeraniae* (Fig. 84), *Hydroides bandaensis* (Fig. 25)
15b Internal spinules present, funnel tips pointed 16
16a Each spine with a small basal spinule each only .... *Hydroides arnoldi* (Fig. 22)
16b Each spine with a basal small spinule and a mid-level one pointing down ....*Hydroides augeneri* (Fig. 23)
16c Each spine with conspicuous subterminal internal spinule, and row of smaller ones....................................... .... *Hydroides sinensis* (Fig. 106)
17a Verticil tips with laterally positioned projections (tips T-shaped or anchor-shaped) 19
17b Verticil tips with radially positioned projections........ 18
18a Basal spinules absent, funnel radii with T-shaped tips... *Hydroides qiui* (Fig. 95)
18b Basal spinules present; funnel radii with pointed tips.......................................*Hydroides dirampha* (Fig. 42)
19a Verticil tips with a terminal dark hook pointing inwards and a long whip-like filament pointing outwards.................................. *Hydroides bisecta* (Fig. 29)
19b Radially positioned projections of verticil spine tips otherwise 20
20a Pointed spine tips with external distal knobs .... *Hydroides recurvispina* (Fig. 97)
20b Pickaxe-shaped modified spine tips................................ .... *Hydroides dolabrus* (Fig. 43)
20c Triangular modifications of spine tips............................ .... *Hydroides panamensis* (Fig. 87)
21a Tips of verticil spines strongly incurved into hooks....... 22
21b Tips of verticil spines straight or slightly curved 24
22a Funnel with T-shaped radii. Verticil spines without basal spinules. Bases of verticil spines touching, no central disk .... *Hydroides gairacensis* (Fig. 53)
22b Funnel with pointed or blunt-tipped radii; verticil spines with basal spinules, central disk present 23
23a Verticil with black spines ending in centrally incurved hooks, twisted anticlockwise; funnel radii with long, sharply pointed tips; long basal spinules almost meeting over small central disk .... *Hydroides fusca* (Fig. 51)
23b Verticil with brown spines abruptly and strongly curving inwards; funnel with blunt tipped radii; small basal spinule above small central disk. .... *Hydroides rzhavskyi* nov. spec. (Fig. 99)
23c Verticil with light incurving spines, with sharp tips twisted clockwise; funnel with pointed radii; small knob-like basal spinule over small central disk .... *Hydroides trompi* (Fig. 114)
24a Verticil spines straight or slightly curved outwards.. 25
24b Verticil spines only slightly curved inwards 30
25a Verticil larger than funnel.. 26
25b Verticil smaller than funnel... 27
26a Verticil with about 30 spines, fused basally for 1/3–2/3 of their length; each spine with a row of 6–8 inner spinules; central disk not elevated dome-like. ....*Hydroides ezoensis* (Fig. 48)
26b Verticil with 14 slender spines not fused basally; rows of inner spinules absent; central disk elevated dome-like.......................... *Hydroides huanghaiensis* (Fig. 117A)
27a Funnel radii blunt; verticil with outcurving spines tapering to blunt tips.. .... *Hydroides sceptrifera* nov. spec (Fig. 102)
27b Funnel radii pointed; verticil with pointed tips 28
28a Verticil spines ending in outcurving acute tips, sometimes lightly twisted clockwise; each spine with knob-shaped inner basal spinule .... *Hydroides protulicola* (Fig. 91)
28b Verticil spines with spines curving outwards or straight, not twisted; no internal spinules.................... 29
29a Funnel with short base .... *Hydroides humilis* (Fig. 62)
29b Funnel with distinctly long base ....*Hydroides longistylaris* (Fig. 70)
30a Basal part of verticil spines flattened into wings 31
30b Basal part of verticil spines not flattened into wings 34
31a Funnel with 45–61 of radii ending in very sharp elongated tips................. *Hydroides spongicola* (Fig. 107)
31b Funnel with 26–38 radii ending in triangular tips 32
32a Basal spinule short, never crossing mid-point of central disk 33
32b Basal spinule long, crossing mid-point of central disk .... *Hydroides alatalateralis* (Fig. 18)
33a Wings occupying less than half of the spine length.................................. *Hydroides floridana* (Fig. 49)
33b Wings occupying more than half of the spine length.................................*Hydroides elegantula* (Fig. 45)
34a Funnel radii tips elongated pointed with or without an additional internal spinule each... .... *Hydroides pseudouncinata* (Figs 93, 94)
34b Funnel radii tips almost triangular pointed without an additional internal spinule each.. .... *Hydroides gracilis* (Fig. 57)
35a Verticil spines with both external and lateral spinules 36
35b Verticil spines with external or lateral spinules 37
36a Verticil spines with groups of 2–4 yellow small external spinules halfway and 3–5 pairs of black lateral spinules in basal-middle position...... *Hydroides lambecki* (Fig. 67)
36b. Verticil spines with groups of yellow 5–13 tiny external spinules halfway; 4 or 5 pairs of black lateral spinules in basal-middle position.. ....*Hydroides mongeslopezi* (Fig. 74)
37a Spines of verticil with external spinules only 38
37b Spines of verticil with lateral spinules only.................. 39
38a Verticil spines with simple pointed tips. Each verticil spine with a single pointed external spinule at mid-length .... *Hydroides sanctaecrucis* (Fig. 101), *Hydroides vizagensis* (Fig. 117C)
38b Verticil spines with radially positioned modified tips. Each verticil spine with a single blunt external spinule at mid-length..................... *Hydroides calopoma* (Fig. 33)

39a Verticil spines with one pair of lateral spinules mid-length 40
39b Verticil spines with more than one pair of lateral spinules 47
40a Verticil spines straight with sharp or blunt tips 41
40b Verticil spines incurved or with incurved tips only 42
41a Verticil spines with sharp tips and a pair of outwardly curved lateral spinules and an internal spinule at mid-length .... *Hydroides tambalagamensis* (Fig. 110)
41b Verticil spines with straight blunt tips and a pair of straight lateral spinules about midway, some with one internal spinule .... *Hydroides mucronata* (Fig. 76)
42a Verticil with central tooth, straight spines with tips curved inwards, each with a pair of straight lateral spinules at about mid-length and a basal spinule .... *Hydroides nodosa* (Fig. 82)
42b Verticil without central tooth 43
43a Funnel with rounded, not chitinised radii tips 44
43b Funnel with chitinised pointed or anchor-shaped tips 45
44a Verticil with lightly incurved spines, without basal spinules .... *Hydroides stoichadon* (Fig. 109)
44b Verticil with strongly incurved spines, each with a pointed basal spinule .... *Hydroides bispinosa* (Fig. 30)
45a Funnel with chitinised anchor-shaped radii. Verticil with slender incurving spines; each with a pointed tip .... *Hydroides homoceros* (Fig. 61)
45b Funnel with chitinised pointed tips 46
46a Verticil with incurved spines with pointed tips, but without an external knob on the 'knee' .... *Hydroides crucigera* (Fig. 37)
46b Verticil with spines abruptly incurving with a pointed tips and an external knob on the 'knee' .... *Hydroides parva* (Fig. 88)
47a Verticil spines with up to 20 pairs of lateral spinules .... *Hydroides ochoterena* (Fig. 85)
47b Verticil spines with 2–6 pairs of lateral spinules 48
48a Central tooth absent 49
48b Central tooth present 50
49a Verticil with 9–10 spines, almost straight to lightly curving outwards, each with 6–8 pairs of lateral spinules and 5–8 internal spinules, including distinct basal one .... *Hydroides nanhaiensis* (Fig. 79)
49b Verticil with 5–9 spines smoothly curving inwards then downwards, each with 2–4 lateral spinules and a sharply pointed basal spinule .... *Hydroides caribensis* nov spec. (Fig. 35)
50a Central tooth long .... *Hydroides longispinosa* (Fig. 69)
50b Central tooth short 51
51a Each verticil spine with 2–4 pairs of lateral spinules and 1–3 internal spinules, verticil spines fused basally, forming a wide shallow cup 52
51b Each verticil spine with 4–8 pairs of lateral spinules, verticil spines not fused basally, not forming a wide shallow cup 53
52a Special collar chaetae with proximal rasp only .... *Hydroides elegans* (Fig. 44)
52b Collar chaetae bayonet, with two knobs .... *Hydroides norvegica* (Fig. 83)
53a Incurved verticil spines with lateral spinules at basal half of each spine, single dark internal spinule pointing toward verticil centre at base of each spine .... *Hydroides capensis* (Fig. 34)
53b Straight to outcurved verticil spines with lateral spinules throughout each spine length, multiple basal spinules pointing upwards 54
54a Spines with 4–5 pairs of lateral spinules .... *Hydroides azorica* (Fig. 24)
54b Spines with 6–8 pairs of lateral spinules .... *Hydroides multispinosa* (Fig. 78)
55a Two or more curved dorsal verticil spines larger than the rest 56
55b One dorsal verticil spines larger or somehow different from the rest 59
56a. External and/or lateral spinules present, 2 larger incurved spines 57
56b External and/or lateral spinules absent, 2–4 larger spines incurved over the rest 58
57a Verticil spines (except dorsal) with a pair of dark lateral spinules and with an external distal spinule .... *Hydroides externispina* (Fig. 47), *Hydroides ralumiana* (Fig. 117B)
57b Verticil spines (except dorsal) without a pair of dark lateral spinules, but with an external spinule .... *Hydroides glasbyi* (Fig. 55)
58a 26–39 verticil spines .... *Hydroides similoides* (Fig. 104)
58b 12–22 verticil spines .... *Hydroides similis* (Fig. 103)
59a Both longer dorsal spine and rest of verticil spines bifurcated at mid-length .... *Hydroides heterofurcata* (Fig. 60)
59b None of verticil spines bifurcated at mid-length 60
60a Dorsal spine narrow with pointed tip and small sub-distal internal spinule, straight or outcurved, only slightly longer than other spines; remaining verticil spines outcurved, without sub-distal spinules .... *Hydroides recta* (Fig. 96)
60b Dorsal spine either distinctly larger than other spines, raised or curved over remaining spines, or as large bulbous structure 61
61a Dorsal spine slender, raised over the rest and ending in a hook, remaining spines the same in size and shape 62
61b Dorsal spine large vesicular structure, accompanied by much smaller spines or smaller spines absent 71
61c Dorsal spine curving over the rest or below the rest, remaining spines the same or their spine sizes may change gradually 79
62a Dorsal-most spine more than twice the length of the others, its acute tip with two lateral spinules 63
62b Dorsal-most spine about or less than twice the length of the others, its acute tip lacking lateral spinules 64

63a Remaining spines without lateral spinules; internal spinules present ....*Hydroides lirs* (Fig. 68)
63b Remaining spines without lateral spinules; internal spinules absent ....*Hydroides minax* (Fig. 73)
64a Remaining spines with 1–3 pairs of lateral spinules.. 65
64b Remaining spines always without lateral spinules........ 66
65a Spines thick, almost conical, with 2 or 3 pairs of short, acute lateral spinules .... *Hydroides ibarzabalae* nov. spec. (Fig. 63)
65b Spines thin, elongated, not conical, with one pair of lateral spinules ....*Hydroides heterocera* (Fig. 59)
66a Remaining spines with T-shaped or bifurcate tips......... ....*Hydroides bifurcata* (Fig. 28)
66b Remaining spines with pointed tips............................. 67
67a Remaining spines with outcurved tips......................... 68
67b Remaining spines with incurved tips........................... 70
68a Funnel radii with T-shaped tips .... *Hydroides pseudexaltata* (Fig. 92)
68b Funnel radii with pointed tips 69
69a Verticil spines without pointed basal spinule ....*Hydroides dipoma* (Fig. 41)
69a Verticil spines with pointed basal spinule each.............. .... *Hydroides exaltata* (Fig. 46)
70a Verticil black, funnel with black well-defined, darkly chitinised constriction. Spines either curving smoothly, or at most with a recognisable external knee sometimes accentuated by clear external knob ....*Hydroides nigra* (Fig. 80)
70b Verticil light-coloured, black constriction absent ....*Hydroides inornata* (Fig. 65), *Hydroides operculata* (Fig. 86)
71a Verticil consists of large bulbous dorsal spine with 6 longitudinal groves, smaller spines absent....................... .... *Hydroides steinitzi* (Fig. 108)
71b Verticil consists of large bulbous dorsal spine and 4–14 much smaller spines.. 72
72a Dorsal bulbous verticil spine more than five times longer than the smaller verticil spines 73
72b Dorsal bulbous verticil spine less than five times longer than the smaller verticil spines 77
73a Dorsal spine erect with strongly recurved acute tip; four smaller spines, median two of which fused to base of dorsal spine, two lateral ones positioned above median ones...................... *Hydroides glandifera* (Fig. 54)
73b Dorsal spine either distinctly globular, incurving beak-like or consisting of bulbous median part and two dorso-lateral processes; small spines pressed against base of large dorsal spine...74
74a Dorsal spine consists of bulbous median part and two dorso-lateral processes; funnel tips truncated............ 75
74b Dorsal spine distinctly globular, incurving beak-like; funnel tips pointed 76
75a Dorsal spine elongated, trilobed in cross-section, with 4–7 outwardly curved spines .... *Hydroides trivesiculosa* (Fig. 113)
75b Dorsal spine bulbous, with 7–14 longitudinal groves corresponding to number of spines inserted at base ....*Hydroides bulbosa* (Fig. 32)
76a All verticil spines without a pronounced external spinule on the knee-like bend. Small spines thin, with pointed tips ....*Hydroides simplidentata* (Fig. 105)
76b All verticil spines with a pronounced external spinule on the knee-like bend. Small spines thick, with rounded tips....................... *Hydroides gottfriedi* (Fig. 56)
77a Funnel radii and verticil spines with distinctly anchor-shaped tips .... *Hydroides ancorispina* (Fig. 21)
77b Funnel radii and verticil spines with club-shaped tips.. 78
78a Dorsal spine with beak-like medial extension and bifid tip .... *Hydroides malleolaspina* (Fig. 71)
78b Dorsal spine without beak-like medial extension and bifid tip .... *Hydroides albiceps* (Fig. 19)
79a Dorsal spine curving strongly inward (its tip almost in position of an internal spinule) 80
79b Dorsal spine curving over the rest 81
80a Verticil with 5–8 spines, large dorsal incurving spine with a distal knob, spines straight or directed outward with blunt tips .... *Hydroides salazarvallejoi* (Fig. 100)
80b Verticil with 9–14 spines, large dorsal spine and two smaller adjacent ones partially fused, each with a distal knob. Other spines directed outward with sharp tips.......................................*Hydroides tenhovei* (Fig. 111)
81a Enlarged dorsal spine incurved, other spines straight or directed outwards 82
81b All spines directed invards.. 83
82a Enlarged dorsal spine overhangs the others like a helmet, with a pointed tip and an accessory external knob; the rest stubby and digitate, slightly outcurved...... ....*Hydroides helmata* (Fig. 58)
82b Enlarged dorsal spine laterally flattened, with a pointed tip horizontally directed inward, and without an accessory external knob; the rest irregularly tetrahedral, vertex pointed upwards .... *Hydroides monroi* (Fig. 75)
83a Verticil with all bulbous spines 84
83b Verticil with more or less elongated spines with pointed tips or blunt, all strongly curving inwards 86
84a Funnel radii with T-shaped tips, 6 diamond-shaped spines with rounded tips .... *Hydroides adamaformis* (Fig. 17)
84b Funnel radii with pointed tips, 4–7 equal bulbous spines, more or less triangular in cros-section, dorsal spine with a sharp point horizontally directed inward 85
85a Small tubercles on outer surface on bend of verticil spines present .... *Hydroides tuberculata*-complex (Fig. 115).
85b Small tubercles on outer surface on bend of verticil spines absent .... *Hydroides perezi* (Fig. 89), *H. dafnii* (Fig. 38)
86a Large dorsal spine covering the central disk and equally sized remaining spines...................................... 87

86b Large dorsal spine covering the central disk and remaining spines decreasing in size ventrally 89

87a Large dorsal spine with blunt tip, remaining spines wide, almost valvular............*Hydroides inermis* (Fig. 64)

87b Large dorsal spine with pointed tip, remaining spines narrow 88

88a Smaller verticil spines with external tubercles .... *Hydroides brachyacantha* (Fig. 31)

88b Smaller verticil spines without external tubercles .... *Hydroides deleoni* (Fig. 39)

89a Verticil spines without external tubercles, not geniculate .... *Hydroides amri* (Fig. 20)

89b Verticil spines, especially ventral ones, may be geniculate ....*Hydroides zibrowii* nov. spec. (Fig 118)

90a Elongated slender spines pointing in one direction (dorsal inside, ventral outside), becoming gradually smaller from dorsal to ventral side .... *Hydroides dianthus* (Fig. 40)

90b Spines slender or short, curved inwards, becoming smaller gradually from dorsal to ventral side 91

91a Spines short, pair of lateral spinules absent ....*Hydroides nikae* (Fig. 81)

91b Spines long, pair of lateral spinules present ....*Hydroides chilensis* (Fig. 36)

12 Taxonomic account

GENUS *HYDROIDES* GUNNERUS, 1768

Type species. *Hydroides norvegica* Gunnerus, 1768 (original binomen)

Codonytes Quatrefages, 1866, erected for *Eupomatus dipoma* Schmarda, 1861

Eupomatus Philippi, 1844, type species *Eupomatus uncinatus* Philippi, 1844 (by subsequent designation)

Eucarphus Mörch, 1863 as subgenus of *Hydroides*, erected for taxa with (sub)terminal pair of side spinules to verticil spines such as *Hydroides* (*Eucarphus*) *crucigera* Mörch, 1863; however, confused

Glossopsis Bush, 1905 erected for *Serpula minax* Grube, 1878b

Olgaharmania Rioja, 1941b, erected for *Hydroides glandiferum* Rioja, 1941a

Polyphragma Quatrefages, 1866, erected for *Eupomatus uncinatus* Philippi, 1844

Protohydroides Uchida, 1978, erected for *Hydroides elegans* (Haswell, 1883)

Schizocraspedon Bush, 1905, erected for *Serpula furcifera* Grube, 1878b

Diagnosis (after ten Hove & Kupriyanova 2009): Tube white (sometimes bluish), more or less circular to trapezoidal (with flattened upper surface) in cross-section, peristomes and shallow longitudinal ridges may be present, no distinct keels. A granular overlay may be present. Operculum two-tiered, composed of basal funnel of fused radii and distal verticil (crown) of chitinised spines. Peduncle cylindrical, smooth, without wings, may or may not be separated from opercular funnel by a constriction; formed from second dorsal radiole on one side. Pseudoperculum present. Arrangement of radioles in semicircles, usually about 15, exceptionally up to 33 per lobe. Radiolar eyes absent. Inter-radiolar membrane generally absent, rarely present (Bastida-Zavala & ten Hove 2003b mention only one species, *H. trompi*, where it is present). Stylodes absent. Mouth palps absent. Seven thoracic chaetigerous segments, exceptionally more (9 in *H. bisecta* Imajima & ten Hove 1989 and *Hydroides* sp. 2 Bastida-Zavala & ten Hove 2003a, described herein as *H. sceptrifera* ten Hove nov. spec.; 7–9 in *H. bannerorum* Bailey-Brock, 1991). Collar trilobed, tonguelets absent. Thoracic membranes long, forming ventral apron. Collar chaetae bayonet-type and limbate. *Apomatus* chaetae absent. All uncini saw-shaped with relatively few (up to 7) teeth; anterior fang simple pointed. Triangular depression present. Abdominal chaetae flat trumpet-shaped with denticulate edge. Achaetous anterior abdominal zone absent. Posterior capillary chaetae present. Posterior glandular pad absent.

SPECIES INFORMATION

1. *Hydroides adamaformis* Pillai, 2009

Figure 17

Type information: Lucas Island, Kimberley, Western Australia, 15°13′S, 124°31′E, holotype AM W.21457.

Etymology: Refers to the diamond-shaped external appearance of the verticil spines (Latin *adamas* = diamond, *forma* = shape), although their distal ends are curved inwards (cf. Pillai 2009: 114).

Selected synonymy

Hydroides adamaformis Pillai, 2009: 113–115, fig 11A–J (description [original]; Western Australia); Sun *et al.* 2015: 8–10, fig. 1, map 1 (diagnosis, discussion; WA); 2018, figs 2, 3, 5, 6 (DNA, phylogeny; Clerke Reef, WA).

Diagnosis: Peduncle with well-defined, chitinised constriction. Funnel with 30 T-shaped radii with chitinised tips. Verticil on short stalk, bilateral symmetrical, with 6 rhomboidal spines including one dorsal, bulbous hook. All spines incurving with rounded tips, without accessory spinules or central disk.

Remarks: Pillai (2009: 113) mentioned a similarity between *Hydroides. adamaformis* and, for example, *H. tuberculata*. However, *H. adamaformis* can be distinguished from all other species with few (3–8) valvular verticil spines by its funnel radii with crescent-shaped tips as opposed to pointed or triangular ones in *H. dafnii*, *H. kimberleyensis*, *H. perezi*, *H. rhombobula*, *H. trilobula*, *H. tuberculata*-complex, *H. uniformis* and *H. xishaensis*. Phylogenetically it is distant from the other taxa mentioned, as far as their DNA is known (*kimberleyensis*, *tuberculata*-complex, *uniformis*; Sun *et al.* 2018). Pillai did not mention the similarity with taxa like *H. albiceps* and *H. ancorispina*, where crescent-tipped funnel radii are common. In the DNA analysis by Sun *et al.* (2018), however, *H. adamaformis* indeed is sister to the clade including the latter two species. Differences with *H. gottfriedi*, *H. inermis* and *H. monroi* are discussed under *H. helmata*.

Distribution: Kimberley, Western Australia.

Fig. 17. *Hydroides adamaformis*, various views of the same operculum, Kimberley, Western Australia, AM W.21457 (photo E. Wong).

2. *Hydroides alatalateralis* (Jones, 1962)

Figure 18

Type information: Port Royal, Jamaica, holotype and paratypes AMNH 3617, 3618.

Etymology: Jones (1962: 207) says, 'The trivial name refers to the limbations that are to be found on the sides of the spines of the distal opercular circlet' (Latin *alatus* = winged, *latus -eris* = side).

Selected synonymy

Eupomatus alatalateralis Jones, 1962: 205–207, figs 139–146 (description [original]; Jamaica).

Eupomatus floridanus **sensu** Augener 1925a: 16 [**non** Bush, 1910] (diagnosis; St Croix, Lesser Antilles).

Hydroides alatalateralis – Bastida-Zavala & ten Hove 2003a: 115–118, fig. 5, table 2 (redescription, synonymy; Cuba to Colombian Atlantic, Colombian Pacific); 2003b: 71–72, fig. 2H–I (diagnosis; Bahía Malaga, Pacific Colombia); Sun *et al.* 2018, fig. 1 (phylogeny, after Pillai 1972).

Diagnosis: Peduncle without constriction. Funnel with 34 (25–44) chitinised, finely pointed radii. Verticil sitting in funnel, radially symmetrical, with 13 (8–16) slender incurving spines. All spines similar in shape, some dorsal spines may be slightly longer, proximally flattened into lateral wings for up to half length; each with long basal spinule, knobbed in the middle, and crossing small central disk. Without central tooth.

Remarks: This taxon is very similar to *H. elegantula*, *H. floridana* and slightly less so *H. spongicola* in general opercular shape with slender, incurving, winged verticil spines. Differences between *H. alatalateralis*, *H. floridana* and *H. elegantula* are very subtle and found in the length of these wings and/or shape of basal spinules (below) and morphometrics (see Bastida-Zavala & ten Hove 2003a, fig. 9), notably in the larger *H. spongicola*. Wings in *H. floridana* extending less than half of the spine length, in *H. alatalateralis* and *H. elegantula* up to more than half of the spine length, in *H. spongicola* variable to absent. A further difference is found in the basal spinules, meeting each other across the central disk in *H. alatalateralis* and *H. spongicola*, not meeting in *H. floridana* and even shorter in *H. elegantula*. All three have fewer numbers of radii and radioles than *H. spongicola*; moreover, *spongicola* exclusively lives in sponges. Bastida-Zavala & ten Hove (2003a: 118) conclude that more and better material is needed to decide whether or not *H. elegantula* should have precedence over *H. alatalateralis*. Mainly on biogeographical grounds they decided to treat the three nominal taxa as full species, an approach we have followed here. Bastida-Zavala & ten Hove (2003a, fig. 5e) document the rare occurrence of a bifid funnel radius, otherwise only known to us from a paratype of *Hydroides recta* (QM 3817; HAtH unpublished).

Distribution: Jamaica, Caribbean, Colombian Pacific.

Fig. 18. *Hydroides alatalateralis*, various views of the same operculum, Piscadera Baai, Curaçao, AM W.42845 (photo E. Wong).

3. *Hydroides albiceps* (Grube, 1870)-complex

Figure 19, 120C

Type information: Tor (Gulf of Suez, Red Sea, Egypt), holotype ZMB 533.

Etymology: Not given by Grube (1870: 521), but he mentions a white (Latin = *albus -a*) operculum.

Selected synonymy

Serpula (*Eupomatus*) *albiceps* Grube, 1870: 520–521 (Latin description [original]; Red Sea).

Hydroides multispinosa var. *ternatensis* Fischli, 1903: 130–131, pl. 5, figs 32, 33, pl. 7 figs 76, 78, pl. 8, figs 89, 90 (description; synonymised by Imajima & ten Hove 1984: 43; Ternate, Moluccas, Indonesia but see remarks below).

Hydroides exaltatus var. *vesiculosus* Fauvel, 1919: 342–343, fig. 1i (diagnosis, but see Remarks below; Gambier Islands, French Polynesia); Monro 1937: 316 (diagnosis; Zanzibar); Mesnil & Fauvel 1939: 30–31 (diagnosis; Java); Kumaraswamy Achari 1969: 40 (catalogue; Palk Bay, Sri Lanka); Amoureux *et al.* 1978: 146–147, fig. 14Hv (diagnosis, material studied; Elat, Israel, Red Sea); Bleeker & van der Spoel 1992: 137 (Mesnil & Fauvel's material in ZMA V.Pol. 1646, 2627).

Hydroides minax **sensu** Gibbs 1971: 203 [**non** Grube, 1878b] (Solomon Islands).

Hydroides albiceps – Imajima 1976a: 133–135, fig. 8a–v (description; south-west Japan); 1987: 9 (distribution in Japan); Zibrowius 1979: 132–133 (ship's hull in Toulon, France); Fiege & Sun 1999: 112–114, fig. 2D–F (description, differential diagnosis, discussion; Hainan, China); Bailey-Brock & Magalhães 2013: 1–3, figs 1, 2 (description, opercular development and variations, discussion; Guam, Mariana Islands; Sun & Yang 2014: 206–208, figs 105A–I, 106 (diagnosis in Chinese; Xiamen, Fujian; Hong Kong; Zhanjiang, Guangdong; Bailongwei, Guangxi; Hainan; South China Sea); Kupriyanova *et al.* 2015: 280–282, fig. 2 (redescription, synonymy; Lizard Island, Queensland, Australia); Sun *et al.* 2015: 10–14, fig. 2 a, b; map 2 (opercular variation, diagnosis, synonymy; widely distributed in the Indo-West Pacific); 2018: 738–739, 741–742, figs 1, 2, 3–6 (DNA, phylogeny; Darwin, Northern Territory; Heron Island, Dunk Island and Mission Beach, Qld, Australia; Kavieng, Papua New Guinea; Gizo, Solomon Islands; Okinawa, Japan; Ratnagiri and Sindhuburg, Maharashtra, India).

Hydroides exaltatus vesiculosus – Uchida 1978: 26–27, pls VII, VIII (taxonomic remarks; Sabiura, Japan).

Hydroides spiratubus Pillai, 2009: 128–131, figs 22A–F, 23A–J (description, discussion; synonymised by Kupriyanova *et al.* 2015; Kimberley, Western Australia).

Diagnosis: Peduncle with ill-defined to lightly chitinised constriction. Funnel with 27 (14–35) chitinised, clavate to bottle-shaped radii. Verticil sitting in funnel, bilateral symmetrical, with 11 (5–14) spines, curving outwards, with rectangular to clavate tips, without any accessory spinules and meeting each other basally (no central disk). Dorsal hook vesicular, consisting of a bulbous median part and 2 latero-dorsal more or less triangular (seen from dorsal side) outpockets; the three parts vary considerably in relative size in various specimens. Bayonet of collar chaetae with 2 teeth (exceptionally 3, or with 1 or 2 accessory ones).

Remarks: Specimens of what we presently perceive as *H. albiceps* show a high intraspecific variability in number, shape, and relative size of verticil spines (see Sun *et al.* 2015). This has been the cause of much debate and confusion. In 1986, during a stay in the Australian Museum, one of us (HAtH unpublished) had a two-way analysis done between the opercula (about 70) of what he then perceived as *H. albiceps* respectively *H. trivesiculosa*. Based on numbers of verticil spines *vs* funnel radii, the 'populations' were statistically different, the number of verticil spines is larger in *H. trivesiculosa*. He then had the impression that *H. albiceps* was better represented in mainland localities of Australia, while *H. trivesiculosa* prevailed offshore, in reefs. Ten Hove (1990: 119, figs 7–22) gave a graphic comparison between *H. albiceps*, *H. bulbosa*, *H. steinitzi* and *H. trivesiculosa*.

Fiege & Sun (1999) compared *H. albiceps*, *H. malleolaspina* and *H. trivesiculosa* in detail, and showed an overlap in the number of spines and funnel radii, along with differences in the shapes of dorsal verticil spine and tubes. In a comparison of the holotypes, the nominal species *H. albiceps* differs from *H. trivesiculosa* notably in the size and shape of the dorsal verticil spine. *Hydroides trivesiculosa* has an exceptionally large dorsal vesicular verticil spine which is more than five times longer than the smaller verticil spines. The lateral outpockets of the dorsal verticil spine are straight in *H. albiceps*, while those of *H. trivesiculosa* are larger and curved ventrally.

Though Imajima & ten Hove (1984) synonymised *Hydroides multispinosa* var. *ternatensis* with *H. albiceps*, it should be noted its huge dorsal vesicular spine not only approaches that as figured by Imajima & ten Hove for *H. albiceps*, but also that of *H. exaltata vesiculosa* (synonymised herein, see Remarks below).

Fig. 19. *Hydroides albiceps*-complex, various views of opercula. Top row: North-east of Aride, Seychelles, AM W.28296; middle row: Lizard Island, Queensland, Australia, AM W.20155; bottom row: Banacon Island, Phillipines, AM W.201660 (photo E. Wong).

The name *Hydroides exaltata* var. *vesiculosa* Fauvel, 1919 (his figure reproduced herein as Fig. 120C) had been discussed as being similar to *H. albiceps*, but never been formally synonymised and therefore regarded to be a *species inquirenda* by Read *et al.* (2017: 17). We have carefully reviewed the original description (and one additional specimen) and compared counts of funnel radii (24–26) and verticil spines (12 +1; however, in Mesnil & Fauvel 7–8 +1) with those of *H. albiceps* (14–31, 5–12 +1; including '*H. multispinosa* var. *ternatensis*': 26, 12 +1). Shape and numbers fall within the ranges of variation of the other nominal taxa, as well as its reportedly wide distribution. We therefore have formally synonymised the taxon here.

Hydroides albiceps is identical to the nominal species *H. spiratuba* except for the roughly spirally coiled tube in the holotype of the latter (AM W.21443). However, spirally coiled tubes are very common in various species of *Hydroides*, and Sun *et al.* 2015 synonymised *H. spiratuba* with *H. albiceps*.

Liu (2008: 408) includes *Hydroides albiceps* **sensu** Sun & Yang (2000, figs 1 A–G) in *H. recta*. In our opinion, however, their fig. 1A–D show typical *H. albiceps*; their figs 1E–G, showing an internal spinule on the dorsal hook, are regarded to be varieties of *H. albiceps* by Sun & Yang (2000). This internal spinule makes the hook somewhat similar to that of *H. recta*, and might be the reason that Liu attributed all these figures to *H. recta*. In absence of molecular data, we regard all forms in these figures for the time being to belong to *H. albiceps*. For an additional discussion see remarks under *H. adamaformis* as well. A so-called *H. 'prisca'*-type operculum, developmental stage, has been recorded for *H. albiceps* (ten Hove & Ben-Eliahu 2005: 18, 19).

On the basis of molecular data of 45 morphospecies, Sun *et al.* (2018: 142) conclude: 'Our phylogenetic results, however, indicate that there are at least four genetically highly divergent lineages of *H. albiceps* morphotypes that potentially constitute separate species.' A clear pattern according to geographical areas was not retrieved. The nominal taxon is a complex of species. A much needed revision of all populations should be underpinned by DNA, the morphology minimally be based on statistically testable counts of verticil spines *v* funnel radii of well-developed opercula.

In view of the above, we should realise that names attributed to specimens in collections, certainly for members of difficult complexes as *H. albiceps*, are hypotheses only, based on the knowledge at the time of identification, and may turn out to be a wrong choice if scrutinised against the contemporary insights.

Distribution: Supposedly widely distributed in the Indo-West Pacific, from Zanzibar and Red Sea to South Japan and Polynesia, Australia: NT, Qld and WA; an as yet unresolved species complex.

4. *Hydroides amri* Sun, Wong, ten Hove, Hutchings, Williamson & Kupriyanova, 2015

Figure 20

Type information: South Bass Point, New South Wales, Australia, 34°36′12″S, 150°53′43″E, holotype AM W.46591.

Etymology: The species is named *'amri'* in honour of Australian Museum Research Institute (AMRI).

Selected synonymy

Eupomatus ralumianus **sensu** Poore *et al.* 1975: 30 [**non** Augener, 1927] (name in list only; Port Phillip Bay, Victoria. Material studied by ten Hove in 1986, reidentified as probable juvenile of *H.* cf. *brachyacantha*, nowadays named *H. amri*).

Hydroides amri Sun *et al.* 2015: 14–20, fig. 3a, b, 4a–h, map 3 (description [original], synonymy; Forster NSW, Port Lincoln SA, to Bremer Bay WA, Australia); 2016: 41–59, fig. 2 (DNA, within *H. brachyacantha*-complex); 2018: 738, figs 2, 3, 5, 6 (DNA, phylogeny; same localities).

Hydroides brachyacantha **sensu** Dew 1959: 28, fig. 7A–G [**non** Rioja, 1941] (diagnosis; in part, New South Wales, but the Queensland part of the material belongs to *H. tuberculata*, **fide** Imajima & ten Hove (1984: 44, 50); Kupriyanova *et al.* 2009: 391 (SA, Australia; DNA).

Diagnosis: Peduncle without constriction. Funnel with 33 (27–42) chitinised, sharply tipped radii. Verticil inserted with short basis into funnel, bilaterally symmetrical, with 9 (7–10) incurving spines with pointed tips. Three to 5 dorsal spines larger than others, covering central disk. Other verticil spines with a sharp to round knob on their bend. All verticil spines with short basal spinule. Central tooth absent.

Remarks: A phylogenetic analysis of material identified as *Hydroides brachyacantha*, from the Pacific side of Mexico and from Australia, showed that the latter was genetically quite distinct, not even recovered as a neighbouring clade in the phylogram. Sun *et al.* (2015) described the Australian material as *H. amri*. Nevertheless, morphologically the similarity is sufficient to have the taxon identified as '*H. brachyacantha*' by Dew (1959) and 20 other following papers on Australian and other non-Mexican Pacific material. However, its operculum shows a more gradual reduction in size from the dorsal hook(s) to the smooth latero-dorsal verticil spines while in *H. brachyacantha* there is a single large dorsal hook covering the central disk and the (almost) equally sized remaining verticil spines are knobbed. Further, *H. amri* resembles *H. similis* (Treadwell 1929) in having multiple larger dorsal hooks and an external knob on smaller verticil spines. However, *H. amri* differs from the latter species in the number of verticil spines, which is 7–10 in *H. amri* but 12–16 in *H. similis* (Bastida-Zavala & ten Hove 2003b). For a full discussion see Remarks under *H. brachyacantha*.

Distribution: Australia: ACT, NSW, SA, southern Qld, WA, Vic.

Fig. 20. *Hydroides amri*, various views of the same operculum, Bass Point, New South Wales, Australia, AM W.46591 (photo E. Wong).

5. *Hydroides ancorispina* Pillai, 1971

Figure 21

Type information: Wellawate, Colombo, Sri Lanka, holotype BMNH 1968.152.

Etymology: Not given by Pillai (1971: 108–110) but refers to the fact that both radii and verticil spines have anchor-shaped tips (Latin *ancora* = anchor, *spina* = thorn, spine).

Selected synonymy

Hydroides ancorispina Pillai, 1971: 108–110, figs 6A–E (description [original]; Sri Lanka); 2009: 132–133, fig. 25D–F (new figures of holotype, differential diagnosis); Sun & Yang 2000: 118, fig. 1H (diagnosis; Dongshan, Fujian; Longya Bay, Sanya, Hainan Island); Sun *et al.* 2018, figs 1–3, 5, 6 (DNA, phylogeny; Clerke and Imperieuse Reefs, Western Australia).

Pro parte *Hydroides ancorispina/us/malleolaspina/us*-complex – ten Hove 1994: 108 (discussion; Seychelles, Malaysian Peninsula); Fiege & Sun 1999: 114–116, figs 5A–C (description, differential diagnosis, discussion; Hainan, China).

New material: Malacca Straits, 05°09′08″N, 99°48′10″E, sand, mud, 'Store Nordiske' 25 December 1935, ZMC (11 spec.), ZMA V.Pol. 3878 (4 spec.), AM W.42851 (1 spec., Fig. 21).

Diagnosis: Peduncle with well-defined chitinised constriction. Funnel with 29 (19–40) chitinised anchor-shaped radii. Verticil inserted on a short stalk into funnel, bilaterally symmetrical, with stout dorsal spine and 6 (–9) subequal outcurving spines. Dorsal spine columnar, somewhat laterally flattened, lacking any accessory processes. Remaining spines simple, smooth, gradually increasing in size ventrally, with anchor-shaped tips similar to those of funnel. Central tooth absent.

Remarks: The taxon is similar to *Hydroides albiceps*, *H. malleospina* and *H. trivesiculosa*, as all four taxa have T-shaped or anchor-shaped tips to verticil spines and funnel radii. However, the dorsal spine is vesicular in *H. albiceps*, vesicular and trilobed in *H. trivesiculosa*, hammer-like in *H. malleospina*, while it is like a large flattened column in the type of *H. ancorispina*. Specimens of the latter nominal species from Malacca Straits and from Hainan show faint lateral dilations (Fig. 21 here), or even a blunt incurving knob-like tip (Fiege & Sun 1999, fig. 5A, B). Sun & Yang (2000, fig. 1H, 5A–B) clearly depict two different forms under the names *H. ancorispina* and *H. malleospina* and no transitional forms between them, consequently regarding both as different species. Morphologically their *H. ancorispina* indeed resembles Pillai's figure, but Sun & Yang's *H. malleospina* approaches the form identified as *H. novaepommeraniae* by Imajima & ten Hove (1984, figs 4a–b). After studying altogether almost 40 specimens of the nominal taxa *H. ancorispina* and *H. malleolaspina*, juvenile as well as large, adults including the types, one of us (HAtH unpublished) concluded that if the opercula of both taxa could be separated on morphology at all, such only could be done in fully grown specimens.

In the DNA analysis by Sun *et al.* (2018), *H. ancorispina* (from Western Australia), however, is recovered as sister to *H. adamaformis*, with anchor-shaped funnel radii, but

Fig. 21. *Hydroides ancorispina*, various views of the same operculum, Malacca Straits, AM W.42851 (photo E. Wong).

also to the morphologically quite dissimilar *H. lirs*, but not closely associated with *H. trivesiculosa* and/or *H. albiceps*. In view of the reported opercular variability, convoluted synonymy and wide distribution, we regard the taxon as a complex of a species, awaiting further analysis with help of DNA.

Ten Hove & Ben-Eliahu (2005: 19) mentioned a *H. 'prisca'*-type operculum, a developmental stage, for the *H. ancorispina/malleolaspina*-complex.

Distribution: Sri Lanka, Malacca Straits to South China Sea, Hainan and Western Australia (all Indo-Polynesian Province); Seychelles (Western Indian Ocean).

6. *Hydroides arnoldi* Augener, 1918

Figure 22

Type information: Annobón, Gulf of Guinea, holotype ZMH V.8724.

Etymology: Named after its collector, Arnold Schultze.

Selected synonymy

Hydroides arnoldi Augener, 1918: 595–598, fig. 107, pl. 6, figs 151, 152 (description [original]; in part, Island of Annobón; Zibrowius (1973a: 26) referred a second specimen, from Togo, to his new species *H. augeneri*; Zibrowius 1973: 24–26, fig. 1a–k (diagnosis, opercular variability, synonymy, discussion; Liberia, Ghana, Togo, Annobón, São Tomé, Congo, Zaire).

Hydroides uncinata **sensu** Monro, 1939: 223–224 (diagnosis; Congo) [**non** Philippi, 1844, and renamed by Zibrowius (1973, above)]; **in part only,** other parts of Monro's material belong to *H. augeneri*, *H. dipoma* and *H. monroi*.

Not *Hydroides arnoldi* **sensu** Uschakov 1970: 445 [**non** Augener, 1918] (Guinea; see *H. augeneri*).

Diagnosis: Peduncle without, or at most with ill-defined constriction. Funnel with 28 (17–35) lightly chitinised, triangularly tipped radii. Verticil inserted with short basis into funnel, radially symmetrical, with 7 (5–9) spines, almost straight to slightly curving outwards, stubby, with bluntly rounded to faintly club-shaped tips. Spines without lateral spinules, with tiny basal knob or spinule above small central disk. Central tooth absent.

Remarks: As evidenced by the synonymy above, the species has been misidentified for or mixed up with other taxa, even in Augener's original description (see Zibrowius 1973a: 26), especially with *H. augeneri*. With hindsight, however, the internal spinules halfway the verticil spines of the latter clearly differentiate this taxon from *H. arnoldi*. By its rather undifferentiated verticil spines, it also shows a certain similarity to *H. bandaensis*, *H. microtis*, *H. novaepommeraniae* and *H. sinensis*. However, these four taxa all have verticil spines with dilated, truncated tips, as opposed to the rounded tips in *H. arnoldi*.

Distribution: Liberia, Ghana, Annobón, São Tomé, Congo, Zaire.

Fig. 22. *Hydroides arnoldi*, various views of the same operculum, Annobón, Golfe de Guinée, ZMA V.Pol. 8724 (photo E. Wong).

7. *Hydroides augeneri* Zibrowius, 1973

Figure 23

Type information: Zaire, Malembe, holotype and four paratypes MRAC 34.951, 23.736, 23.764–6.

Etymology: Named after Hermann Augener.

Selected synonymy

Hydroides augeneri Zibrowius, 1973: 26–29, fig. 2a–l (diagnosis [original], opercular variability, synonymy, discussion; Liberia, Ghana, Togo, Guinea, Congo, Zaire).

Hydroides uncinata **sensu** Monro 1939: 223–224 (diagnosis; Congo) [**non** Philippi, 1844, and renamed by Zibrowius (1973, above)]; **in part only**, other parts of Monro's material belong to *H. arnoldi*, *H. dipoma* and *H. monroi*.

Diagnosis: Peduncle with ill-defined constriction. Funnel with 35 (25–43) chitinised, triangularly tipped radii. Verticil inserted with short basis into funnel, radially symmetrical, with 8 (7–9) spines, almost straight to curving inward, slender with bluntly rounded tips. Spines without lateral spinules, but with well-developed medial hook, knobbed in the middle, and short basal spinule above small central disk. Central tooth absent.

Remarks: In his description of *Hydroides arnoldi*, Augener (1918: 597) diagnosed a second specimen from Lomé (Togo), with opercular spines longer and thinner than those of the holotype, moreover incurving and with a medial 'tooth'. He expressed his doubts if this specimen belonged to the same species. Zibrowius (1973: 24–26) did not find this second specimen with the holotype in the Zoological Museum Hamburg, but its diagnosis was a perfect fit with his newly described *H. augeneri*. By its rather undifferentiated verticil spines, *H. augeneri* also shows a certain similarity to *H. bandaensis*, *H. microtis*, *H. novaepommeraniae* and *H. sinensis*. However, these four taxa all have verticil spines with dilated, truncated tips, as opposed to the rounded tips in *H. augeneri*.

Distribution: West Africa: Congo, Togo, Zaire, Ghana, Liberia, Guinea.

Fig. 23. *Hydroides augeneri*, various views of the same operculum, Vista, Democratic Republic Congo, NHML 1939.7.17-19 (photo E. Wong).

8. *Hydroides azorica* Zibrowius, 1972b

Figure 24

Type information: Ponta Delgada, São Miguel, Azores, holotype MNHN, five paratypes USNM 44069. On shipwreck *Doria* east of Ponta Delgada port, Ilha de Sao Miguel, Açores (Azores).

Etymology: Named after its type locality.

Selected synonymy

Hydroides azorica Zibrowius, 1972b: 435–438, fig. 1a–c (description [original], opercular variability, synonymy, discussion; Azores).

?*Hydroides elegans* **sensu** Morton & Salvador 2009: 145, 150, 153, fig. 4b, d [**non** Haswell, 1883] (off São Miguel, Azores; see Remarks).

Diagnosis: Peduncle without or with ill-defined constriction. Funnel with 30+ (24–40) radii, with rounded, not chitinised tips. Verticil sitting in funnel, with 15 (12–21) spines, straight or slightly curving outwards, radially symmetrical. Spines each with 6–10 irregularly spaced lateral spinules (at both sides), as well as up to 6 internal spinules over about 2/3 of their length. Central disk slightly elevated, with well-developed smooth central tooth. Bayonet of collar chaetae with 2 teeth, without proximal rasp.

Remarks: *Hydroides azorica* is similar to *H. elegans* and *H. norvegica*, both occurring in the Atlantic as well, and to the Indo-West Pacific *H. longispinosa*, *H. multispinosa*, *H. nanhaiensis* and, to a lesser extent, the Eastern Pacific *H. ochotereana*. *H. azorica*, *H. norvegica* and *H. ochotereana* have essentially collar chaetae with 2(–3) large teeth, without a proximal rasp (although there may be some accessory teeth in *H. norvegica*, and the endings of the fibres making up the shaft of the chaetae may give the impression of a rasp). All four other taxa are characterised by a well-developed proximal rasp of small teeth, merging into 2–5 larger teeth just proximal from the distal blade. Moreover, *H. azorica* has 6–10 irregularly spaced lateral spinules as opposed to 2–4 in *H. elegans*, 3–5 in *H. norvegica* and 6–18 in *H. ochotereana*; a central tooth is missing in *H. norvegica* and *H. ochotereana*, but generally present in *H. azorica* and *H. elegans*. See Remarks under the other similar taxa as well.

By its depth distribution (100–250 m) and shape of tube, *Hydroides elegans* **sensu** Morton & Salvador (2009) would belong to *H. azorica* rather than to the fouling *H. elegans*, reported from shallow depths in the Azores (Morton & Britton 2000).

Distribution: Azores.

Fig. 24. *Hydroides azorica*, various views of the same operculum, Sao Miguel, Azores, SMF 21570 (photo E. Wong).

9. *Hydroides bandaensis* Zibrowius, 1972a

Figure 25

Type information: Banda Island, Indonesia, holotype NHMW Evert. Varia 2039.

Etymology: Named after its type locality.

Selected synonymy

Hydroides bandaensis Zibrowius, 1972a: 92– 93, fig. 2a–g (diagnosis [original]; Banda, Indonesia).

?*Hydroides* cf. *bandaensis* Sun & Yang 2014: 208–209, figs 107A–B, 106 (Dongshan, Fujian; 10–15 m; the specimen from Hainan is probably a juvenile of an undescribed species; compare Remarks under the rejected name *H. prisca*, below); Sun *et al.* 2018: 742, figs 2, 3, 5, 6 (DNA, phylogeny; Kavieng, Papua-New Guinea; Gizo, Solomon Islands).

Diagnosis: Peduncle with ill-defined to lightly chitinised constriction. Funnel with 16 radii (21–23 in non-type material), long, with truncated to slightly swollen T-shaped tips. Verticil inserted with short basis into funnel, radially symmetrical, with 9 (6–11 in non-types) equal spines, almost straight to slightly outcurving, with bluntly dilated, truncated tips. Spines simple, smooth, joining at centre of verticil, without central disk. Bayonet of collar chaetae with 2 teeth and 1 or 2 accessory ones.

Remarks: Differences with similar taxa *Hydroides dirampha*, *H. microtis*, *H. sinensis* and *H. qiui* are listed in the Remarks of the last. By its opercular morphology, rather undifferentiated verticil spines, *H. bandaensis* shows a certain similarity to the south-west African *H. arnoldi* and *H. augeneri*, but these have spines with rounded tips. A summary of similar forms, even if just faintly so, is found in Tovar-Hernández *et al.* (2016, fig. 4).

Hydroides bandaensis was regarded to be a questionable synonym of *H. novaepommeraniae* by Imajima & ten Hove (1984: 47), who compared these nominal taxa further with the morphologically somewhat similar taxa *H. malleolaspina* and *H. recta*. The material mentioned by Sun & Yang (2000) differs from the type by a slightly higher number of verticil spines, 11 as opposed to 9, and might rather belong to *H. novaepommeraniae*, and see discussion there as well.

Two populations of the nominal taxon *H. bandaensis*, from Kavieng, Latangai Island, and Gizo, Solomon Islands respectively, were retrieved as separate phylogenetic lineages by Sun *et al.* (2018); in view of the wide geographical distances between these two lineages, it is not unlikely that they form a complex of species, which also may include material from three other localities, Sumbawa (with only 6 verticil spines, our fig. 25, as opposed to the 9 in the type) and Banda, Indonesia, and the north South China Sea).

Distribution: North South China Sea; Banda Sea, Indonesia; Kavieng, Latangai Island; Gizo, Solomon Islands. See Remarks.

Fig. 25. *Hydroides bandaensis*, various views of the same operculum, Dangar Besar, Indonesia, *Siboga* Expedition sta. 313, ZMA V.Pol. 1854.1 (photo E. Wong).

10. *Hydroides bannerorum* Bailey-Brock, 1991

Figure 26

Type information: 70 m, near Sand Island and Barbers Point outfalls off Oahu, Hawaii, USA, holotype BPBM R 2153, paratypes BMHN ZB 1990.16, ZB 1990.17, 18; BPBM R2154, USNM 136581.

Etymology: Named in honour of the biologists Albert H. (Hank) and Dora May (Dee) Banner.

Selected synonymy

Hydroides bannerorum Bailey-Brock, 1991: 199, fig. la–r (description [original]; Hawaii, USA); Bastida-Zavala & ten Hove 2003b: 72–73, fig. 2A–G (redescription, discussion; Oahu, Hawaii, USA); Bailey-Brock & Magalhães 2013: 3–4, fig. 3 (description, discussion; Guam, Mariana Islands; Sun 2017: 17, 151, 161, fig. 1A (diagnosis; same).

Diagnosis: Peduncle merging smoothly into or separated by constriction from verticil, no funnel. Verticil with 17 (16–19) dark spines, straight to slightly curving outwards and gradually narrowing to fine tip. Each spine with basal internal knob adjacent to central disk; small and irregular tubercles in middle sections of some verticil spines, but without further adornments. Bayonet of collar chaetae with 2–3 large teeth and proximal rasp. Seven to 9 thoracic chaetigers.

Remarks: A very aberrant species by its one-tiered operculum. Bastida-Zavala & ten Hove (2003b: 73) hypothesise that of the two tiers it is the funnel which has been lost in this species, and what is left is the verticil. A further difference with all known *Hydroides* species is the presence of rasp-shaped thoracic uncini in all chaetigers. *Hydroides bannerorum* is one of three *Hydroides* species known with more than the usual 7 thoracic chaetigers; the others are *H. bisecta* from Japan (Imajima & ten Hove 1989) and *H. sceptrifera* ten Hove nov. spec., described herein.

Distribution: Hawaii, Mariana Islands.

Fig. 26. *Hydroides bannerorum*, various views of the same operculum, Oahu, Hawaii, USA, AM W.42847 (photo E. Wong).

11. *Hydroides basispinosa* Straughan, 1967a

Figure 27

Type information: Ross River, Townsville, Queensland, Australia. Holotype AM W.4012.

Etymology: Not given by Straughan (1967a: 230) but named after the basal spinules of the verticil spines.

Selected synonymy

Hydroides basispinosus Straughan, 1967a: 230–231, fig. 12a–f (diagnosis [original]; material studied; Townsville, Qld, Australia).

Hydroides gradata Straughan, 1967a: 225–227, fig. 9a–g (diagnosis [original]; holotype AM W.4000, Pretty Beach, 40 km north of Cairns, Queensland, Australia).

Not *Hydroides basispinosus* **sensu** Sun & Yang 2014: 209–211, figs 108A–H, 117 (diagnosis, in Chinese; Dalian, Liaoning; Dongshan, Fujian; Mirs Bay, Hong Kong) [**non** Straughan, 1967] referred herein to *H. inornata* conform Sun *et al.* (2017b).

Pro parte *Hydroides operculata* Sun *et al.* 2015: 62–65, map 17 (description, remarks, but not the synonymy; only the material from Queensland).

Hydroides basispinosa Sun *et al.* 2017b: 39–46, fig. 1C–D, fig. 2 (discussion, DNA supported raise from the synonymy with *H. operculata*, where it was placed by Zibrowius 1971a: 695; Mission Beach, Qld, Australia); 2018: 741, figs 2, 3, 5, 6 (DNA, phylogeny; Qld, Australia).

Diagnosis: Peduncle with constriction (mostly ill-defined). Funnel with 34 ((25)29–41) lightly chitinised, pointed radii. Verticil sitting in funnel, radially or more rarely bilaterally symmetrical, with 7 ((4)6–7) slender incurving spines; with blunt basal spinule above central disk. All spines of the same shape and size, less frequently one or two dorsal spines larger than the rest. Bayonet of collar chaetae with 2 teeth.

Remarks: Species complex with the African/Arabian *H. operculata* **sensu stricto** and the Indian-Chinese *H. inornata* (see Sun *et al.* 2017b). For some time, it was considered to be a synonym of *H. operculata*, first proposed by Zibrowius (1971a: 695; respectively *in litteris* 1973), which was still followed by Sun *et al.* (2015: 63), but a molecular study by Sun *et al.* (2017b) revealed three well-supported clades, suggesting a natural phylogeographical pattern. For a full discussion see Remarks under *H. operculata*.

Distribution: Qld, Australia.

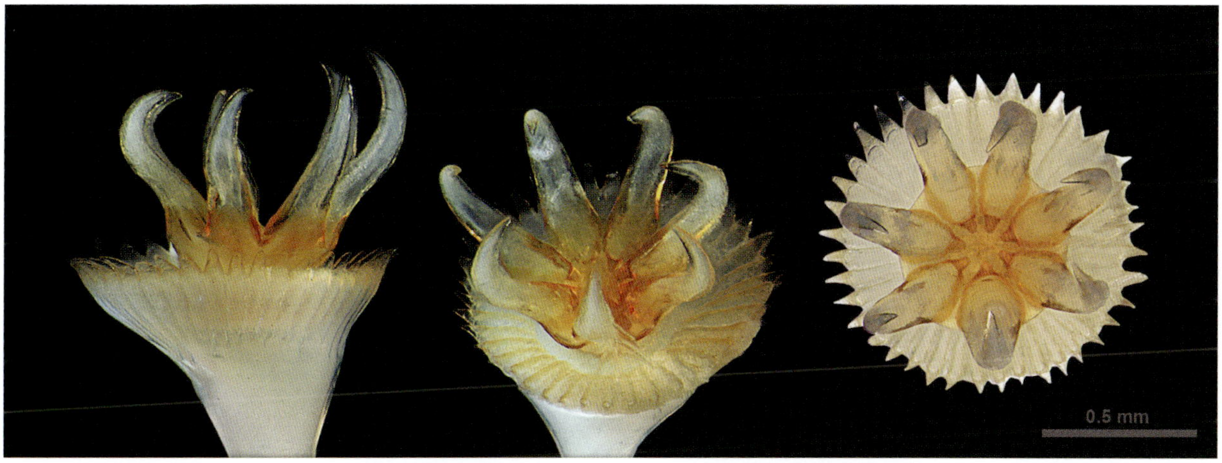

Fig. 27. *Hydroides basispinosa*, various views of the same operculum, Townsville, Queensland, Australia, AM W.42360 (photo E. Wong).

12. *Hydroides bifurcata* Pixell, 1913

Figure 28

Type information: Minikoi, Maldive Islands, holotype 'Prof. Stanley Gardiner's collection', holotype NHML PO01 1972.117.

Etymology: Not given by Pixell (1913: 76) but apparently refers to the bifid verticil spines (Latin *bis* = twice, *furca* = fork).

Selected synonymy

Hydroides bifurcatus Pixell, 1913: 76, pl. 8, fig. 3 (diagnosis [original]; Maldive Islands); Day 1951: 64, fig. 8f–g (diagnosis; Umhlali, Natal, South Africa); Kumaraswamy Achari 1969: 40 (name in Catalogue; Hare Island, Gulf of Mannar, India); Sun *et al.* 2018 fig. 1 (phylogeny, after Pillai 1972).

Hydroides bifurcata Day 1967: 808, fig. 38.4q–r (diagnosis; Natal, Mozambique); Rullier 1972: 154 (diagnosis; Baie de Saint Vincent, New Caledonia).

Additional material: Oman, Masirah Island, Ra's al Ya, low tide, under stones & beach (16 spec. ZMA V.Pol 3839, 4 spec. AM W.42848; see Remarks).

Diagnosis: Peduncle without or with ill-defined constriction. Funnel with 33 (25–40) hardly chitinised, pointed radii. Verticil bilaterally symmetrical, inserted with short basis into funnel, with 7 (7–9) almost straight spines including a large dorsal, slender hook, smoothly incurving. Other verticil spines equal in size, bifid, their 2 tips slightly pointing outwards, and a short internal spinule below; all verticil spines with basal spinule curving away from smooth central disk. Bayonet of collar chaetae with 2 teeth and 2 accessory ones.

Remarks: At first sight the operculum of *Hydroides bifurcata* is not unlike that of *H. dipoma*, both with a slender dorsal incurving hook, and verticil spines with a bifid tip. However, in *H. bifurcata* the bifurcation is sideways, and in *H. dipoma* one tooth is directed outwards, the other inwards. Although Day (1951: 64, fig. 8f, g) clearly diagnosed *H. bifurcata*, he erroneously used the name *H. spinosus* for his text figures. Our additional material from Oman is missing the internal spinules subdistally from the bifid tip and therefore labelled *Hydroides* aff. *bifurcata* (see our Fig. 28). More material, preferably supported by DNA, is needed for definite identification. Rullier's material should be checked; his diagnosis conforms ours (with as minor difference 10 verticil spines, the material might belong to *H. externispina*), but its locality is in a different ocean than the other finds.

Distribution: Natal, South Africa; Maldives, ?Oman and southern India; ?New Caledonia.

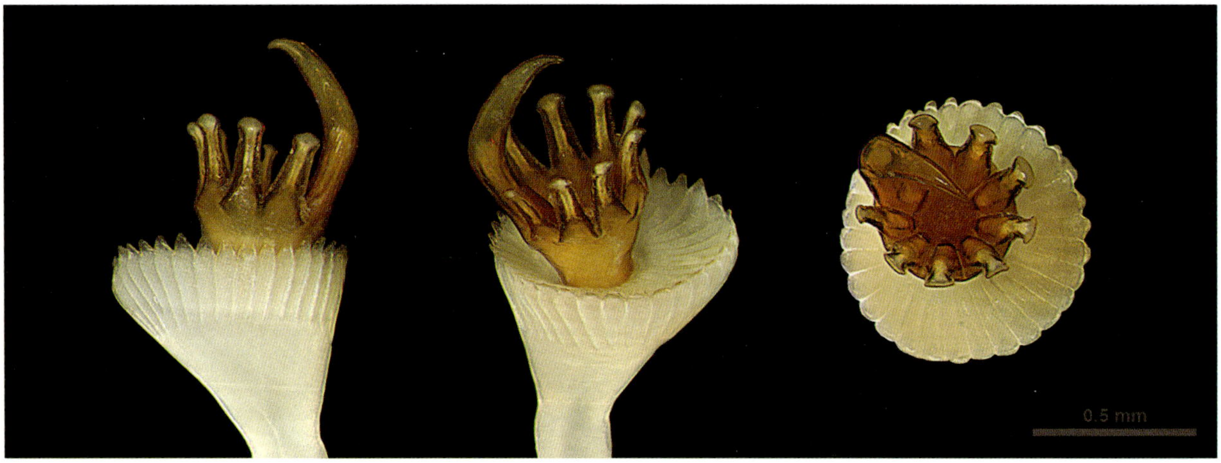

Fig. 28. *Hydroides* aff. *bifurcata*, various views of the same operculum, Masirah Island, Oman, AM W.42848 (photo E. Wong).

13. *Hydroides bisecta* Imajima & ten Hove, 1989

Figure 29

Type information: Off Sesoko Marine Station, Okinawa, Japan, holotype in NSMT–Pol. 256, figured herein.

Etymology: Not given by Imajima & ten Hove (1989) but refers to the bifid verticil spines (Latin *bis* = twice; *sectus -a* = cut).

Selected synonymy

Hydroides bisectus Imajima & ten Hove, 1989: 13–15, fig. 2a–l (description [original]; Japan).

Diagnosis: Peduncle with well-defined constriction. Funnel with 27 hardly chitinised, sharply tipped radii. Verticil radially symmetrical, inserted with short basis into funnel; with 7 spines, slender, all sharply incurving with dark hook. Subterminally the spines bear an outer, irregularly shaped whip-like appendage. Spines without further appendages, but slightly swollen internally above small, unadorned central disk. Nine thoracic chaetigers.

Remarks: Known from the holotype only (Fig. 29). By its outer, whip-like appendages of the verticil spines quite distinct from all known species of the genus.

Distribution: Okinawa, Japan.

Fig. 29. *Hydroides bisecta*, various views of the same operculum, Sesoko, Okinawa, Japan, holotype NSMT H-256 (photo E. Wong).

14. *Hydroides bispinosa* Bush, 1910

Figure 30

Type information: Bermuda, holotype unknown to us. The Yale Peabody Museum type (syntype? YPM IZ 001367.AN), from the Verrill Bermuda Expedition in 1898, evidently has no further location data, but Castle Harbour is a collection site mentioned by Bush (1910).

Etymology: Not given by Bush (1910: 496) but evidently refers to the pair of lateral spinules of the verticil spines (Latin *bis* = twice): Bush compared *bispinosa* with *multispinosa* [= 'many spined'].

Selected synonymy

Hydroides bispinosa Bush, 1910: 496 (diagnosis [original]; Bermuda); Bastida-Zavala *et al.* 2017: 23–24, figs 4A, 5 (diagnosis, synonymy).

Hydroides crucigera **sensu auctores**, e.g. Rioja 1958: 250–251 [**non** Mörch, 1863] (Verde Island and Santiaguillo Island, Veracruz, eastern Mexico) (for Caribbean records under this name, see Remarks); Vittor & Johnson 1977: 167, Johnson & Vittor 1982: 137 (name only, material studied; Florida, USA).

Hydroides bispinosus Bastida-Zavala & ten Hove 2003a: 125–127 (redescription, synonymy; Bermuda to Aruba).

Hydroides sanctaecrucis **sensu** Díaz-Díaz & Liñero-Arana 2001: 11, 13–14, figs 2a–f, 3g–l [**non** Krøyer in Mörch, 1863] (Cariaco Gulf, Venezuela) **fide** Bastida-Zavala & ten Hove (2003a).

Diagnosis: Peduncle with ill-defined constriction. Funnel with 16 (12–24) not chitinised, rounded radii. Verticil inserted on very short basis in funnel, faintly bilateral symmetrical, with 8 (7–10) spines strongly curving inwards, each with pair of lateral spinules at mid-length and a sharply pointed basal spinule above small central disk. One to 3 dorsal spines may be somewhat larger than the others; spines mostly smoothly curving inwards, rarely with 'knee', but never with external knob on the curve.

Remarks: The opercular verticil of *Hydroides bispinosa* is very similar to that of *H. parva*, and has been confused with that taxon in routine identifications. However, the funnel radii are pointed, chitinised in the latter; moreover, the verticil spines are hooked with a knob on the hook, absent in *H. bispinosa*. At least 12 authors reported *H. crucigera* (*-us*) from Florida and vicinity. However, this is an Eastern Pacific species, easily differentiated from both *H. bispinosa* and *H. parva* by the combination of gradually curving slender spines and sharply pointed radii. See Remarks under the similar *H. caribensis* as well. Western Atlantic records of the nominal species *H. crucigera* should thus be referred to one of the other taxa, generally only possible after study of the material. The pair of lateral spinules give *H. bispinosa*, and certainly *H. parva*, a certain similarity to the verticil of the Pacific *H. panamensis*; however, in *panamensis* these shift to a more external position in the ventral spines, even to the extent that the pair is merging into a single spine, while they are always in lateral position in the other mentioned taxa. The operculum of *H. bispinosa* is similar to that of *H. stoichadon*, the main difference being the tips of the verticil spines, sharply pointed in the former, bluntly rounded and mostly abraded in the latter. See the remark on parasitism under *H. dianthus* as well.

Distribution: Bermuda, Gulf of Mexico, Eastern USA and Caribbean.

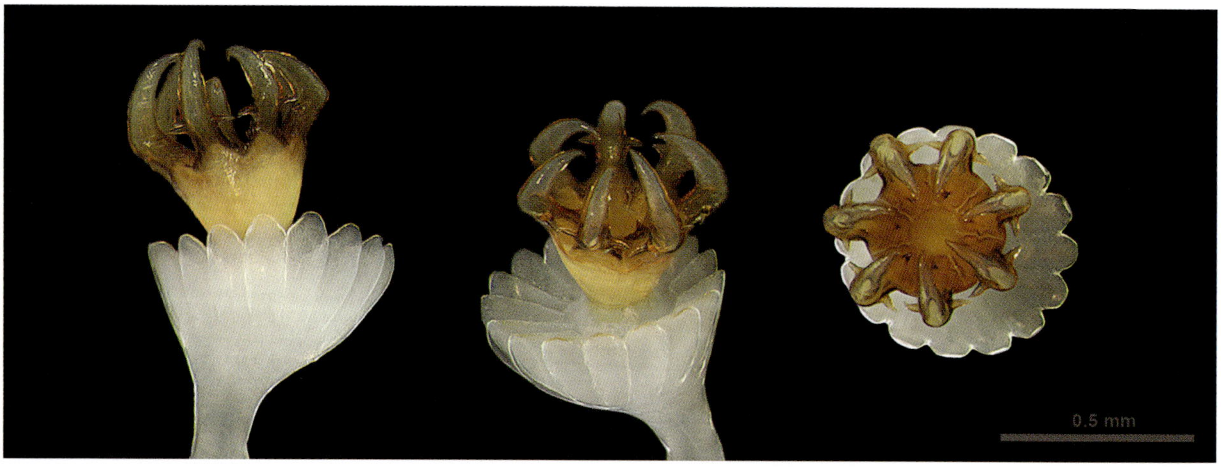

Fig. 30. *Hydroides bispinosa*, various views of the same operculum, Fuikbaai, Curaçao, AM W.42844 (photo E. Wong).

15. *Hydroides brachyacantha* Rioja, 1941a

Figure 31

Type information: Gulf of California, Sinaloa, Mazatlán, Marina Mazatlán, Mexico, 23°16′47″N, 106°27′40″W, neotype AM W.46899 (see Remarks, below).

Etymology: Not given by Rioja (1941a: 169–172) and his description gives no clue why he called it 'short-spined' (Greek βραχυ = short; ακανθα = spine).

Selected synonymy

Hydroides brachyacantha Rioja, 1941a: 169–172, pl. 3, fig. 2, pl. 4, figs 1–9; Sun *et al.* 2018: 733–734, figs 1–3, 5, 6 (name, DNA, phylogeny; Mazatlán, Pacific Mexico).

Hydroides brachyacanthus – Bastida-Zavala & ten Hove 2003b: 73–76, figs 3a–m, 7a–f (redescription, Mexican Pacific; they regarded records from Hawaii as questionable); (?) Lewis *et al.* 2006: 669 (probably ship-transported from Mexico; see Remarks); Bastida-Zavala 2008: 22–23, fig. 6c (discussion of complex; several localities Mexico); Tovar-Hernández *et al.* 2009: 328–330, figs 3j, 8d–f (diagnosis, discussion; Mazatlán, Mexico); Sun *et al.* 2016, table 2, fig. 6 (name, DNA, phylogeny; Mazatlán, Mexico).

Not *Hydroides brachyacantha* **sensu** Dew 1959: 28, fig. 7A–G [**non** Rioja, 1941] the material from New South Wales, Australia belongs to *H. amri* **fide** Sun *et al.* (2015: 14), and that from Queensland, Australia to *H. tuberculata* **fide** Imajima & ten Hove (1984: 44, 50).

Not *Hydroides brachyacantha* **sensu** Bastida-Zavala 1993: 35 (Bahîa de La Paz, Gulf of California) [**non** Rioja, 1941], see *H. similis* **fide** Bastida-Zavala & ten Hove (2003b: 74).

Diagnosis: Peduncle with shallow to well-defined constriction. Funnel radii with 30 (14)26–35(39 in Neotype) chitinised bluntly tipped radii. Verticil inserted with short basis into funnel, bilaterally symmetrical, with 8 (7–12) incurving spines with pointed tips. Dorsal hook broad, curved, bigger than all other spines, covering central disk. First and second pair of dorsal spines (lateral to dorsal hook) with tips and trunks wider than remaining ones, all with a pronounced knob on their bend. All verticil spines with short basal spinule. Central tooth absent.

Remarks: As Rioja (1941) did not leave any collections and thus the type material of *H. brachyacantha* is not available, and in accordance with the International Code of Zoological Nomenclature (ICZN 1999), Article 75.3, given the absence of a holotype of *H. brachyacantha*, Sun *et al.* (2016) designated a neotype collected from the original type locality (Mazatlán, Mexico). The complexity of the synonymic list above, and certainly that of the existence of a second genetic clade in the type locality (Sun *et al.* 2016, fig. 2), clearly demonstrates the necessity of this action. The status of this second clade needs further research. The selected synonymy, above, has been restricted to records from the type locality and surrounding Panamian province, but does not differentiate between the two genetic clades. We thus explicitly have refrained from including here, for instance, Eastern Mediterranean records (e.g. Grosse *et al.* 2021), which, in the light of two genetic lineages in the type locality alone, should be checked for genetic support.

Examination of topotypical specimens of *H. brachyacantha* showed that the first and second pairs of dorsal verticil spines (lateral to the very large dorsal hook) are wider than lateral and ventral spines (not explicitly mentioned in descriptions by Bastida-Zavala & ten Hove 2003b and

Fig. 31. *Hydroides brachyacantha*, various views of the same operculum, Mazatlań, Mexico, neotype AM W.46899 (photo E. Wong).

Tovar-Hernández *et al.* 2009, but conform Rioja's original description and figures by Bastida-Zavala & ten Hove 2003b). In *H. amri* and *H. nikae* the dorsal hook is far less pronounced than in *H. brachyacantha*, and the dorso-ventral gradient in size of verticil spines is more gradual. On the possession of a large dorsal hook and relatively few rather simple, incurving verticil spines, 11 taxa, now regarded to be valid species in their own right, have been confused with the nominal species *Hydroides brachyacantha*, although already for a couple of decades *H.* cf. *brachyacantha* has been in use for non-Mexican records, the 'cf.' indicating the researchers' doubt as to the identity of their material. For instance, what in the meantime has been described as *H. amri*, for decades was named *H. brachyacantha* (see Remarks with *H. amri*). Differentiating characters with other species confused under the name *H. brachyacantha* will be found in the Remarks with the following names: *H. deleoni*, *H. inornata*, *H. monroi*, *H. nikae*, *H. ralumiana*, *H. similis*, *H. similoides*, *H. tuberculata* and *H. zibrowii* nov. spec. Yet another similar taxon is *H. gottfriedi*.

Some fouling specimens on the German barque *Gorch Fock* (AM W.20166), arriving in Sydney after the Australian Bicentennial Tall Ships race, were tentatively identified as *H. monroi* (by HAtH unpublished) and reidentified as *H. brachyacantha* in Lewis *et al.* (2006: 669). In view of our recent reappraisal of *H. brachyacantha* (e.g. Sun *et al.* 2016), we re-examined specimen AM W.20166, and compared it with *H. amri*, *H. brachyacantha* **sensu stricto**, *H. monroi* and *H. nikae*. The number of verticil spines (12) occurs in *H. brachyacantha* as well, but not in *H. amri*, nor *H. monroi* nor *H. nikae* (7–10, 7–9 and 6–9 respectively). In some specimens from the type locality of *H. brachyacantha* there seems to be one dorsal hook only (as in AM W.20166), though close inspection may show a pair of slightly larger laterodorsal spines. The number of funnel radii (estimated at 24–26, not completely visible in photographs of AM W.20166) is similar to that in *H. nikae* (18–26), but low as compared to the other taxa ((30(14–35), 33(26–42), 35(31–39)); it should be mentioned that the funnel in specimen AM W.20166 appears to be not fully developed by its unchitinised rounded radii (as opposed to clearly chitinised tips in the other four taxa) and by being narrower than the verticil (only occasionally seen in *H. nikae*). Altogether, an identification as *H. brachyacantha* for this ship-hull fouling specimen is still the most likely.

Distribution: Mexican Pacific, records from other areas (e.g. Eastern Mediterranean, Moçambique, southern Japan, Polynesia, Hawaii), if not already covered in the list of mistaken identifications above, should be carefully checked; see Remarks above.

16. *Hydroides bulbosa* ten Hove, 1990

Figure 32

Type information: Khor Ghubb Ali, Oman, holotype BMNH ZB 1972.42, paratypes BMNH ZB 1980.548, ZMA 3704, AM W.19862, USNM 123384.

Etymology: Not given by ten Hove (1990: 116–119) but evidently refers to the bulbous dorsal verticil spine.

Selected synonymy

Hydroides bulbosus ten Hove, 1990: 116–119, figs 2–9 (description [original]; Oman).

Hydroides bulbosa – Sun *et al.* 2018: 742, figs 2, 3, 5, 6 (DNA, phylogeny; Al-Doha, Kuwait).

Diagnosis: Peduncle with well-defined lightly chitinised constriction. Funnel chitinised, with 23 (20–29) chitinised, truncated radii. Verticil sitting in funnel, bilaterally symmetrical, with 9 (7–14) spines, curving outwards, with rectangular tips, without any accessory spinules, basally attached in grooves of dorsal bulb (no central disk). Dorsal spine consists of a bulbous median part and 2 latero-dorsal processes, triangular in cross-section, almost like a cross when seen from the strictly dorsal side. Median bulb with 7–14 longitudinal grooves.

Remarks: Morphologically, *Hydroides bulbosa* belongs in a complex with species like *H. albiceps*, as evidenced by the fact that Zibrowius (unpublished) identified the material from Oman under the latter name. In a DNA analysis (Sun *et al.* 2018, figs 2, 3; Fig. 13 here), however, the position of *Hydroides bulbosa* is problematical, between two lineages of *H. bandaensis* in one clade, as a sister group of a large clade including species complexes like *H. albiceps* and *H. exaltata*. DNA information of *H. steinitzi* has not been obtained.

The bulbous median part of the dorsal spine in *Hydroides bulbosa* is always smoothly rounded with furrows, to which the other verticil spines are fused at 1/3 their length. In both *H. albiceps* and *H. trivesiculosa* the remaining verticil spines are separate from the dorsal bulbous one, at most the dorsal-most of these are fused to the base of the bulb. Other verticil spines are absent in *H. steinitzi*, which thus is characterised by a verticil consisting of one lobed bulb only.

Distribution. Kuwait to Strait of Hormuz, Arabian (Persian) Gulf.

Fig. 32. *Hydroides bulbosa*, various views of the same operculum, Khor Ghubb Ali, Arabian (Persian) Gulf, paratype AM W.19862 (photo E. Wong).

17. *Hydroides calopoma* Zibrowius, 1973a

Figure 33

Type information: Isla Tortuga, Annobón, Gulf of Guinea, holotype MRAC 34.950.

Etymology: Refers to the operculum (Latinised Greek καλος = beautiful, πωμα = cap, stopper).

Selected synonymy

Hydroides calopoma Zibrowius, 1973a: 36–38, fig. 5a–j (diagnosis [original]; Gulf of Guinea); Sun 2017: 154, 163, fig. 2P (diagnosis; same).

Diagnosis: Peduncle with well-defined chitinised constriction. Funnel with 29–32 chitinised radii (counted on photograph of Zibrowius's specimen), with long, slender tips slightly bending outwards and a subapical denticle. Verticil almost radially symmetrical, sitting in funnel; with 8 or 9 spines, slender, all sharply incurving with a sharp hook and an additional blunt hook on their bend, directed outwards; this bifold end resembles a pickaxe. In addition, each verticil spine with a blunt external spinule at 2/3 of its length and a long, upright basal spinule above unadorned central disk. The dorsal spine slightly larger than others, only different by a longer basal spinule and a larger, slightly bifurcate hook on its bend.

Remarks: The species is known from a single, large specimen, length 40 mm, about 145 abdominal segments, the diameters of its two opercula being 1.2 mm and 1.5 mm. These are very typical and cannot be mistaken for any other known species even though the pickaxe-like tips of the verticil spines are similar to those in *H. dolabrus*.

Distribution: Annobón, Equatorial Guinea.

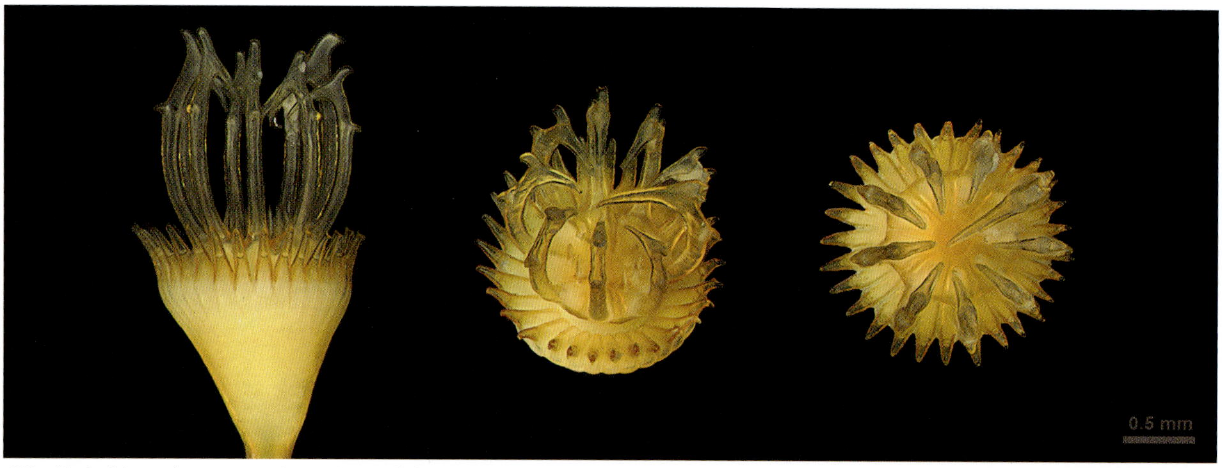

Fig. 33. *Hydroides calopoma*, various views of the same operculum, Annobón, Golfe de Guinée, holotype MRAC 34.950 (photo E. Wong).

18. *Hydroides capensis* Zibrowius, 1972b

Figure 34

Type information: Lambert's Bay, 32°05′S, 17°56′E, South Africa, holotype and paratypes BMNH 1961.9.819–833.

Etymology: Named after the Cape Province of South Africa.

Selected synonymy

Hydroides capensis Zibrowius, 1972b: 438–443, fig. 2a–f (description [original], opercular variability, synonymy, discussion; Western Sahara, Sénégal, Gambia, Guinée, Côte d'Ivoire, Ghana, Nigeria, São Tomé, Gabon, Congo, Angola, Cape Province, South Africa); Sun 2017: 149, 155, 163, fig. 3H (diagnosis; same).

Hydroides uncinata (*norvegica*) var. *multispinosa* – **sensu** McIntosh 1925: 89–91, pl. 9, fig, 7, pl. 10, fig. 13 [**non** Gunnerus, 1768] (description, discussion; South Africa).

Hydroides norvegica – **sensu** e.g. Monro 1930: 208 (very short diagnosis; South Africa) [**non** Gunnerus, 1768]; Day 1961: 549 (South Africa).

Diagnosis: Peduncle without constriction. Funnel not chitinised, with 28 (20–38) blunt, yellowish (very lightly chitinised) radii. Verticil on short stalk in funnel, with 10 (8–15) spines, distally curving inwards, radially symmetrical. All verticil spines similar in shape and size, smooth and pointed for their distal half, but proximally each with 2–3 (4) irregularly spaced lateral spinules (at both sides) as well as up to 3 internal spinules. The most distal of these spinules may be hooked and pointing to the centre of the verticil. Central disk small, flat, generally with central tooth of varying size (up to a third of peripheral verticil spines). Bayonet of collar chaetae with 2 teeth only, without proximal rasp.

Remarks: The confusion within the nominal taxon *Hydroides norvegica* was untangled by Zibrowius (1971a; 1972b); for a full discussion see Remarks under *H. azorica* and *H. norvegica*. One of the species from this complex, *H. capensis*, can be distinguished from *H. norvegica* **sensu stricto** by its operculum, with fewer and incurving spines.

Distribution: Western Sahara to Cape Province, South Africa.

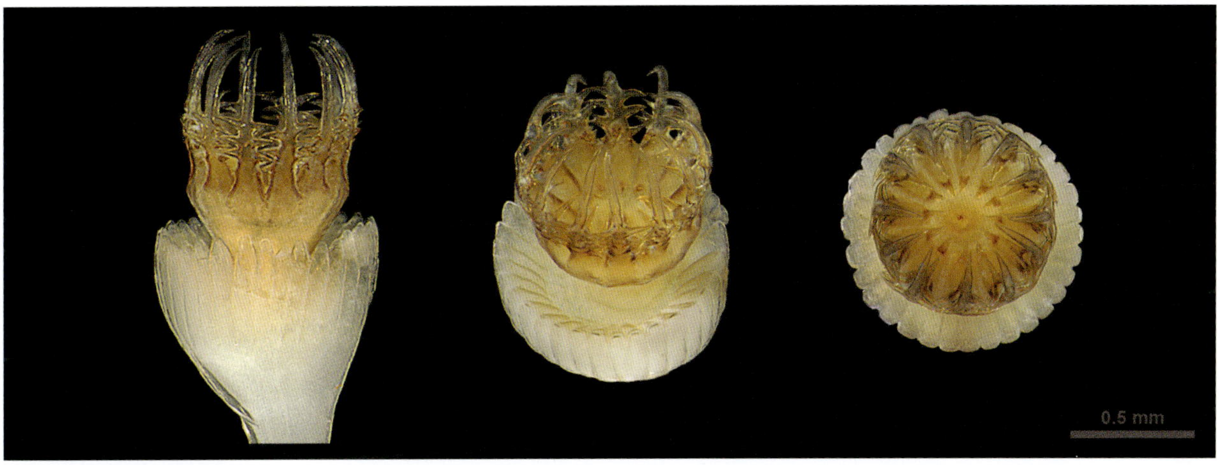

Fig. 34. *Hydroides capensis*, various views of the same operculum, Abidjan, Ivory Coast, SMF 21001 (photo E. Wong).

19. *Hydroides caribensis* nov. spec. ten Hove

Figure 35

Type information: West of Sanibel Island Light, RV *Hourglass* stn M, 26°24′N, 83°43′W, 73.2 m, Florida, eastern Gulf of Mexico. Holotype USNM 1143384, and nine paratypes: USNM 1143385, ZMA V.Pol. 3688, Ver. P1277, 1278 (microscopic slides), NHML 1992-282-283, FSBC I 39205, AM W.20154, NSMT-Pol P-231.

Etymology: Named after its distribution in the Caribbean Sea.

Synonymy

Hydroides sp. A – ten Hove & Wolf 1984: 55–17, 19, figs 55–11, 12a–p (diagnosis; off Florida, Gulf of Mexico; off Klein Curaçao, Netherlands Antilles; Shelf of Surinam); ten Hove & San Martín 1995: 16 (name; Canarreos Archipelago, Cuba).

Hydroides norvegica **sensu** Vittor 1975: 79, Vittor & Johnson 1977: 167 (both name in list only, material studied; Florida) [**non** Gunnerus, 1768].

Hydroides californicus **sensu** Nonato & Luna 1970: 99–100, figs 106–109 (diagnosis, discussion, see remarks; Southern Brazil, off Sergipe) [**non** Treadwell].

Hydroides sp. 1 – Bastida-Zavala & ten Hove 2003a: 167–179, figs 37A–O, 40 (description, discussion; Northern Gulf of Mexico, Mexican Caribbean and Cuba).

Additional material: Eastern Gulf of Mexico, 29°43′N, 85°15′W, 67 m, RV *Columbus Iselin*, BLM stn 2533, ident. B.A. Vittor *H. crucigera* (1 spec., USNM 56045); BLM 2645C, 2/78, 106 m, det. P.G. Johnson *H. norvegica* (1 spec., MESC); 28°24.0′N, 84°57′15″W, 85–95 m, BLM cruise 11, stn 22 (1 spec., MESC); 27°37′–26°24′N, 84°13′–83°43′W, 55–73 m, RV *Hourglass* stn E, L, M (37 spec., USNM 1143387–1143398, ZMA V.Pol. 3689, NHML 1992.284–289, FSBC I 39203–4, HUJ, NSMT-Pol P-231); approximately 200 km west of Venice, Florida, 27°07′N, 84°49′W, 239–247 m (1 spec., FSBC I 39810); approximately 219 km west of Sanibel Island, Florida, 26°28′N, 84°42′W, 236 m (2 spec., FSBC I 40226); White Shoal, Dry Tortugas, Florida, 24°38.2′N, 82°53.9′W, 5.2 m (1 spec., FSBC I 45909); Straits of Florida, 24°28.4′–32′N, 81°05′–17.6′W, 135–146 m, RV *Gerda* stn 377, 378, 1028 (3 spec., UMML 22:279, USNM 1143411, FSBC I 33239).

Cuba, Isla de la Juventud, Punta Pedernales, on gorgonian, 50 m, legit G. San Martín, 26 April 1984 (2 spec., UAM 19-0.3.02.05).

Jamaica, Discovery Bay, experimental asbestos caves, 40 m, legit J.B.C. Jackson 1974–5 (11 spec., ZMA V.Pol. 3227, 3693, ZMA VER.P.1263 (microscopic slide), ZMA V.Pol. 5076 (the last lot originally identified as *H. mongeslopezi* by H.J.L. Lambeck 1976–1977, later recognised by him as a new species and redet. H.A. ten Hove, USNM 1143386).

Puerto Rico, Mayagüez Bay, 4 km off Añasco River, soft bottom, sandy mud, 10 m, legit J.E. Randall, 14 September, 1963, det. C. E. van Eekhout 1968 *H. crucigera*; redet H.A. ten Hove 2003 *H. caribensis* (1 spec., ZMA V.Pol. 5131).

Netherlands Antilles, Klein Curaçao, north-west, from concrete block submerged for 1 month on sandy reef, 35 m, legit H.A. ten Hove, stn 2102, 13 June 1970 (6 spec., ZMA V.Pol. 3690).

Barbados, 0.8 km off Holetown, shell debris on muddy sponge bottom, 100 m, legit P. Wagenaar Hummelinck stn 1442, 19 February 1964 (1 spec., ZMA V.Pol. 3691).

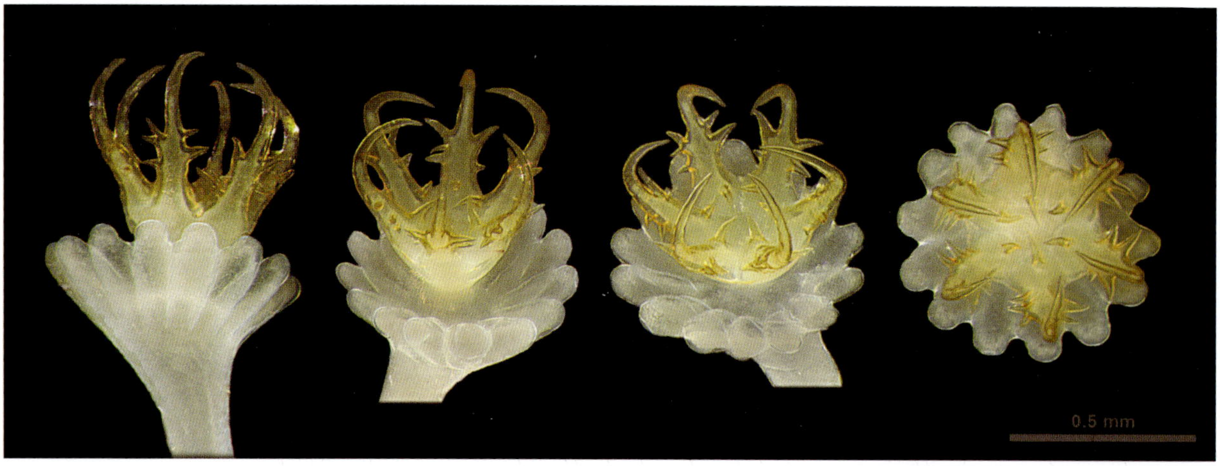

Fig. 35. *Hydroides caribensis* nov. spec., various views of the same operculum, Eastern Gulf of Mexico, paratype AM W.20154 (photo E. Wong).

Surinam, shelf, 07°35′–06°38′N, 53°48′–56°52′W, 38–130 m, HNLMS *Snellius* and *Luymes* stn C21, B23, B24, D33, F40, G54, H57, M97, K101B, J112 (together 22 specimens, RMNH 18179–86, FSBC I 39194, ZMA V.Pol. 3692).

Diagnosis: Peduncle without, at most with ill-defined constriction. Funnel with 17 (11–25) not chitinised, rounded radii. Verticil inserted on short basis in funnel, radially to bilaterally symmetrical, with 7 (5–9) spines smoothly curving inwards then downwards, each with 2 or 3 (exceptionally 4) lateral spinules and a sharply pointed basal spinule above small central disk. Rarely two dorsal spines may be larger than the others. Bayonet of collar chaetae with 2 teeth, sometimes 1 accessory.

Description: Tube: White, about 1.3 (up to 2.0) mm wide, with lumen of about 0.8 (up to 1.4) mm. Subcircular to trapezoidal in cross-section, often with coarse irregularly placed growth rings. Outer layer of tube semi-transparent, granular. Three longitudinal ridges may be present, a small median and 2 large lateral ones.

Radiolar lobes: With up to 13 radioles each, generally 11. Terminal filament of radioles up to 4 times longer than longest pinnules.

Peduncle: Smooth, circular in cross-section, except for shallow dorsal groove in some larger specimens; inserted just below first and second normal radiole. Generally, no distinct constriction below funnel, but faint joint sometimes present.

Operculum: With distal verticil inserted on short stalk into proximal slightly oblique funnel. Verticil radially to bilaterally symmetrical; with 7 (5–9) spines, curving inward then downward, each having 2 to 3 pairs (exceptionally 4) of lateral spinules and usually 1 sharply pointed basal tubercle. Spines with pointed tips. Spinules not always placed regularly, not infrequently with 3 lateral spinules at one side of a spine and 2 at the other. Spines almost meeting in centre; small central disk present, rarely absent; central tooth absent. Funnel not chitinous, with 17 (11–25) rounded radii. Grooves separating radii extend 1/3–2/3 of funnel length. Length of operculum about 1.2 (up to 1.7) mm, width 0.7 (0.9) mm.

Collar and thoracic membranes: Collar low, with entire edge; continuous with thoracic membranes, which form apron ventrally across anterior abdominal chaetigers. A small wart-like protuberance of collar chaetiger is present on each side between the ventral and lateral collar parts.

Thorax: With collar chaetiger and 6 uncinigerous chaetigers. Collar chaetae bayonet with 2 (–3) large conical teeth and distal striated 'limba' (Bastida-Zavala & ten Hove 2003a, fig. 37e–h) and narrowly limbate. Subsequent chaetae limbate, of 2 sizes (Bastida-Zavala & ten Hove 2003a, fig. 37j). Uncini saw-shaped, with fang and 5 teeth (Bastida-Zavala & ten Hove 2003a, fig. 37k–l). Prostomial ocellar clusters not observed.

Abdomen: Usually with about 50 (up to 70) chaetigers. Uncini saw-shaped anteriorly, with fang and 4 teeth; rasp-shaped with 5 rows of teeth (Bastida-Zavala & ten Hove 2003a, fig. 37n–o), fang and 8 teeth in profile view posteriorly. Chaetae flat trumpet-shaped with denticulate edge (Bastida-Zavala & ten Hove 2003a, fig. 37m), each with slightly enlarged, blunt anterior tooth about 3 times width of normal denticle. 'Capillary' chaetae present posteriorly. Pygidium with 2 small rounded lobes.

Size: Length up to 19 mm, usually about 13 mm. Width of thorax 0.5–0.9 (1.1) mm ($x = 0.7$, $SD = 0.1$, $n = 13$). Radioles (including operculum) accounting for 1/4–1/3 of entire length.

Colour: No records.

Remarks: Bastida-Zavala & ten Hove (2003a) gave a description of this taxon, as *Hydroides* sp. 1, without naming it because at that time ten Hove still hoped to include this new species, already known to him for a couple of decades, in a planned overview of all Floridean *Hydroides* species, together with Tom Perkins. This project was never finished, so a more complete description and list of material have been added here.

There are slight differences between opercula from the populations in the Eastern Gulf of Mexico, Cuba, Jamaica and Surinam (Table 3). Whereas the first three have almost radially symmetrical opercula, those of Surinam specimens show a tendency towards zygomorphy, in which two or three of the dorsal opercular spines are larger than the ventral ones. Eastern Gulf, Cuban and Surinam specimens have 2(–3) pairs of lateral spinules to the verticil spines; in the Jamaican population it is mostly only the distal pair that is well developed, the proximal pair being very small. However, opercula like that of the paratype (Fig. 35) occur in all populations.

Because proximal spinules of *H. caribensis* nov. spec. may be so small that they are easily overlooked, such specimens might be (and have been) mistaken for *H. bispinosa*, with one pair of lateral spinules on each verticil spine, as well as blunt, unchitinised radii. However, the verticil spines are more massive and generally more abruptly bent (almost geniculate) in *H. bispinosa*. An additional difference may be found in the abdominal chaetae, which have

Table 3. Opercular characters of three populations of *Hydroides caribensis* nov. spec.

	Number of verticil teeth	*x*	*SD*	Number of funnel denticles	*x*	*SD*	*n*
Types and topotypical material	6–8	7	0.7	14–21	17.6	2.1	17
Jamaica	5–9	6.6	1.5	11–17	14	2.1	9
Surinam	6–7	6.5	0.5	14–20	17.2	2.2	10

blunt anterior teeth in *H. caribensis*, but have sharp teeth and a thickened anterior tooth in *H. bispinosa*.

Hydroides caribensis resembles *H. capensis* from tropical Western Africa regarding the incurving verticil spines with up to 3 pairs of lateral spinules and a basal tubercle. Whereas all spinules are equally well developed in *H. capensis*, the proximal 1 or 2 pairs in *H. caribensis* are generally smaller than the distal ones. Moreover, subapical medial teeth and a central tooth are present in *H. capensis*, absent in *H. caribensis*. Furthermore, numbers of verticil spines in the 2 species differ, there being 5–9 in *H. caribensis* and 8–15 in *H. capensis*.

The record of *Hydroides californica* from Southern Brazil (Nonato & Luna 1970) must be incorrect. *Hydroides californica* is a junior synonym of *H. crucigera*, a Pacific species, which has sharp funnel radii and slightly geniculate verticil spines with lateral spinules at mid-length. The figure of the operculum given by Nonato & Luna (1970) shows rounded funnel radii and smoothly curving verticil spines with lateral spinules implanted proximally, characters typical for *H. caribensis*. They did not mention a second pair of lateral spinules, nor basal spinules, as found on *H. caribensis*, but these characters are easily overlooked. In our opinion their specimen most probably belongs to *H. caribensis*. In the Gulf of Mexico, this species has been misidentified routinely as *H. crucigera* or *H. norvegica*. More often, however, Caribbean records of *H. crucigera* should be referred to *H. bispinosa*, and those of *H. norvegica* should mostly be referred to *H. elegans*.

Ecology: So far, the species has been recorded only from 35–247 m depths. These records probably define the shallower margin of the depth range only; quite extensive collecting has been done in scuba diving depths in its area of distribution, while deeper samples are under-represented.

Distribution: Tropical West Atlantic.

20. *Hydroides chilensis* Hartmann-Schröder, 1962

Figure 36

Type information: Arica, Chile, type series ZMH V.15179, 15180 and ZMA V.Pol. 3687, ZMA VER.P.1264 (microscopic slide).

Etymology: Named after its type locality.

Selected synonymy

Hydroides chilensis Hartmann-Schröder, 1962: 165–167, figs 225–228 (description [original]; Arica, northern Chile); Bastida-Zavala & ten Hove 2003b: 76–78, fig. 4 (redescription; Pacific side of the Americas, from Colombia to Chile); Sun *et al.* 2018, fig. 1 (phylogeny, after Pillai 1972).

Hydroides crucigera **sensu** Laverde-Castillo 1988: 88–89 [**non** Mörch, 1863] (diagnosis; Bahía Málaga, Pacific side of Colombia); ?Knight-Jones & Knight-Jones 1991: 584 (name; Lima, Peru).

Diagnosis: Peduncle with shallowly to well-defined constriction. Funnel with 40 (21–58) lightly chitinised, short radii with pointed tips. Verticil sitting in funnel, bilaterally symmetrical, with 8 ((6)7–11) yellow spines, curving inwards, each with pair of lateral spinules at mid-length and pointed tip; all spines similar in shape but increasing in size towards dorsal side. Spines with a short, wart-like to long, downwards curving basal spinule above small central disk, without central tooth. Spines without external spinules or wings.

Remarks: This species is similar to *H. crucigera*, and for some time has been confused with it (for a detailed review see Bastida-Zavala & ten Hove 2003b: 78). The main difference is the asymmetric arrangement of the verticil spines (larger dorsally) in *H. chilensis*, as opposed to the radial symmetry in *H. cruciger.* Further differences can be found in body size, and in numbers and shape of funnel radii: 40 (21–58) short, hardly chitinised ones in *H. chilensis*, 26 (22–32) longer and well-chitinised in *H. cruciger.* By its incurving verticil spines with one pair of lateral spinules it also shows some resemblance to the Caribbean *H. bispinosa* and *H. parva*; however, in the latter two taxa the verticil spines are less slender, and more abruptly curving inwards, recurving, often with a 'knee' in the curve.

Distribution: Pacific, Colombia to Arica (Chile).

Fig. 36. *Hydroides chilensis*, various views of the same operculum, Arica, Chile, ZMA V.Pol. 3687 (photo E. Wong).

21. *Hydroides crucigera* Mörch, 1863

Figure 37

Type information: Punta Arenas, western Costa Rica, type material neither found in Copenhagen by HAtH, nor by Mary Petersen (pers. comm. to HAtH).

Etymology: Cross bearing (Latin *crux*, genitive *crucis* = cross, *gero* = carry), refers to the verticil spines.

Selected synonymy

Hydroides (*Eucarphus*) *crucigera* Mörch, 1863: 378, pl. 11, fig. 8 (diagnosis [original]; Western Costa Rica).

Hydroides cruciger – Bastida-Zavala & ten Hove 2003b: 78–80, fig. 5 (redescription, synonymy; Pacific side of the Americas, from Baja California to Bahia Malaga (Colombia), questionably Hawaii); Bastida-Zavala 2008: 23, fig. 6E (name; Pacific Mexico and Hawaii, ship fouling); Tovar-Hernández *et al.* 2009: 12–13, figs 3k, 6b–c, 7g–i (diagnosis, synonymy, ecology; Mazatlán, Pacific Mexico); Sun *et al.* 2016, table 2, fig. 6 (name, DNA, phylogeny; Mazatlán, Mexico).

Hydroides crucigera – Sun *et al.* 2018, figs 1–3, 5, 6 (name, DNA, phylogeny; Sinaloa, Pacific Mexico and Panama Bay).

Diagnosis: Peduncle with shallowly defined constriction. Funnel with 26 ((18)22–32) radii with well-chitinised, pointed tips. Verticil sitting in funnel, bilaterally symmetrical with 8 (7–10) yellow spines, curving inwards; each with pair of lateral spinules at 1/3 to half their length and pointed tip. All spines similar in shape and size; all with a short, wart-like to sharply pointed downward-curving basal spinule above small central disk, without central tooth. Spines without external spinules or wings.

Remarks: Most similar to *H. chilensis*; see above.

Distribution: Pacific, Punta Arenas, Baja California to Colombia, questionably Hawaii.

Fig. 37. *Hydroides crucigera*, various views of the same operculum, Mazatlań, Mexico, AM W.40538 (photo E. Wong).

22. *Hydroides dafnii* (Amoureux, Rullier & Fishelson 1978)

Figure 38

Type information: Elat, Israel, Red Sea, holotype MNHN A526 UB348 though not mentioned in Solis-Weiss *et al.* (2004); incomplete paratype in HUJ, Jerusalem.

Etymology: Named after the collector, Yaacob Dafni (Amoureux, Rullier & Fishelson, 1978: 60, 148).

Selected synonymy

Eupomatus dafnii Amoureux, Rullier & Fishelson, 1978: 148–149, fig. 11A–F (description [original]; Elat, Israel, Red Sea); Wehe & Fiege 2002: 128 (name only as synonym of *H. perezi*).

Hydroides perezi **sensu** Ben-Eliahu 1976: 106, 110, fig. 2 (diagnosis; Gulf of Elat, Israel; restudied by HAtH 1999, and reidentified herein to *H. dafnii*) [**non** Fauvel, 1918].

Eupomatus daphnii [*sic*] – Goren 1979: 22 (fouling asbestos panels, name in list, Rullier det.; material mentioned by Amoureux *et al.* 1978, Elat).

Fig. 38. *Hydroides dafnii*, holotype removed from its tube and close-up view of the operculum, Elat, Israel, MNHN A526 UB348 (photo S. Salazar).

Diagnosis: Peduncle without (in the original fig. 15B) or with ill-defined constriction. Funnel with 16–21 chitinised, sharply tipped radii. Verticil sitting in funnel, bilaterally symmetrical, with 5 spines enlarged at their base and tapering towards their tips, with smooth border. Dorsal spine larger, with sharp subapical tip pointing inwards (easily overlooked); with high power microscope all verticil spines show a small external knob below the tip. No lateral or internal spinules. No central tooth.

Remarks: Amoureux *et al.* (1978) mention one specimen collected by Dafni (originally deposited in Angers, now in Paris; and one by Goren, deposited in Tel Aviv, now in Jerusalem. Both specimens were restudied by Ben-Eliahu, one (in the meantime incomplete) in Jerusalem by ten Hove as well (both unpublished). There is one more known specimen, tentatively identified as *H. dafnii*, but published as *H. perezi* by Ben-Eliahu (1976); see above. The narrow tips of the verticil spines are unlike the broadly rounded ones in *H. perezi*, and also different from the more triangular spines in *H. tuberculata*; the outside knee/knob might be mistaken for a 'point' under a not-too-good stereomicroscope and be a reason for confusion. However, the latter has accessory teeth in its bayonet collar chaetae (Imajima 1976a, fig. 2d), a character not figured by Amoureux, nor in unpublished sketches of type material by Ben-Eliahu, neither by ten Hove. The synonymy with *H. perezi*, as given in the checklist of Wehe & Fiege (2002: 128), was based on a personal communication by Ben-Eliahu and ten Hove, an opinion that in view of our present knowledge – such as the multiple genetic lineages in *H. tuberculata* – may have been slightly premature. Additional studies are needed to determine whether *H. dafnii* and *H. perezi* are separate species, but unfortunately material for a DNA analysis was unavailable. See Remarks under *H. perezi* as well.

Distribution: Elat, Israel, Red Sea.

23. *Hydroides deleoni* Bastida-Zavala & ten Hove, 2003b

Figure 39

Type information: Punta San Juanico, western coast of Baja California Sur, Mexico, holotype and 20 paratypes ZMA V.Pol.3784.

Etymology: Named after Dr Jesús A. de León-González.

Selected synonymy

Hydroides deleoni Bastida-Zavala & ten Hove, 2003b: 80–83, fig. 7G–T (description, synonymy [original]; several localities West Mexico, partly fouling); Bastida-Zavala *et al.* 2016: 417, figs 4, 11D (name, new localities; South Baja California to North Peru); Sun *et al.* 2018, figs 2, 3, 5, 6 (DNA, phylogeny; Sinaloa, Mexico).

Diagnosis: Peduncle with ill-defined constriction. Funnel with 36 (30–43) lightly chitinised radii with pointed tips. Verticil sitting in funnel, bilaterally symmetrical, with 9 (7–11) spines, tips curving inwards. The most dorsal spine decidedly larger than the others, which are similar in shape and size, with pointed tip. Spines with short basal spinule; without external knob or lateral spinules and/or wings. Small central disk without tooth.

Remarks: *Hydroides deleoni* is rather similar to *H. brachyacantha* and has been routinely identified as *H.* cf. *brachyacantha* by students supervised by ten Hove (see Bastida-Zavala & ten Hove 2003b: 83), and it is even more similar to *H. amri* or *H. nikae*. However, these three taxa all have an external knob on the spines, lacking in *H. deleoni*. For a comparison with the Caribbean similar taxa *H. similoides* and *H. zibrowii* nov. spec. see Remarks under the latter.

Distribution: Baja California Sur to Ecuador.

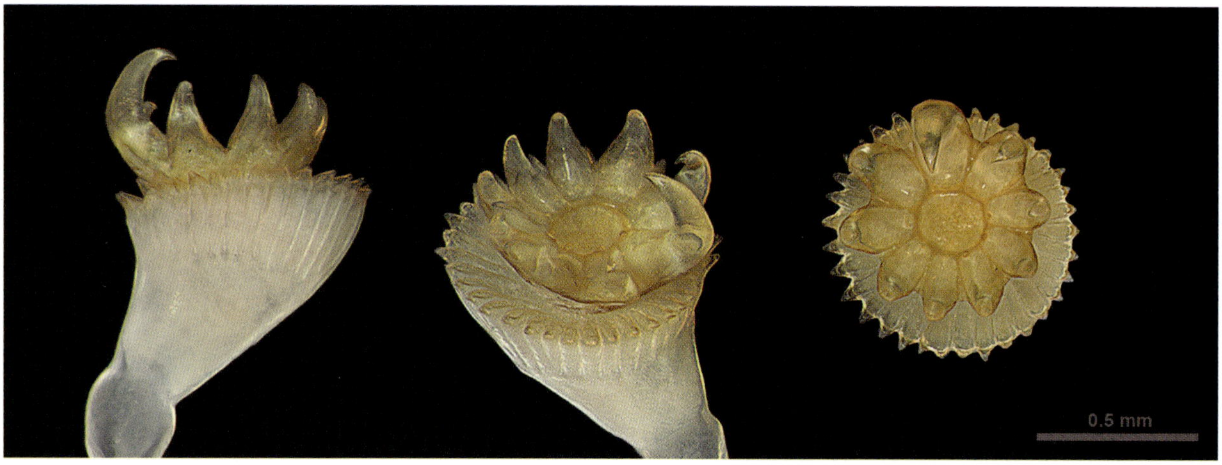

Fig. 39. *Hydroides deleoni*, various views of the same operculum, Miraflores Locks, Panama, AM W.42849 (photo E. Wong).

24. *Hydroides dianthus* (Verrill, 1873)

Figure 40

Type information: Great Egg Harbor to New Haven and Cape Cod, New Jersey to Massachusetts, USA, syntypes YPM 2698, 2699.

Etymology: Verrill (1873: 28) says, 'its elegant branchiae ... different kinds of pinks (as in the flowers of *Dianthus*). The name ... alludes to this resemblance'.

Selected synonymy (but see Remarks, below)

Serpula dianthus Verrill, 1873: 28, 620, no figures (description [original], material studied; New Jersey to Massachusetts, USA).

Serpula dianthus var. *citrina* Verrill, 1873: 620–621 (diagnosis, colour morph of preceding and co-occurring with it).

Hydroides hexagonus **sensu** Pratt 1916: 302, fig. 477 (name in handbook; Cape Cod to Florida, very common) [**non** Bosc, 1802]; see e.g. discussions in Zibrowius (1971a: 697) and Read *et al.* (2017: 13, 22–23) and *H. hexagona*, a *species inquirenda*, below; Colwin & Colwin 1961a, b, c (structure of spermatozoa).

Hydroides (*Eupomatus*) *dianthoides* Augener, 1922: 49–50 (diagnosis; in part *H. dianthus*, in part *H. sanctaecrucis*; Veracruz, Mexico, material studied).

Eupomatus dianthus – Hartman 1945: 45, 48, pl. 10, fig. 1 (diagnosis; Beaufort, North Carolina).

Hydroides dianthus – Zibrowius 1971a: 697–705, figs 1–5 (revision, synonymy; Cape Cod to Gulf of Mexico, USA; European and African Atlantic, Ile de Ré France to La Coruña Spain, Senegal, ?Nigeria; Mediterranean); Bastida-Zavala & ten Hove 2003a: 143–147, figs 23A–M, 24A–K, 28 (redescription, synonymy; off New Haven, Connecticut, USA south to Quintana Roo, Mexico); Kitsos *et al.* 2005: 258 (epibionthic on loggerhead turtles; Greece); Link *et al.* 2009: 1–5, figs 1A–G, 2 (description, discussion, distribution; introduced in Japan); Ben-Eliahu & ten Hove 2011: 14–15 (name, discussion; extensive Mediterranean synonymy); Boltachova *et al.* 2011: 34–38, figs 1–3 (name, fouling, new introduction; Sevastopol, Black Sea); Sun & Yang 2014: 212–213, figs 110A–F, 119 (diagnosis in Chinese; Dalian, Liaoning; Qingdao, Shandong, Yellow Sea); Bastida-Zavala *et al.* 2017: 25–28, figs 4C, 5 (diagnosis, remarks, very extensive synonymy; east coast USA to south Gulf of Mexico); Sun *et al.* 2017a: 1–12, figs 1–3 (DNA, phylogeny; several localities on western and southern coasts of USA, Brasil, Italy, Turkey, Ukraine, Yellow Sea, Japan); 2018 figs 1–3, 5, 6 (DNA, phylogeny; Woods Hole Massachusetts, St Petersburg, Florida, USA, and Sevastopol, Russia); Dong *et al.* 2018: 86–92, figs 1–3 (similar to clade A from Sun *et al.* 2017a; massively fouling, reef-like structures; Yellow Sea); Rodrigues *et al.* 2020: 6–8, figs 5, 6 (description; Rio de Janeiro, Brazil); Grosse *et al.* 2021: 15–28 (fouling, DNA; Mallorca, Spain); Palmer *et al.* 2022 (name, reefs, salinity tolerance 1–51.7, temperature 3–31.2°C).

Hydroides cf. *uncinata* **sensu** Sun & Yang 2014: 259–260, figs 119, 144A–H (diagnosis in Chinese; Dalian, Liaoning) [**non** Philippi, 1844], synonymised herein; see Remarks.

Diagnosis: Peduncle without or with ill-defined constriction. Funnel with 28 (24–33 (42)) chitinised, pointed radii. Verticil bilaterally symmetrical, sitting or on short basis in funnel, with 10 (8–13) stubby to more slender spines. Three to 5 dorsal spines curving inwards, slightly

Fig. 40. *Hydroides dianthus*, various views of the same operculum, Woods Hole, Massachusetts, USA, AM W.42071 (photo E. Wong).

larger and gradually decreasing in size towards remaining spines, which all point ventrally, thus outwards. Verticil spines with basal spinule, without external or lateral spinules, ending in pointed tips. Central disk without tooth. Bayonet of collar chaetae with 2 teeth, exceptionally 1 or 3.

Remarks: With well over 130 records in the literature this century alone, *Hydroides dianthus* is the second most frequently mentioned taxon in the genus (after *H. elegans*, with 270). In the older experimental literature and in several handbooks (starting with Pratt 1916) the taxon often has been named *Hydroides hexagonus* (see 'Selected synonymy' above). *Hydroides sanctaecrucis* and *H. dianthus* can be distinguished by the presence of external spinules on the main spines of the verticil, present in *H. sanctaecrucis* but absent in *H. dianthus*. Genetically these two taxa are well separated (Sun *et al.* 2018, figs 1, 2, 5, Fig. 13 here). A genetic study by Sun *et al.* (2017a), based on 112 specimens collected from 17 localities worldwide confirmed that *H. dianthus* **sensu stricto** is indeed a global invader. The genetic diversity in populations of the Mediterranean was higher than that in populations of eastern USA (Sun *et al.* 2017a: 1, fig. 1), but this seems to contradict the currently accepted native range of *H. dianthus* **sensu stricto** (clade A) in the USA. Moreover, this study revealed the existence of a potential cryptic species (clade B) in Galveston, Texas, for the time being indicated as *H.* cf. *dianthus*, with a genetic distance from the other specimens of 5.6%, but without obvious morphological distinguishing characters. This cryptic lineage was evidently introduced to the Black Sea only recently. Grosse *et al.* (2021) mentioned both lineages of *H. dianthus* from Mallorca (Spain) for the first time, but see Remarks under *H. nigra* as well.

Hydroides cf. *uncinata* **sensu** Sun & Yang (2014 fig. 119) co-occurs with *H. dianthus* in Dalian (China). A careful comparison of their figures (110 *H. dianthus*, 144 *H.* cf. *uncinatus*) with those given in the literature for *H. dianthus* (especially in Bianchi 1981: 61, Thorp *et al.* 1987: 866, Bastida-Zavala & ten Hove 2003a: 146, Link *et al.* 2009: 2) makes it extremely likely that both nominal taxa fall in the range of variability as shown in other localities.

Hydroides dianthus has been reported to be parasitised by the gastropod *Odostomia dianthophila* Wells & Wells, 1961: 149–157. Later this snail (nowadays known as *Fargoa dianthophila*) was reported from other species of *Hydroides* in south-eastern USA too: *H. microtis*, *H. floridana*, *H. bispinosa* (as *H. crucigera*) (e.g. Wells & Wells 1969: 109–110). Robertson & Mau-Lastovicka (1979) mentioned a second parasitic snail, *F. bartschi* (Winkley 1909), sometimes competing for food with its congener.

Distribution: Originally described from off New England, USA, reported along the east coast of North America down to Florida and Grand Caribbean, and reportedly introduced to Europe (Atlantic as well as Mediterranean, but see Remarks), China, Japan, and West Africa and Brazil. However, the full pattern of haplotype diversity (Sun *et al.* 2017a) indicative of the possible existence of different (cryptic) species is not yet clear enough to draw firm conclusions.

25. *Hydroides dipoma* (Schmarda, 1861)

Figure 41

Type information: 'Vorgebirge der Guten Hoffnung' (Cape of Good Hope, South Africa, south-eastern Atlantic).

Type material: Types apparently lost from NHMW.

Etymology: 'Das Thier hat zwei Deckel' (Schmarda, 1861: 29), thus refers to 2 opercula (Greek δις = twice, πωμα = cap, stopper).

Selected synonymy

Eupomatus dipoma Schmarda, 1861: 29: figs a–d, pl. 61, fig. 177 (diagnosis [original]; South Africa).

Hydroides uncinata var. *macronyx* Ehlers, 1913: 582–583, pl. 46, figs 1, 2 (diagnosis, discussion; South Africa, Simonstown).

Eupomatus spinosus Pixell, 1913: 78, pl. 8, fig. 5 (diagnosis; Suez).

Hydroides dipoma – Zibrowius 1973a: 33–35, fig. 4f–i (redescription, synonymy; from Río de Oro, Morocco to Cape of Good Hope, intertidal and subtidal; Suez); Ben-Eliahu & ten Hove 2011: 15–17 (name, synonymy, discussion; Gulf of Suez, Egypt and Cape Peninsula, South Africa).

Hydroides spinosus – Monro 1933: 508 (Still Bay, South Africa); Sun *et al.* 2018 fig. 1 (phylogeny, after Pillai 1972).

Hydroides uncinata **sensu** Monro 1939: 223–224 (diagnosis; Zaïre) [**non** Philippi, 1844, and renamed by Zibrowius (1973a, above)]; **in part only,** other parts of Monro's material belong to *H. arnoldi*, *H. augeneri* and *H. monroi*.

Not *Hydroides spinosus* **sensu** Day 1951: 56, fig. 8f, g [**non** Pixell, 1913] (Natal, South Africa; *lapsus calami* for *H. bifurcata* (see Day 1951: 64, and *H. bifurcata*, above).

Diagnosis: Peduncle with ill- to well-defined, chitinised constriction. Funnel with 39 (25–53) well-chitinised radii, long, pointed. Verticil bilaterally symmetrical, inserted with short basis into funnel, with 11 (8–14) spines curving inwards then outwards, including a large, slender, dorsal hook, smoothly incurving. Other verticil spines equal in size, almost bifid, the main tip pointing outwards, and a short, subapical medial spinule (incidentally missing in specimens). All verticil spines with a blunt basal spinule above smooth central disk.

Remarks: *Hydroides dipoma* has been described independently under three different names: *Eupomatus dipoma*, *Hydroides uncinata* var. *macronyx* and *Eupomatus spinosus*. Schmarda's diagnosis does not allow identification of the species, but that by Ehlers, also from South African material, is more detailed. Augener (1918: 595) examined Schmarda's material (which now no longer can be found in the collections of the NHMW), and confirmed the synonymy of the taxa described by Schmarda and Ehlers. Monro (1933) consecutively compared his South African specimen with a syntype of *E. spinosus* from Suez (Zibrowius 1973a, fig. 4), and decided that they were the same, a conclusion confirmed by Zibrowius. The species has a very disjunctive distribution: along the length of the south-east Atlantic coast, with a single Indo-West-Pacific location, Suez, without intermediate records from either the Mediterranean and Red Sea or from the Indian Ocean (Zibrowius 1973a: 34). The record from Suez is probably due to a shipborne introduction (Zibrowius 1973a: 36; Ben-Eliahu & ten Hove 2011: 17). The superficial similarity of the operculum with that of *H. bifurcata* is explained there. Remarkable is that in the material of Rio de Oro subapical internal spinules are missing (Zibrowius 1973, fig. 4h–i, Fig. 41 here; both from same locality and collector), a feature shared with a population of the latter taxon (see above).

Distribution: Tropical West Africa, South Africa and Suez (Red Sea).

Fig. 41. *Hydroides dipoma*, various views of the same operculum, Rio d'Oro, Western Sahara, ZMA V.Pol. 5064 (photo E. Wong).

26. *Hydroides dirampha* Mörch, 1863

Figure 42

Type information: St Thomas, Lesser Antilles ('*in portu urbis St. Thomae Antillarum*'); although some other types of Mörch (e.g. of *Hydroides (Eucarphus)* Mörch, 1863, synonymised with *H. dirampha*), and other old material is present in the ZMC, even from St Thomas, none of this fits Mörch's (1863: 379) remarks.

Etymology: Refers to the 2 sharp lateral spinules of the tip of the verticil spines (Greek δις = twice, ραμφος = beak, especially that of a bird of prey; Latinised to *dirampha -us*).

Selected synonymy

Hydroides (Eucarphus) dirampha Mörch, 1863: 379, fig. 10 (diagnosis [original]; Antilles).

Eupomatus lunulifer Claparède, 1870: 181–182, pl. 31, fig. 3 (description; Italy, Naples, on ship's hull).

Hydroides lunulifera – Fauvel 1927: 358–359, fig. 122p–s (diagnosis; Spain, Valencia; Italy, Naples; ship fouling); Sun *et al.* 2018, fig. 1 (phylogeny after Pillai 1972).

Not *Hydroides lunulifera* **sensu** Uschakov & Wu 1963: 158 [**non** (Claparède, 1870)]; see *H. sinensis* **fide** Zibrowius (1972a: 89).

Hydroides malleophorus Rioja, 1942: 126–130, figs 7–14 (description; Mazatlán, Gulf of California).

Hydroides dirampha – Imajima 1978: 54–56, fig. 3 (description, synonymy; S. Japan); 1987: 10 (distribution in Japan); Bastida-Zavala *et al.* 2017: 28–30, figs 4D, 5 (diagnosis, remarks, very extensive synonymy; Florida, Texas, California, Hawaii); Sun *et al.* 2018: 733, figs 1–6 (name, DNA, phylogeny; Sydney Harbour NSW, Bunburry WA, Australia; Colon, Panama; São Sebastião, Brazil); Grosse *et al.* 2021: 15–28 (fouling, DNA; Mallorca, Spain).

Hydroides diramphus – Bastida-Zavala & ten Hove 2003a: 161–164, figs 34A–P, 36 (redescription, synonymy, (sub) tropical Western Atlantic); Bastida-Zavala & ten Hove 2003b: 83–86, fig. 10A–L (description, synonymy, discussion, California and Hawaiian Islands); Ben-Eliahu & ten Hove 2011: 16–19, figs 4, 5A–C (discussion; Suez Canal); Sun *et al.* 2012b: 7–11, figs 2A, 5A–I (Hong Kong); Sun & Yang 2014: 214–216, figs 111A–I, 112 (diagnosis in Chinese; Xiamen, Fujian; Mirs Bay, Hong Kong; Xisha Islands, South China Sea); Tovar-Hernández *et al.* 2016: 694, 696, 700, 702, table 2, fig. 4 (name, DNA, phylogeny; Hong Kong, China).

Diagnosis: Peduncle with ill- to well-defined constriction. Funnel with 30 ((16)25–42) chitinised, pointed radii. Verticil radially symmetrical, inserted on short basis into funnel; with 14 ((9)11–18) spines similar in size and shape, rather straight, slightly curving outwards and ending in flattened T to crescent-shaped tips. All spines with blunt basal spinule above small central disk. Bayonet of collar chaetae with 2 teeth only, without proximal rasp.

Remarks: *Hydroides dirampha* shows low morphological variation in literature records worldwide (Day 1967; Zibrowius 1971a; Imajima 1978; Hayes *et al.* 2005; Sun *et al.* 2012b; 2015). Even populations from the Caribbean and the Indo-Pacific Panamian province were not essentially different in ranges of opercular radii and spines (Bastida-Zavala & ten Hove 2003a, b). Differences with similar taxa such as *H. bandaensis*, *H. microtis*, *H. sinensis* and *H. qiui* are enumerated in the Remarks after *H. qiui*. A summary of similar forms, even if just faintly

Fig. 42. *Hydroides dirampha*, various views of the same operculum, Port Jackson, New South Wales, Australia, AM W.27696 (photo E. Wong).

similar, is found in Tovar-Hernández *et al.* (2016, fig. 4). A so called *H. 'prisca'*-type operculum, developmental stage, has been recorded (ten Hove & Ben-Eliahu 2005: 8). In phylogenetic analyses it forms a clade with two Panamian Pacific, not even vaguely similar taxa: *H. crucigera* and *H. brachyacantha* (see Sun *et al.* 2018). This biogeographical discrepancy with the ecology based hypothesis one line below needs further study.

The species is fairly common in reasonably undisturbed lagoonal habitats in the Caribbean (Bastida-Zavala & ten Hove 2003a) where its presence appears to be 'natural' (probably its native distribution). Many literature records of *H. dirampha* (e.g. Day 1967; Imajima 1978; Zibrowius 1979, 1994; Ben-Eliahu & ten Hove 2011; Sun *et al.* 2012b, 2015) are from harbours and ships' hulls, suggesting that wide distribution of this species is due to accidental transfer by shipping.

Distribution: Mediterranean, Indo-West Pacific, tropical America, West Atlantic, East Atlantic, South Africa, New Zealand, Australia: NSW, Qld, Tas, WA.

27. *Hydroides dolabrus* Tovar-Hernández, Villalobos-Guerrero, Kupriyanova & Sun, 2016

Figure 43

Type information: Mexico, Gulf of California, Sinaloa, Mazatlán, Mazatlán Marina, 23°16′47.42′′N, 106°27′39.94′′W, holotype AM W.47012, about 150 paratypes AM, ECOSUR

Etymology: 'The name *dolabrus* is from the Latin *dolabra*, a sort of pickaxe that resembles the shape of the verticil spines'; however, *dolabrus* should be regarded as invariant noun in apposistion.

Selected synonymy

Hydroides dolabrus Tovar-Hernández, Villalobos-Guerrero, Kupriyanova & Sun, 2016: 696–703, figs 1–3, 4I, 5B, D, F (description [original], DNA, phylogeny, extensive discussion; Gulf of California, Mexico); Sun *et al.* 2018: 733, 739, 742, figs 1–3, 5, 6 (DNA, phylogeny; Sinaloa, Mexico).

Hydroides recurvispina **sensu** Tovar-Hernández *et al.* 2009: 13–15, figs 3m, 8d–f [**non** Rioja, 1941] (diagnosis; Mazatlán, Sinaloa, Mexico; synonymy **fide** Tovar-Hernández *et al.* 2016, above).

Diagnosis: Peduncle without constriction. Funnel with 26 (18–30) rounded radii. Verticil bilaterally symmetrical, inserted on short basis into funnel; with 9 (8–11) spines similar in shape and size, with short, blunt basal spinule above small central disk. Spine tips pickaxe-shaped, with two projections in the medial plane: internal ones short and pointing inwards, external projections longer than internal, pointing outwards. One dorsal verticil spine, however, with trifid tip, though easily overlooked. Verticil without central tooth.

Remarks: The original description is somewhat confusing in the statement (p. 701) 'all distal projections of verticil spines in *H. dolabrus* sp. identical', whereas the insert in their fig. 4 (Tovar-Hernández *et al.* 2016) clearly shows a trifid tip of 1, most dorsal, verticil spine. The authors give differential diagnoses of their new species *v H. bisecta*, *H. panamensis*, *H. recurvispina* and *H. trompi*, and additionally give photographs of the opercula of *H. bandaensis*, *H. dirampha*, *H. microtis* and *H. sinensis* as more or less similar forms (Tovar-Hernández *et al.* 2016, fig. 4).

In phylogenetic analyses *Hydroides dolabrus* forms a clade with *H. panamensis* (see Tovar-Hernández *et al.* 2016, fig. 6; Sun *et al.* 2018, Fig. 13 here), with which taxon it easily can be confused (Tovar-Hernández *et al.* 2016: 200). The main differences between *H. dolabrus* and *H. panamensis* are verticil spines mostly with bifid tip *v* mostly trifid; basal spinules short *v* long; funnel radii stubby, almost rounded *v* thin and with long, pointed tips (Tovar-Hernández *et al.* 2016: 700–701, fig. 5).

Distribution: Southern Gulf of California.

Fig. 43. *Hydroides dolabrus*, various views of the same operculum, Mazatlań, Mexico, AM W.40547 (photo E. Wong).

28. *Hydroides elegans* (Haswell, 1883), *nomen protectum*

Figure 44

Type information: Port Jackson, NSW, Australia, deposition of material is not given by Haswell (1883: 633); material of this taxon and this age was not encountered by HAtH in either the AM or the BMNH.

Etymology: Not given by Haswell, but Latin *elegans*, *-antis* = elegant.

Selected synonymy

Eupomatus elegans Haswell, 1883: 633, PL 32, fig. 11 (description [original]; Port Jackson, NSW).

Hydroides multispinosa [**non** Marenzeller, 1884] **sensu** Augener 1914: 139–142 (description, discussion; Shark Bay, Swan River, WA); Allen & Wood 1950: 100 (name, fouling; Eden, NSW to Moreton Bay, Qld).

Hydroides norvegica [**non** Gunnerus, 1768] **sensu** Allen 1953: 310–312 (ship-fouling; Port Jackson, NSW; New Zealand); Dew 1958: 52–54, figs 1–4 (opercular variations, aquarium at Taronga Zoological Park, Sydney); 1959: 23–25, fig. 4A–I (Townsville (Qld) to Geelong (Vic), Port Lincoln (SA), Fremantle (WA); Straughan 1967a: 219–220 (diagnosis; Qld, NSW); Kitsos *et al.* 2005: 258 (epibionthic on loggerhead turtles, Greece).

Hydroides spinalateralis – Straughan 1967a: 230, fig. 11a–g (diagnosis; Qld; material studied); Sun *et al.* 2018 fig. 1 (phylogeny after Pillai 1972).

Hydroides elegans – Zibrowius 1971a: 721–725, figs 56–64 (extensive description, discussion, synonymy; Mediterranean, from Marseille, France to the Suez Canal, Egypt; around the world in tropical to subtropical harbours); Imajima 1976b: 237–238, fig. 3a–n (description; various harbours in Japan); 1987: 10 (distribution in Japan); Moran 1984: 487–492 (ontogenetic and teratological variability of opercula; Wollongong and Port Kembla harbors, NSW); Bastida-Zavala & ten Hove 2003a: 164–165, figs 35A–J, 36 (description, synonymy, discussion, Florida to Curaçao); 2003b: 86–87, fig. 11A–S (description, synonymy, California and Hawaiian Islands); Sun & Yang 2014: 216–219, figs 113A–L, 114 (diagnosis, in Chinese; Dalian, Liaoning; Qindao, Shandong; Dongshan, Fujian; Shantou; Guangdong; Hong Kong; Hainan; Xisha Islands); Tovar-Hernández *et al.* 2016: table 2, fig. 6 (name, DNA, phylogeny; Sydney, NSW, Australia); Bastida-Zavala *et al.* 2017: 30–34, figs 4E, 5 (diagnosis, remarks, very extensive synonymy; Florida, California, Hawaii; and occasionally from Biscayne Bay, Florida and California, USA); Sun *et al.* 2017a: 2–3, fig. 2, table 1 (DNA, phylogeny; Sydney, NSW, Australia); 2018: 733–734, 738, 742, figs 1–3, 5 (DNA, phylogeny; Sydney, NSW, Bunbury, WA, Australia; Venice, Italy; Colon, Panama; Rio de Janeiro, Brazil; Los Angeles, USA); Grosse *et al.* 2021: 15–28 (fouling, DNA; Mallorca, Spain).

Protohydroides elegans – Uchida 1978: 27–30, pls IX–X (new combination, description; Seto, Shirahama, Japan).

Diagnosis: Peduncle without or with ill-defined constriction. Funnel not chitinised, with 24 (15–35) blunt radii. Verticil sitting in funnel, with 11–18 spines, straight, radially symmetrical. All verticil spines similar in shape and size, each with pointed tip, (1)2–4 irregularly spaced lateral spinules (at both sides), sometimes with up to 3 pointed internal spinules. Central disk distinctly elevated, generally with central tooth, simple, but sometimes with numerous teeth. Bayonet of collar chaetae with 2–5 teeth and proximal rasp.

Fig. 44. *Hydroides elegans*, various views of the same operculum, Sydney, New South Wales, Australia, AM W.40541 (photo E. Wong).

Remarks: Because the species is a notorious fouler, as well as easily cultured in the lab, the name *H. elegans* has been mentioned in the literature at least 500 times, of which 270 records in this century alone, and our synonymy indeed only can be a minimal selection. In our list we want to explicitly draw the attention to the different names attributed to what we now perceive as *H. elegans*, clearly reflecting that names are hypotheses only, based on the knowledge at the time of identification.

For several decades *Hydroides elegans* was synonymised with the nominal taxon *H. norvegica*, but Zibrowius (1971a: 717–725), on ecological as well as morphological grounds, correctly described them again as separate species (see Remarks under *H. norvegica*), further elucidated by ten Hove (1974: 45–48). Furthermore, it shows similarity with *H. azorica*, *H. longispinosa*, *H. multispinosa*, *H. nanhaiensis* and, to a lesser extent, *H. ochotereana* (see Remarks under *H. azorica*). Of the eight morphologically similar taxa only four were available for genetic analysis. In both Bayesian inference trees *H. elegans*, *H. longispinosa* (though polyphyletic) and *H. norvegica* formed a reasonably well-defined clade, while the position of *H. ochotereana* is less clear (Sun *et al.* 2018, figs 2, 3, 7, Fig. 13 here).

The taxon was originally described from Port Jackson, NSW, Australia, and in the meantime is regarded to be cryptogenic and probably ship-transported to a circum (sub)tropical distribution. *Hydroides elegans* is one of the most common foulers in (sub)tropical harbours, even in artificially heated docks and harbours in temperate climates (ten Hove 1974). It can stand fluctuations in salinity and some specimens even may survive 50 h in 15 ppt seawater (Mak & Huang 1982).

The first Mediterranean record of *H. elegans* dates back to 1888, from a fouling community in the harbour of Naples (Zibrowius 1992: 91). In the Mediterranean, *H. elegans* is restricted to lagoonal and harbour communities, and is absent from the open coasts. The same is true for tropical Curaçao, where *H. elegans* was found for the first time in the main harbour Schottegat by HAtH in 1970, and later (1975–85) spread into some other semi-enclosed lagoons, where it is replacing the local lagoonal species *H. sanctaecrucis* and *H. alatalateralis*. According to Allen (1953), at the time of its discovery in 1883, *H. elegans* (under the name *H. norvegica*) occurred rarely other than on certain seaweeds at a depth of about 20 m. This suggests natural occurrence rather than introduction by ship fouling. Moreover, Allen (1953) mentions observations made by docking officers that *H. elegans* was seldom, if ever, seen on ships fouled in the 1930s, whereas it was the most important fouling organism in Port Jackson in the 1950s (e.g. Allen & Wood 1950, as *H. multispinosa*; see also Wood & Allen 1958, as *H. norvegica*). He hypothesises its rise in importance to be due to the increasing pollution in the harbour, a thesis supported by Moran & Grant (1984). These ecological observations, together with the fact that *H. elegans* is part of a complex of six taxa, see above, of which three (the probably polyphyletic *Hydroides longispinosa*, *H. multispinosa* **sensu stricto** and *H. nanhaiensis*) occur in the Indo-Polynesian province adjacent to the type-locality of *H. elegans*, is making an Australian origin for the latter the most likely.

Distribution: Subtropical locations around the world (anthropogenic); Mediterranean-Atlantic, Indo-West Pacific including all around Australia, tropical Pacific America, West Atlantic, East Atlantic, South Africa.

29. *Hydroides elegantula* (Bush, 1910)

Figure 45

Type information: Bermuda, paratype YPM 1323

Etymology: Not given by Bush (1910: 497–498), but the Latin *elegans*, *-antis* = elegant and the suffix *-ulus* is the male diminutive form.

Selected synonymy

Eupomatus elegantulus Bush, 1910: 497–498 (description [original]; Bermuda).

Hydroides elegantulus – Bastida-Zavala & ten Hove 2003a: 118, figs 6A–J, 9A–F, 10 (description, synonymy; Bermuda); Sun 2017: 153, 166, Fig. 2G (diagnosis, figure of operculum).

Diagnosis: Peduncle without constriction. Funnel with 32–36 chitinised, finely pointed radii. Verticil sitting in funnel or on short basis, almost radially symmetrical, with 11–13 slender incurving spines. All spines similar in shape, dorsal spines slightly longer, proximally flattened into lateral wings for up to more than half length; each with short basal spinule, without knob in the middle, reaching 1/3 of small central disk. Without central tooth.

Remarks: This species is very similar to *H. alatalateralis*, *H. floridana* and *H. spongicola* in having wings along its slender, incurving verticil spines. For a full discussion see Remarks for *H. alatalateralis*.

Distribution: Bermuda Islands.

Fig. 45. *Hydroides elegantula*, various views of the same operculum, St George Bay, Bermuda, ZMA V.Pol. 4536 (photo E. Wong).

30. *Hydroides exaltata* (Marenzeller, 1884)-complex

Figure 46

Type information: East Coast of Island of Enoshima, Honshu, Japan ('Ostküste der Insel Eno-sima'), holotype unknown.

Etymology: Not given by Marenzeller (1884: 217) but Latin *exaltatus* = (up)lifted, referring to the elevated opercular verticil (Marenzeller 'auf einer … Säule', on a pillar).

Selected synonymy

Eupomatus exaltatus Marenzeller, 1884: 217, pl. IV, fig. 3A–C (diagnosis [original]; Enoshima Island, Japan); Willey 1905: 312–313, pl. VII, fig. 182 (diagnosis; off Chilaw, Gulf of Mannar); Pixell 1913: 77–78, pl. 8, fig. 4 (description; ship bottom, Wasin Basin, Kenya; Dungonab Harbour, Red Sea, Sudan); Mohammad 1975: 7 (name, experimental fouling; Kuwait).

Hydroides exaltatus – Dew 1959: 27–28, fig. 6 (diagnosis; Townsville, Queensland, Australia); Fiege & Sun 1999: 116, fig. 5D–F (description; Hainan, China); Pillai 2009: 115–117 (description, discussion; Kimberley, WA); Bailey-Brock *et al.* 2012: 969–972, fig. 3B–C (description; Rongelap, Marshall Islands); Sun *et al.* 2012b: 13–14, fig. 7A–I (description; Hong Kong); Sun & Yang 2014: 220–221, figs 115A–I, 119 (diagnosis, in Chinese; Dongshan, Fujian; Mirs Bay, Hong Kong; Longya Bay, Hainan; Xisha Islands).

Hydroides (*Eupomatus*) *exaltatus* – Augener 1914: 142–144 (diagnosis, discussion; Shark Bay, Western Australia).

Hydroides exaltata – Imajima 1976a: 127–128, fig. 4a–j (description, partial synonymy; south-west Japan); 1987: 10 (distribution in Japan); Imajima & ten Hove 1984: 48 (name only; Ponape Island, Micronesia; Okinawa, Japan, experimental fouling; Shark Bay, WA); Sun *et al.* 2015:

Fig. 46. *Hydroides exaltata*–complex, various views of opercula. Top row: New Caledonia, AM W.201332; bottom row: Kimberley, Western Australia, AM W.202945 (photo E. Wong).

28–32, fig. 7, map 6) (diagnosis; NT, Qld, WA); 2018: 738–740, figs 1–3, 5, 6 (DNA, phylogeny; Ashmore and Mermaid Reefs, Bremer Bay, WA, Australia; Gizo, Solomon Islands; Okinawa, Japan).

Diagnosis: Peduncle with well-defined chitinised constriction. Funnel with 24 (18–28) chitinised, sharply pointed radii; base of funnel chitinised, ending in long constriction. Verticil on columnar basis in funnel, bilaterally symmetrical, with 8 (6–10) spines. Dorsal spine large hook, somewhat swollen on its knee-like bend, strongly incurving with acute tip; remaining verticil spines much shorter, tapering, ending in simple outcurving acute tips. Spines almost meeting each other basally, all with basal spinule slightly elevated above meeting point. Bayonet of collar chaetae generally with 2 teeth, sometimes with an accessory tooth; see Remarks.

Remarks: Three genetic lineages were detected (Sun *et al.* 2018) in *Hydroides exaltata*, one in Japan (from Okinawa, within the same Sino-Japanese biogeographical province as its type locality Enoshima, but fairly distant), as sister to a cluster with the other two lineages, from Gizo, Solomon Islands and from the Kimberley, Western Australia. The nominal species *H. exaltata* clearly is a complex of cryptic species. We carefully compared all available figures from Red Sea, Sri Lanka/South India, Japan, China Seas, Polynesia, the Kimberley, WA and Queensland, Australia, and found no evident morphological differences in the opercula, nor in the counts of funnel radii (18–28) and verticil spines (6–10), which show a complete overlap. Sample sizes did not allow statistical comparisons.

Members of the *Hydroides exaltata*-complex might be confused with *H. lirs*, *H. minax* and *H. pseudexaltata*, and a differential diagnosis is given in the remarks under *H. lirs*. Pillai (2009, fig. 13C–I) figures bayonet collar chaetae with a range of 2–6 teeth, or 2 with up to 4 accessory ones for his material from the Kimberley. Marenzeller (1884, pl. IV, fig. 3) does not mention the number, figures only 1 tooth and Imajima (1976a: 128, fig. 3d) mentions 2, but figures a small accessory as well. Whether or not such differences are interspecific variations, or simply overlooked in other material should be kept in mind in future analyses.

Distribution: The complex has a wide distribution in the Indo-West Pacific from Zanzibar, Red Sea, and Arabian (Persian) Gulf to South Japan, Polynesia, and tropical Australia. More DNA analyses probably will reveal more specific lineages.

31. *Hydroides externispina* Straughan, 1967b

Figure 47

Type information: Heron Island, Qld, Australia, holotype AM W.4105.

Etymology: Not given by Straughan (1967b), but apparently refers to the lateral and external (curved outwards) spinules of the verticil.

Selected synonymy

Hydroides (Eupomatus) ralumianus **sensu** Fauvel 1930: 502, 506, 558–559 [**non** Augener, 1927]; Île de Pins, New Caledonia; referred herein to *H. externispina* – see Remarks).

Hydroides externispina Straughan, 1967b: 31–33, fig. 3a–f (diagnosis [original]; Heron Island, Qld; material studied); Imajima 1976a: 126–127, fig. 3a–k (description; Tanegashima, south-west Japan); 1987: 11 (distribution in Japan); Imajima & ten Hove 1986: 4 (name; Solomon Islands); Kupriyanova *et al.* 2015: 281–283, fig. 3 (diagnosis, discussion, synonymy; Lizard Island, Qld, Australia); Sun *et al.* 2015: 32–34, fig. 8, map 7 (redescription, discussion; Qld, WA, Australia); 2018: 733, 738, 742, figs 1–3, 5, 6) (DNA, phylogeny; Imperieuse Reef WA, Lizard Island Qld, Australia).

Hydroides externispinosa [in error] – Day & Hutchings 1979: 144 (name in checklist; Qld, Australia).

Diagnosis: Peduncle with shallowly to well-defined constriction. Funnel with 28 (24–36) pointed radii with chitinised tips. Verticil sitting in funnel, bilaterally symmetrical, with 9 (8–10) spines ending in sharply pointed tips, curved inwards. The two dorsal verticil spines large, with a pair of small lateral spinules, and hovering over the other spines. Remaining verticil spines all with a pair of outcurving lateral spinules at mid-length as well as one outcurving external spinule at 2/3 of their length (subterminal). All verticil spines with a small, blunt basal spinule above small central disk.

Remarks: *Hydroides externispina* is very similar to the figure of *H. ralumiana* by Augener, but for the fact that the latter has 2 pairs of lateral spinules to the verticil spines (as opposed to 1), as well as a pair of small lateral spinules to the two dorsal hooks, explicitly mentioned as smooth for *H. ralumiana*. However, Ben-Eliahu (unpublished) restudied Fauvel's specimen identified as *H. ralumiana* (in Paris) and only observed verticil spines with 1 pair of lateral spinules, as well as a basal spinule, not mentioned by Fauvel. With low resolution these basal spinules might be mistaken for proximal lateral spinules. Moreover, she mentions that the two dorsal hooks do have a pair of small lateral spinules, overlooked by Fauvel (and possibly by Augener in his specimen too), and reidentified Fauvel's specimen as *H. externispina*. It should be noted that the latter taxon had not yet been described in Fauvel's time, but in the meantime has been collected from New Caledonia (Zibrowius, ten Hove det., unpublished, ZMA V.Pol. 2558). New material from Papua New Guinea is necessary to solve the question if *H. ralumiana* is the senior name, and *H. externispina* its junior synonym.

Hydroides externispina is also very similar to *H. glasbyi* in overall appearance of the operculum with two large hooks and incurving other verticil spines with a subterminal external spinule. In fact, the field identification of the latter was *H.* cf. *externispina*. However, the pair of

Fig. 47. *Hydroides externispina*, various views of the same operculum, Heron Island, Queensland, Australia, holotype AM W.4105 (photo E. Wong).

lateral spinules at mid-length of all verticil spines is missing in *H. glasbyi*. At first sight *H. externispina* might be confused with *H. tambalagamensis* due to the similar dark curved lateral spinules on the verticil spines of both species. However, all spines are alike in *H. tambalagamensis*, while *H. externispina* has 2 large hooks. See Remarks under *H. tambalagamensis* as well.

In their phylogenetic analyses, Sun *et al.* (2018, figs 2, 3, 5) recovered *H. externispina* as a single clade with two lineages, corresponding with material from WA and Qld, Australia respectively, but with a genetic distance between lineages of about 1/3 of the usual inter-specific genetic distance. This clade is sister to that of the *H. tambalagamensis* specimens from various places in Queensland. Genetic data for *H. glasbyi* are not available.

Distribution: Indo-West Pacific: Southwestern Japan, Indonesia, Australia: Qld, WA; Solomon Islands, New Caledonia.

32. *Hydroides ezoensis* Okuda, 1934

Figure 48

Type information: Akkeshi, Muroran and Oshoro, Hokkaido, Japan, types not known to us.

Etymology: Apparently after its type locality as Ezo is a former Japanese name of the island Hokkaido.

Selected synonymy

Hydroides ezoensis Okuda, 1934: 239–242, figs 5–7 (description [original]; Hokkaido, Japan); Imajima 1976b: 236–237, fig. 2a–o (redescription; Peter the Great Bay, USSR, Northern Japan, Tokyo Bay, Wakayama to Hiroshima); 1987: 12–13 (distribution in Japan); Zibrowius 1978: 141–150, fig. 1 (description, discussion; Sea of Japan, imported to Atlantic coast of France); Hayes & Sliwa 2003 (Australia, well-known fouler); Sun *et al.* 2012b: 12 (DNA; remarks); Sun & Yang 2014: 221–224, figs 116A–L, 117 (diagnosis in Chinese; Dalian, Liaoning; Shandong; probably not their record from Dongshan – see Remarks, below); Sun *et al.* 2015: 34–37, fig. 9 (redescription, discussion; Australian distribution); Tovar-Hernández *et al.* 2016: table 2, fig. 6 (name, DNA, phylogeny; Vladivostok, Russia); Miller *et al.* 2018: table A1 (transported on tsunami debris from Japan to North America); Sun *et al.* 2018: 733, 738, 742, figs 1–3, 5, 6 (DNA, phylogeny; Qingdao, China; Tokyo, Japan); Faasse *et al.* 2020: 415–412, fig. 1A–C (diagnosis; recent range extension and most northern Atlantic record, in the Netherlands).

Not ?*Hydroides* cf. *ezoensis* **sensu** Wang & Huang 1993: 7, 10 [**non** Okuda, 1934], from Hong Kong, probably see *H. longistylaris*, **fide** Sun *et al.* 2012a: 15.

Diagnosis: Peduncle without, at best with very ill-defined constriction. Funnel with 45 ((23)30–63) radii ending in pointed, lightly chitinised tips. Verticil sitting in funnel, radially symmetrical, with 30 (16)22–34) spines curved outward and fused basally for 1/3–2/3 of their length; all spines same in shape and size and tapering into a pointed tip, and with a row of 6–8(9) internal spinules. Small, flat central disk with central tooth of varying size. Bayonet of collar chaetae with 2 teeth, exceptionally 1 or 1 with some accessories. Imajima (1976b: 237) found collar chaetae with 1 tooth only more often on the southern coast of Japan.

Remarks: *Hydroides ezoensis* was originally described from temperate Japan and later reported from the east Chinese and Russian coasts of the Sea of Japan, forming large aggregations on intertidal rocky shores (Miura & Kajihara 1984), also Fig. 12C. It has been introduced in Europe (France, the United Kingdom and the Netherlands) and Australia, and recorded as a nuisance fouler (Imajima 1976b; Zibrowius & Thorp 1989; Faasse *et al.* 2020). *Hydroides ezoensis* was first collected by Julie People from Newcastle, NSW in 1997 (as mentioned in Sun *et al.* 2015), next collected in Sydney (Glebe Island Bridge, floating Pontoon, July 1998, legit J. Powell, det. P. Hutchings, AM W.24658) and listed as a medium priority hazard species by CSIRO in 2005 (Hayes *et al.* 2005). The record from Dongshan (tropical North S. China Sea, Sun & Yang 2014) is questionable, temperatures being too high for *H. ezoensis*, and probably too should be referred to *H. longistylaris*, like that by Wang & Huang (1993:7, 10) from nearby Hong Kong.

Fig. 48. *Hydroides ezoensis*, various views of the same operculum Botany Bay, New South Wales, Australia, AM W.29654 (photo E. Wong).

Hydroides diplochone (Grube, 1878a), in the literature sometimes mentioned as a synonym of *H. ezoensis*, is regarded to be a *species inquirenda* by us (see the section below). In the section, '*Nomina inquirenda* and rejected names, occasionally still in use', *Hydroides prisca*, and in the Remarks on *H. fusicola* we made some observations on the possibility that the name of this ontogenetic stage has been used for material in reality belonging to *H. ezoensis*.

Phylogenetic analyses (Sun *et al.* 2018, Fig. 13 here) recovered *Hydroides ezoensis* as sister clade of the sympatric *H. fusicola*; however, its stubby verticil spines with rows of medial spinules make it easily distinguishable from the long and smooth verticil spines the of *H. fusicola*.

Distribution: Sea of Japan and South China; as introduced species in France, southern United Kingdom and the Netherlands, and temperate Australia: NSW, Tas, Vic.

33. *Hydroides floridana* (Bush, 1910)

Figure 49

Type information: Cape Dear Rio, Florida, USA, types not known to us. Neotype ZMA V.Pol.3393.1; type locality: off Cape Lookout Shoals, North Carolina, 10–20 m.

Etymology: Named after its type locality.

Selected synonymy

Eupomatus uncinatus [**non** Philippi, 1844] – Ehlers 1887: 285–286, pl. 58, figs 6–11 (description; 'Cape Rear', see Remarks); McCloskey 1970: 24 (name, material studied; Beaufort, North Carolina, USA).

Eupomatus Floridanus Bush, 1910 (diagnosis [original], new name for *Eupomatus uncinatus* **sensu** Ehlers 1887: 285 [**non** Philippi, 1844]; Florida, USA); Pequegnat & Pequegnat 1968: 23, 44 (name, rare, below 17 m; off Panama City, Florida).

Hydroides rostrata Iroso, 1921: 53 (a second, independently given, new name for *Eupomatus uncinatus* **sensu** Ehlers 1887: 285 [**non** Philippi, 1844]); *H. rostrata* Iroso, 1921 is the senior homonym of *H. rostrata* Pillai, 1971, for which reason the latter has been renamed *Hydroides gottfriedi* Read, ten Hove, Sun & Kupriyanova, 2017: 35–36 (treated below).

Eupomatus decorus Treadwell, 1931: 4–5, fig. 3 (description, synonymised by Bastida-Zavala & ten Hove 2003a: 119–120 (Grand Isle, Louisiana; material studied).

Hydroides decora – Zibrowius 1971a: 695 (new combination in discussion only).

Hydroides uncinata **sensu** Day 1973: 132–134, fig. 18h–i [**non** Philippi, 1844] (diagnosis, material studied; Beaufort, North Carolina); ?partly Johnson & Vittor 1982: 137 (material partly studied, but their remark might refer to other species as well; Florida).

Hydroides floridanus – Bastida-Zavala & ten Hove 2003a: 118–122, figs 6A–J, 9A–F, 10 (redescription, synonymy; Bermuda).

Hydroides floridana – Barnes, 1982: 13–15 (Mississippi Sound, material partly studied); Sun 2017: 153, 167, fig. 2F (diagnosis, figure of operculum); Bastida-Zavala *et al.* 2017: 34–35, figs 4F, 6 (diagnosis, synonymy; Biscayne Bay, Pensacola Bay, Florida, USA).

Not *Eupomatus floridana/us* and *Hydroides* (*Eupomatus*) *floridanus* **sensu** Augener 1922: 52; 1925a: 16; 1927: 80, 1934: 116 [**non** Bush, 1910] from the Caribbean Islands, material mostly studied, all see *H. alatalateralis*.

Diagnosis: Peduncular constriction very ill-defined, if any. Funnel with 33 ((19)27–43) chitinised, finely pointed radii. Verticil sitting in funnel, almost radially symmetrical, with 12 (10–14) slender incurving spines. All spines similar in shape, dorsal spines slightly longer, proximally flattened into lateral wings for less than half length; each with basal spinule, in most specimens knobbed in the middle, and reaching to halfway the small central disk. Without central tooth.

Material additional to that mentioned in the most recent revisions (by Bastida-Zavala & ten Hove 2003a; Bastida-Zavala *et al.* 2017):

North Carolina, off Cape Lookout Shoals, on shells of *Argopecten gibbus*, 10–20 m, *Ensign*, legit W. Kirby-Smith, 15 March 1971 (**neotype** ZMA V.Pol. 3393.1 and 6 spec., ZMA V.Pol. 3393.2, ZMA VER.P.1343 (microscopic slide)); off Beaufort, 10–20 m, legit L.R. McCloskey, det. J. H. Day as *H. uncinata* (3 spec., USNM 51236–7); same, legit E.A. Andrews 1885 (1 spec., USNM 42725); South Carolina, 32°54′N, 79°12′W, 17 November 1977 (1 spec., USNM 57129).

Fig. 49. *Hydroides floridana*, various views of the same operculum, Blue Water, Florida, USA, AM W.43194 (photo E. Wong).

Florida, off St Augustine to Fort Pierce, 10–64 m, Rock Shrimp Project, stn 1A, 3, 4, 5, 178, 194, 206, 220, 263, 298 (170 spec., FSBC I 17413, 23341, 39613–39623, RMNH 10768–70, ZMA V.Pol. 3395); off Jacksonville, 30°20.2′N, 80°23.9′W, 40 m, SEAMAP Project, stn 91–84 (1 spec., FSBC I 45290); 30°20.2′–32.8′N, 79°58.8'–80°07.7′W, on *Oculina varicosa* Lesueur, 39–80 m, RV *Johnson* and *Sea Diver*, stn J021/JSL II 046A, B, 047, J027/JSL II 125, J037/JSL II 163C, J043/JSL II 187B, S010/JSL II 230C (13 spec., HBOM 50: 0841, 0844, 2562–4, FSBC I 33304); Harbor Branch Benthic Ecology Offshore Station, 27°33.2′N, 80°03.2′W, 33 m (2 spec., HBOM 50: 2565, FSBC I 33301); 27°23′–27°21′N, 80°12′–14′W, Hutchinson Island, 10–11 m, stn 2, 3, 4, 5 (67 spec., FSBC I 33306–18, 39450–62, 39582, USNM 54729–30); Fort Lauderdale, U.S. Naval Oceanographic Office bio-fouling test panels, 100 m (1 spec., USNM 1143414).

Eastern Gulf of Mexico, Florida, Apalachicola, on oyster bed, legit A.S. Pearse, 5 June & 17 September 1935, Treadwell det. *Eupomatus uncinatus* (2 spec., AMNH 2292, 2304); 10 km west of Anna Maria Island, on shell of pagurid, 12 m, legit J.K. McNulty, 14 April 1966 (20 spec., ZMA V.Pol. 3394); 27°35′–26°24′N, 82°06′–84°13′W, 6–55 m, RV *Hourglass* stn A, B, C, I, J, K, L (171 spec., ZMA V.Pol. 3670–1, MCZ, FSBC I 33277–33300); Federal Clam Project, northeast of Bell Shoal, Mexico Beach, Gulf County, 6 m, 17 March 1967 (1 spec., FSBC I 33302); Crystal River Power Plant Study, 28 55′42″N, 82 55′24″W, (1 spec., FSBC I 33305); Seahorse Key, near Cedar Key, 29°05′N, 83°05′W, E.L. Pierce & students, 1958–60 (2 spec., USNM 44073, FSBC I 19409); Missouri Key, 30 May 1967 (1 spec., FSBC I 33303); Alabama, 18.5 km south of Perdido Bay, legit B.A. Vittor, 25 June 1973, P.G. Johnson det. *Hydroides uncinata* (4 spec., deposition unknown, not USNM); Mississippi, Mississippi Sound, 8 km south of Horn Island, experimental fouling plates in wreck, 14 m, legit D.H. Barnes (12 spec., ZMA V.Pol. 3440); Louisiana, Grand Isle, legit Anderson (holotype and 8 paratypes of *Eupomatus decorus*, USNM 19624-1, ZMA V.Pol. 3208, ZMA VER.P.1088 (microscopic slide), AMNH 2033).

South-western Gulf of Mexico, Yucatan Strait, 22°03′N, 88°11′W, RV *Hernan Cortez* Cruise 63, Pelagic Fish Project stn 13, 37 m (1 spec., FSBC I 54838).

Description (based upon Atlantic specimens, unless stated otherwise): Tube white, about 1.5 (from Campeche = C: up to 3.1) mm wide, with lumen of about 0.8 (up to 1.0; C: 1.8) mm. Semicircular to subtrapezoidal in cross-section, with faint growth rings, covered by a hyaline granular layer. Three longitudinal ridges may be present.

Radiolar lobes each with (7)8–14 radioles (x = 10.3, SD = 2.3, n = 9; C: up to 20). Terminal filament of radioles as long as longest pinnule.

Peduncle smooth, subtriangular in cross-section, inserted just below first and second normal radiole; border between peduncle and base of funnel hardly distinguishable. Peduncle implanted eccentrically towards the dorsal side of the funnel, in 80% of the specimens. Pseudoperculum present.

Operculum with distal verticil inserted on short stalk into proximal slightly oblique funnel. Verticil with 10–14 spines (x = 11.8, SD = 1.4, n = 10), curved inward; each spine with prong-shaped basal tubercle, in 75% of the specimens with distal knee-like knob. Spines transversely flattened (with 'wing') basally for up to 1/2 of their length, with pointed tips, arranged around central disk partly covered by basal spinules. Central tooth absent, but disk occasionally centrally elevated. Funnel not chitinised, with (19, juvenile from Gulf of Mexico) 24–43 (x = 32.7, SD = 5.2, n = 10) acute chitinised radii. Grooves separating radii extending 1/2 of funnel length. Length of operculum about 1.2 (up to 1.5; C: 3.0) mm, width 7 (up to 1; C: 1.6) mm.

Collar high, with entire edge; continuous with thoracic membranes, forming apron ventrally across anterior abdominal chaetigers. A triangular protuberance of collar chaetiger present on each side just below junction of ventral and lateral collar lobes.

Thorax with collar chaetiger and 6 uncinigerous chaetigers. Collar chaetae stout bayonet with 1 or 2 large, rounded conical teeth, without distal striated zone, and narrowly limbate. Subsequent chaetae limbate, of two sizes. Uncini saw-shaped, with pointed fang and 5 curved teeth. Prostomial ocellar clusters not observed.

Abdomen with (38) 43–68 (x = 56, SD = 8.6, n = 7; C: 72) chaetigers. Uncini saw-shaped anteriorly, with fang and 5 teeth; rasp-shaped with approximately 5 rows of teeth, fang and up to 12 teeth in profile view posteriorly. Chaetae flat trumpet-shaped with denticulate edge, each with anterior non-denticulate zone approximately covering the width of 10 denticles. 'Capillary' chaetae present in about 12 posterior chaetigers. Pygidium bilobed.

Size: Length up to 16 mm (C: 30), usually about 10 mm. Width of thorax (0.4) 0.5–1.0 mm (x = 0.7, SD = 0.2, n = 10;

C: 1.6). Radioles (including operculum) accounting for 1/5 of entire length.

Colour. No records of live specimens.

Ecology: The season of attachment in the Northern Gulf of Mexico is July–October (Pequegnat & Pequegnat 1968: 44), in concord with Barnes's (1982: 14) results with steel fouling plates. Specimens occur as epibionts on calico scallops (Wells *et al.* 1964: 581), and on the coral *Oculina* (McCloskey 1970: 24); we found it on shells inhabited by hermit crabs as well. The known depth distribution is 6–100 m. See the remark on parasitism under *H. dianthus* as well.

Remarks: Ehlers (1887: 286) gave as localities for his specimens 'Cape Rear' and 'Cape Dear Rio', 'inside fishing ground'. Since such localities are not known from Florida, these very likely are transcription errors for Cape Fear and Cape Fear River. His material apparently is lost: it is not present in the MCZ or any other museum that we know of. Ehlers pointed out that his material differed slightly from the Mediterranean *Eupomatus uncinatus* Philippi (1844: 195). Bush (1910: 498) compared the description of *E. uncinatus* by Ehlers with Mediterranean material, which she believed to be *Eupomatus uncinatus* Philippi, and found the differences large enough to justify separation on specific level. She therefore proposed the new name *Eupomatus Floridanus*. Bush's Mediterranean material (YPM 2839) was studied by us and belongs to *Hydroides dianthus*. Iroso (1921: 53) independently came to the same conclusion as Bush, and proposed yet another name, *Hydroides rostrata*, for *E. uncinatus* of Ehlers. However, the Mediterranean material that Iroso had before her is clearly recognisable as what is known now as *Hydroides pseudouncinata* Zibrowius, 1968. The latter author concluded that *E. uncinatus* Philippi was insufficiently described. Because the types were lost, and in the Mediterranean alone at least two, and worldwide about 13, species had been reported under that name, Zibrowius (1971a) determined that *Eupomatus uncinatus* Philippi was a *nomen dubium* (for a full discussion, see '*Nomina inquirenda* and rejected names, occasionally still in use' later in this book). In view of this and the close similarity to *H. alatalateralis* and *H. elegantula* Bush, 1910, a **neotype** is designated herein for *H. floridana*. The description above is based almost entirely on material from the eastern Atlantic coasts of the USA. The neotype has 2 opercula, an older one and a regenerating one (cut longitudinally), respectively with 11 and 10 verticil spines and 27 and 38 funnel radii (Figs 9, –10). The dorsal knob of the basal spinules of the older operculum is about half the length of those figured for the regenerating one.

Hartman's opinion (1959: 574), followed by Perkins & Savage (1975: 59) that *Eupomatus floridanus* is a synonym of *H. sanctaecrucis* is incorrect. We have not been able to find a reason for the statement by Perkins & Savage that Rioja (1959) referred *E. floridanus* to *H. sanctaecrucis*.

Creyghton (unpublished) compared 23 opercula of *H. floridana* from Florida with the same number of the nominal taxon *H. alatalateralis* from Curaçao. The results have been summarised in the taxonomic remarks on *H. alatalateralis* in Bastida-Zavala & ten Hove (2003a: 115, table 2), and give the main differences between the two taxa. Checking an additional 25 specimens of this species complex from various localities (including a paratype of *H. decora* and a specimen of Day's '*uncinata*' from North Carolina) for these character states, it became apparent that all specimens attributed to *H. floridana* were from temperate localities, and *H. alatalateralis* exclusively from tropical Caribbean localities. Thus, morphological criteria for a specific distinction are backed by geographical distribution.

As for *H. elegantula* from Bermuda, we studied the type (YPM 1323) long ago, but only had one specimen for direct comparison at hand for the present comparison (St Georges Bay, between Nonsuch and Castle Island, 0–5 m, '*Président Théodore Tissier*' stn 231, 14 April 1954). All character states of *H. elegantula* are the same as those in *H. alatalateralis*, except for the length of the basal spinules, which in *H. elegantula*, are not quite as long as half the radius of the central disk. However, this condition can be found in 28% of the tropical specimens of *H. alatalateralis* as well. In conclusion, it is not unlikely that *H. alatalateralis* is a junior synonym of *H. elegantula*; a direct comparison of a series of specimens of the two nominal taxa complemented with DNA sequence data is needed before a more definite opinion of this possible synonymy can be formed.

Finally, *H. floridana* is similar to *H. spongicola* in having wings along its slender, incurving verticil spines. For a full discussion see Remarks for *H. alatalateralis*.

Distribution: North-west Atlantic, warm temperate (North Carolina to Florida, Gulf of Mexico).

34. *Hydroides furcifera* (Grube, 1878b)

Figure 50

Type information: Ubay, Pandanon, Philippines, syntypes MNHW 383, ZMB Q 5337.

Etymology: The Latin *furca* = fork is likely to refer to the double row of forked spines in opercular funnel as well as verticil, *fero* = to carry.

Selected synonymy

Serpula furcifera Grube, 1878b: 268–269, pl. 15, fig. 4 (description [original], material studied; Philippines); Wiktor 1980: 281 (two syntypes in Wroclaw MPW. No. 383; Philippines).

Schizocraspedon furcifera – Bush 1905: 179, 225, 287 (new combination only).

Hydroides bifidus Imajima, 1982: 46–48, fig. 4a–o (description; Palau Islands, off Arumonogui; **fide** Imajima & ten Hove 1984: 51).

Hydroides furcifera – Pillai 1972: 7, 11, 21–23, 26 (name only); Imajima & ten Hove 1984: 51 (diagnosis, synonymy; Philippines and Palau Islands); Sun *et al.* 2015: 38–40, fig. 10, map 9 (redescription, synonymy, discussion; Australian distribution); Sun *et al.* 2018 fig. 1 (phylogeny, after Pillai 1972).

Diagnosis: Peduncle with well-defined, chitinised constriction. Funnel with 22 ((14)18–32) thinly chitinised radii with long, slender bifurcated tips. Verticil radially symmetrical, inserted on short basis into funnel; with 18 ((14)16–23) long bifurcated spines with pointed tips. All spines with row of up to 7 internal spinules, irregularly sized and spaced, distal spinule bi- or tri-furcate. Infrequently branches of radii or verticil spines bifurcate again, making them three-tipped. Small, unadorned central disk.

Remarks: The operculum of *Hydroides furcifera* can be easily recognised by its unique long bifurcated verticil spines and funnel radii tips. The species shares bifurcated verticil spines with *H. heterofurcata*, but lacks the larger dorsal bifurcated hook of the latter; moreover, the verticil spines if *H. heterofurcata* are lacking the rows of internal spinules and its funnel radii are simple, pointed.

Distribution: Philippines, Palau Island; Qld and WA, Australia.

Fig. 50. *Hydroides furcifera*, various views of the same operculum, Gulf of Carpentaria, Queensland, Australia, AM W.201312 (photo E. Wong).

35. *Hydroides fusca* Imajima, 1976a

Figure 51

Type information: North-east of Tanega-Shima, south-west Japan, holotype NSMT–Pol. H 119.

Etymology: Not given by Imajima (1976a: 130–131) but Latin *fuscus* = dark, evidently referring to the black verticil spines.

Selected synonymy

Hydroides fusca Imajima, 1976a: 130–131, fig. 8a–i (description [original]; south-west Japan); 1976b: 231 (diagnosis; northern and southern Kyushu, Japan); 1982: 45 (name; Palau Islands); 1987: 13 (distribution in Japan); Sun 2017: 167–168, fig. 1R (diagnosis, figure of operculum.

Hydroides fuscus – Sun & Yang 2014: 224–226, figs 118A–L, 119 (diagnosis, in Chinese; East China Sea; west of South China Sea; Gulf of Tonkin).

Diagnosis: Peduncle with well-defined constriction. Funnel with 22 (20–26) radii with long, sharply pointed, well chitinised tips. Verticil on short basis in funnel, radially symmetrical, with 7(–9) black spines, all same in shape and size; spines ending in centrally recurved hooks, slightly turning anticlockwise; each with long basal spinule, almost meeting over small central disk. No central tooth. Bayonet of collar chaetae with 2 larger and 2 smaller teeth, as well as proximal faintly striated zone.

Remarks: This species is similar to *Hydroides nigra* in dark colour of the verticil, but latter has 1–3 enlarged dorsal spines. Recurved verticil spines occur too in *H. recurvispina* and *H. trompi*, but these turn clockwise.

Distribution: Dogo Island (Japanese Sea), south Japan, south China, Palau Island.

Fig. 51. *Hydroides fusca*, various views of the same operculum, Kagoshima Bay, Japan, ZMA V.Pol. 3244 (photo E. Wong).

36. *Hydroides fusicola* Mörch, 1863

Figure 52

Type information: Japan (not further specified); being in a private collection when described, the type material might have been lost.

Etymology: According to Mörch (1863: 374) on a shell of the bivalve *Fusus*; the Latin *colo* = to live.

Selected synonymy

Hydroides (*Eupomatus*) *fusicola* Mörch, 1863: 374 (diagnosis [original]; Japan).

Hydroides okudai Pillai, 1972: 10, 15, 23, 25, 26 (*nomen novum* for '*H. uncinata*' **sensu** Okuda (1937) and Uschakov ([1955] 1965), the latter already synonymised with *H. fusicola* by Zibrowius (1971a: 694); phylogenetic tree also in Sun *et al.* (2018 fig. 1).

Hydroides uncinata **sensu** Okuda 1937: 63–64, fig. 10 (diagnosis, opercular variation; Onagawa Bay, Honshu, Japan); Uschakov [1955] 1965: 402, fig. 161I–K (name in key, figure of operculum) [**non** Philippi, 1844]; synonymised by Imajima 1976b.

Hydroides fusicola – Imajima, 1976b: 235–236, fig. 1a-k (redescription, synonymy; Hokkaido, Japan to Kuril Islands, Russia); 1987: 14 (distribution in Japan); Nishi 1993: 17, 19 (name only, tube structure; Sesoko, Okinawa, Japan; most southern record, should be checked); Sun & Yang 2014: 228, figs 120A–J, 117 (same; Yellow Sea; Dalian, Liaoning; Qingdao, Shandong); Tovar-Hernández *et al.* 2016: table 2, fig. 6 (name, DNA, phylogeny; Manazuru, Honshu, Japan); Sun *et al.* 2018: 733, 738, figs 1–3, 5, 6 (DNA, phylogeny; Qingdao, China).

Not ?*Hydroides* cf. *fusicola* **sensu** Mak 1982: 603–604 [**non** Mörch, 1863], from Hong Kong, most likely see *H. longistylaris*, **fide** Sun *et al.* (2012a: 15, 17).

Diagnosis: Peduncle without, at best with very ill-defined constriction. Funnel with 30–44 radii ending in pointed tips. Verticil sitting in funnel, radially symmetrical, with 16 ((7)13–20) slender, smooth spines, all same in shape and size and tapering into a long pointed tip (see Remarks as well). Minute basal spinule present. Small central disk, not elevated but with a tiny central tooth. Bayonet of collar chaetae with 1 or 2 teeth.

Remarks: Both Okuda (1937, as *H. uncinata*) and Imajima (1976b) mention and figure two ranges of verticil spines for this taxon, one with 7–9 relatively short and thick spines, the other with about 16 (13–20) longer and slender ones. This is suggesting a '*prisca*'-like stage, the presence of a developmental stage in the opercular ontogeny (see *Hydroides prisca*, rejected name, below), and explain some of the questionable records of *H. prisca-us* as well as *H. uncinata* **sensu auctores**, recorded together with e.g. *H. ezoensis* as well as *H. fusicola*, but since not recognisable without examining the material purposely left out mostly from the selected synonymies. DNA analysis will be necessary to determine whether or not *H. huanghaiensis*, with similar operculum and distribution and only differing in the (not always easily visible) absence of basal spinules in the verticil spines, is a separate taxon. This taxon is also similar to *H. novaepommeraniae*, but the latter has up to 5 truncated dorsal verticil spines instead of all very thinly tapering ones, as well as a tropical distribution, not cold-temperate.

Phylogenetic analyses (Sun *et al.* 2018; Fig. 13) recovered *Hydroides fusicola* as sister clade of the sympatric *H. ezoensis*; however, its long and smooth verticil spines make it easily distinguishable from the stubby spines with rows of medial spinules of *H. ezoensis*.

Distribution: Japan, North China.

Fig. 52. *Hydroides fusicola*, various views of the same operculum Hokkaido, Japan, AM W.42852 (photo E. Wong).

37. *Hydroides gairacensis* Augener, 1934

Figure 53

Type information: Gairaca, Santa Marta, Colombia (Atlantic side), holotype ZMA V.Pol. 0458.

Etymology: After its type locality.

Selected synonymy

Hydroides (*Eupomatus*) *gairacensis* Augener, 1934: 117, fig. 20a–c (description [original]; Colombian Atlantic).

Hydroides uncinata **sensu** Monro 1933: 1082, fig. 25 (recognisable figure; Taboga Island, Pacific Panama) [**non** Philippi, 1844].

Hydroides gairacensis – Zibrowius 1969: 366–376 (revision, Colombia, Mexican Caribbean, Southern Brazil, Pacific of Panama); Bastida-Zavala & ten Hove 2003a: 129–132, figs 13A–P, 15 (description, synonymy, discussion); Bastida-Zavala *et al.* 2016: 420–421, figs 4, 11H (synonymy, Puerto Ángel, Pacific Mexico).

Diagnosis: Peduncle without or with ill-defined constriction. Funnel with 18 (15–24) radii with T-shaped to crescentic tips, lightly chitinised. Verticil sitting in funnel, radially symmetrical, with 6 (5–8) spines, distally strongly incurving, simple, smooth, without basal or lateral spinules. Bases of verticil spines touching, no central disk. Bayonet of collar chaetae with 2 or 3 teeth.

Remarks: The species is similar to *H. trompi* in the shape and number of the verticil spines except for the presence of a basal spinule in the latter; moreover, *H. trompi* has acutely pointed, rarely narrowly truncated radii while these are T- or crescent-shaped in *H. gairacensis*. It would be interesting to see whether or not the Amphiamerican distribution is supported by DNA data.

The split verticil spine as shown in our Fig. 53 is an anomaly, most probably teratological. Split verticil spines (laterally bifid) are the norm in *H. bifurcata*, *H. furcifera* and *H. heterofurcata*, but otherwise extremely rare, and only reported before for *H. protulicola* (by Bastida-Zavala & ten Hove 2003a, fig. 27c).

Distribution: South Florida and Caribbean Sea to South Brazil; Amphiamerican: Panama, Ecuador.

Fig. 53. *Hydroides gairacensis*, various views of the same operculum, Aruba, Netherlands Antilles, AM W.43519 (photo E. Wong).

38. *Hydroides glandifera* Rioja, 1941a

Figure 54

Type information: Acapulco, Guerrero, Mexico, type specimens lost.

Etymology: Though not given by Rioja (1941a: 174), he writes 'una robusta protuberancia … en forma de glande'; evidently it is a functional name for the bulbous dorsal verticil spine: Latin *glans*, *glandis* = acorn; *fero* = to carry.

Selected synonymy

Hydroides glandiferum Rioja, 1941a: 172–174, pl. 4, figs 10–14 (description [original]; Mexico, Pacific side).

Olgaharmania glandifera Rioja, 1941b: 733–734 (same material attributed to new genus).

Hydroides glandifer – Bastida-Zavala & ten Hove 2003b: 89, figs 12A–M, 14 (redescription, synonymy, discussion, Baja California Sur to Acapulco, Pacific Mexico).

Hydroides glandifera – Bastida-Zavala *et al.* 2016: 422, figs 5, 11I (diagnosis, synonymy, discussion; Oaxaca, Mexico); Sun 2017: 117, 157, 168, fig. 4T (diagnosis; Baja California Sur to Acapulco, Mexico).

Diagnosis: Peduncle with well-defined lightly chitinised constriction. Funnel with 15 (12–17) darkly chitinised radii with pointed tips, straight to outcurving. Verticil sitting in funnel, bilaterally symmetrical, with 4 darkly chitinised spines and a dorsal hook. The latter seemingly a large dome, but in reality a strongly recurved hook with acute tip; the other 4 spines fused to the (latero-)ventral side of this swollen hook, at 1/3 to 2/3 of its length, with pointed tips either appressed to the dome or directed outwards; the 2 laterodorsal spines with broad base. Spines smooth, without spinules. No central disk.

Remarks: Rioja did not mention an incurving tip (beak) on the ventral side of the 'bulbous' spine in his description. Bastida-Zavala & ten Hove (2003b) too missed the closely appressed beak in their new material from the Pacific side of Mexico. However, they found a sharply pointed tip coinciding with the medio-ventral V-shaped cleft between the two ventral-most spines. Sometimes it was necessary to squeeze the base of the verticil with a pair of tweezers to open up the beak and verticil spines to make them visible. They assumed that Rioja overlooked this beak, but that it was present in his material. At first sight the opercula of *H. glandifera* and *H. salazarvallejoi* are rather different, but on closer inspection the arrangement of the verticil spines is rather similar: a large dorsal hook with remaining spines closely appressed. However, the distal surface of the dorsal hook is convex in the first, concave in the latter; the remaining verticil spines are without a basal spinule and have pointed tips in *H. glandifera*, have a basal spinule and are blunt in *H. salazarvallejoi*. The verticil spines are not completely fused to the dorsal hook in *H. salazarvallejoi*: they still resemble a funnel. By the dome-like dorsal hook *H. glandifera* is similar to *H. simplidentata* as well; however, in the latter the dorsal hook has a rounded tip, not pointed, and the remaining verticil spines are not fused to, but form a funnel with the hook.

Distribution: Baja California Sur to Acapulco, Guerrero.

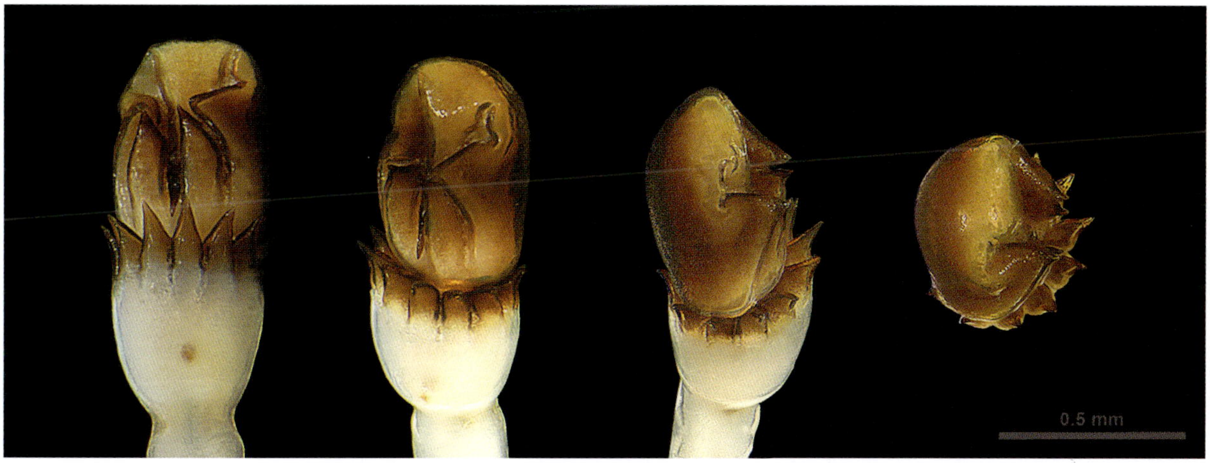

Fig. 54. *Hydroides glandifera*, various views of the same operculum, Huatulco, Oaxaca, Mexico, AM W.43615 (photo E. Wong).

39. *Hydroides glasbyi* Sun, Wong, ten Hove, Hutchings, Williamson & Kupriyanova, 2015

Figure 55

Type information: Northern Territory, Darwin, Fort Hill Wharf, 12°28′17″S, 130°50′48″E, holotype MAGNT W017502.

Etymology: The species is named '*glasbyi*' in honour of Dr Christopher J. Glasby.

Selected synonymy

Hydroides glasbyi Sun, Wong, ten Hove, Hutchings, Williamson & Kupriyanova, 2015: 40–43, figs 11a–b, 12a–f, map 10 (description [original]; Darwin, NT, Australia).

Diagnosis: Peduncle with well-defined constriction. Funnel with 24 (22–26) pointed chitinised radii. Verticil on short basis in funnel, bilaterally symmetrical, with 9 (8–10) spines ending in sharply pointed tips, curved inwards. Remaining verticil spines all with one outcurving external spinule on the bend, at 2/3 of their length (subterminal). All spines with a small basal spinule above small central disk. Bayonet of collar chaetae with 2 large and 2 slightly smaller teeth in between.

Remarks: *Hydroides glasbyi* is very similar to *H. externispina* in overall appearance of the operculum with 2 large hooks and incurving other verticil spines with a subterminal external spinule. In fact, the field identification of the former was *H.* cf. *externispina*. However, a pair of lateral spinules at mid-length in all verticil spines is missing in *H. glasbyi*. Unfortunately, there are no genetic data available.

Distribution: NT, Australia.

Fig. 55. *Hydroides glasbyi*, various views of the same operculum, Darwin, Northern Territory, Australia, MAGNT W21804 (photo E. Wong).

40. *Hydroides gottfriedi* Read, ten Hove, Sun & Kupriyanova, 2017

Figure 56

Type information: Hikkaduwa and Wellawatte, Sri Lanka, holotype NHM, London BM 1968–148 (Hikkaduwa), three paratypes BM 1968–148–151 (Wellawatte); types of *H. rostrata* (see below).

Etymology: The replacement name *H. gottfriedi* is given in memory of T. Gottfried Pillai (1930–2013), author of the original species name.

Selected synonymy

Hydroides gottfriedi Read, ten Hove, Sun & Kupriyanova, 2017: 35–36 [original], as replacement for the homonymous *Hydroides rostrata* Pillai, 1971 (for type material see above).

Hydroides rostrata Pillai, 1971: 107–108, fig. 5B–H (description; Sri Lanka; material studied); recognised as a junior homonym of *H. rostrata* Iroso, 1921: 53 and renamed by Read, ten Hove, Sun & Kupriyanova 2017; the latter has been synonymised with *H. floridana* by Bastida-Zavala & ten Hove (2003a: 119); Pillai 1972: 19, 24, fig. 3N (phylogeny; mistakenly named *H. ancorispina* in fig. 3N, but from text and figure it is evident that this is meant to be *H. rostrata*); Imajima & ten Hove 1984: 50 (comparison with *H. brachyacantha* **sensu auctores**); Sun *et al.* 2018: 734, fig. 1 (name, phylogeny, from Pillai 1972).

Diagnosis: Peduncle with well-defined constriction. Funnel radii with (14)16–19 chitinised pointed radii. Verticil inserted with short basis into funnel, bilaterally symmetrical, with 7 incurving spines with pointed tips. Dorsal hook bigger than all other subequal spines, covering centre of verticil (information on central disk or basal spinules thus inaccessible). All verticil spines with a pronounced external knob (dorsal hook) to pointed spinule (remaining 6) on the knee-like bend.

Remarks: Imajima & ten Hove (1984: 50) compared a number of taxa (including *H. gottfriedi*, still under its homonymic name *H. rostrata*) with what they perceived as *H. brachyacantha* from various localities, Panama to Australia. With help of DNA, this perception has been replaced by the understanding that *H. brachyacantha* **sensu auctores** is a complex of species, partly resolved by Sun *et al.* (2016). Most (all?) members of this *brachyacantha*-complex have a verticil with one large hook, accompanied by one or two pairs of latero-dorsal spines that are visibly larger than the remaining ones as opposed to *H. gottfriedi*, ostensibly with almost imperceptibly larger spines adjacent to the hook (Pillai 1971, fig. 5B). However, Pillai's original freehand drawings are not entirely in proper proportion, as shown by photographs of the paratype (Fig. 56) the shape of the verticil is rather similar to given for *H. brachyacantha* by Bastida-Zavala & ten Hove (2003b, fig. 3a–k). The funnel, however, has fewer radii: 16–19 in *H. gottfriedi*, as opposed to the average of 30 in *H. brachyacantha*. Differences with *H. adamaformis*, *H. inermis* and *H. monroi*, all with few (up to about 8), rather thick spines, including a large dorsal hook are discussed under *H. helmata*.

The original description by Pillai (1971: 107) mentions 16–24 radii, but a study of the holotype and three paratypes revealed a number of (?14)16–19. Tweezing the basis of the verticil slightly revealed that the dorsal hook has a minute basal tubercle, the other verticil spines a rounded basal spinule.

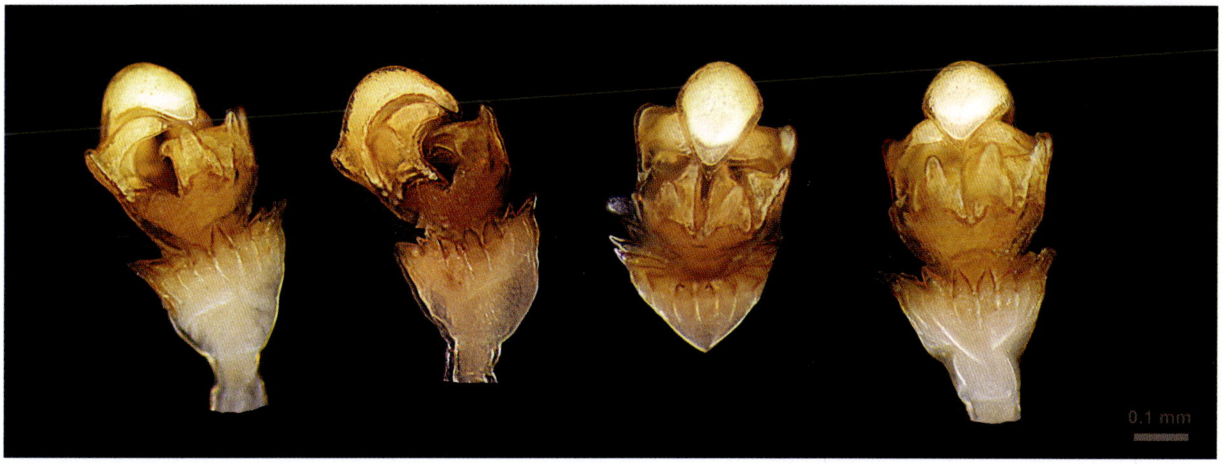

Fig. 56. *Hydroides gottfriedi*, various views of the same operculum, Sri Lanka, paratype of *Hydroides rostrata*, NHML 1968.151 (photo L. Nealova © trustees of the NHM, London).

41. *Hydroides gracilis* (Bush, 1905)

Figure 57

Type information: Pacific Grove, California, USA, two syntypes YPM 2697.

Etymology: Not given by Bush (1905: 234–235), but Latin *gracilis* = slender, thin, apparently referring to the simple verticil spines.

Selected synonymy

Eupomatus gracilis Bush, 1905: 234–235, pl. 27, fig. 9, pl. 34, fig. 25, pl. 37, figs 26–27 (description [original]; California, USA); Pillai 1972: 9 (phylogenetic grouping).

Eupomatus intereans Chamberlin, 1919: 23 (description; Laguna Beach, California, types MCZ Invertebrate Zoology ANNb-2179, synonymised by Zibrowius (1971a: 694).

Hydroides uncinata **sensu** Berkeley & Berkeley 1941: 56 (diagnosis; southern California) [**non** Philippi, 1844, and synonymised by Zibrowius (1971a: 695)].

Not *Eupomatus gracilis* and *Hydroides gracilis* **sensu** Lowenstam 1954: fig. 1; Lowenstam & Weiner 1989, Smith *et al.* 2013 (all mineralogy of tube; Bermuda) indeterminable, but this Californian taxon is not occurring in the Bermudas.

Hydroides gracilis – Bastida-Zavala & ten Hove 2003b: 89–92, figs 13A–V, 14 (redescription, synonymy, discussion; Monterey Bay, California to Bahia Tortugas, Baja California Sur, Pacific); Bastida-Zavala *et al.* 2017: 35–36, figs 4G, 6 (diagnosis, synonymy; San Francisco to Baja California Sur); Sun *et al.* 2018, figs 2, 3, 5, 6 (DNA, phylogeny; San Francisco and Port Hueneme, California).

Hydroides intereans – Bastida-Zavala & ten Hove 2003b: 1, 70 (synonymised with *H. gracilis*).

Diagnosis: Peduncle with shallow to well-defined constriction. Funnel with 30 (16–42) pointed radii. Verticil sitting or at most on short basis in funnel, radially to lightly bilaterally symmetrical, with 11 ((8)9–12) spines, curving inwards. All spines similar in shape and size, with pointed tip; dorsal spines maybe slightly larger. Spines without internal, external or lateral spinules, or wings. Small central disk.

Remarks: Most *Hydroides gracilis* specimens have verticil spines of the same size, but some specimens have 2 or 3 dorsal spines slightly bigger than the others, like in *H. similis*; however, *H. gracilis* has ventral spines without knob and without basal spinule, while *H. similis* shows a conspicuous knob in the ventral spines and a basal spinule. *Hydroides gracilis* shows some similarity with members of the *H. operculata*-complex, but these all show basal spinules above the central disk. *Hydroides gracilis* is similar to *Hydroides pseudouncinata*, but both verticil spines and funnel radii in the latter are more slender and longer. It also might be mistaken for *H. dianthus*, but in the latter taxon all verticil spines are curving in ventral direction.

Distribution: Monterey Bay, California to Bahia Tortugas, Baja California Sur.

Fig. 57. *Hydroides gracilis*, various views of the same operculum, Baja California, Mexico, ZMA V.Pol. 3701 (photo E. Wong).

42. *Hydroides helmata* (Iroso, 1921)

Figure 58

Type information: Gulf of Naples, Italy, holotype unknown; Iroso (1921: 58–59) did not mention where she deposited her material.

Etymology: Named after helmet-like dorsal verticil spine (Iroso 1921: 54, 'che rocade sugli altri ad elmo').

Selected synonymy

Eupomatus helmatus Iroso, 1921: 53–54, figs 16, 17 (description [original]; Naples, Italy).

Hydroides helmata – Zibrowius 1971a: 713–714, figs 31–37 (description, synonymy [excluding that of the *nomen inquirendum H. affinis*]; Marseille, Naples. Rovinj, Gulf of Gabès, all Mediterranean); Bianchi 1981: 68–70, fig. 24a–d (diagnosis; W. Mediterranean, Adriatic Sea); Sun 2017: 158, 169, fig. 5G (diagnosis; Island Rab, Croatia, though not mentioned with photograph).

Hydroides helmatus – Sun *et al.* 2018, fig. 1 (phylogeny after Pillai 1972).

Diagnosis: Peduncle without constriction. Funnel radii with 15–20 rounded, unchitinised radii. Verticil inserted with short basis into funnel, bilaterally symmetrical, with 6 (4–7) spines including dorsal enlarged spine, incurving, shaped like a wide bill or even a helmet (overhanging the others like a helmet), with a pointed tip and a rounded to knee-like bend; the remaining spines stubby, straight, without external or basal spinules. Spines meeting in centre of verticil, no central tooth.

Remarks: The fact that *Hydroides affinis* (*nomen inquirendum*) was regarded to be a synonym of *H. helmata* by Zibrowius (1971a: 713) indicates a certain similarity, and has been discussed below. By its verticil with few (up to about 8), rather thick spines, including a large dorsal hook, *H. helmata* resembles *H. adamaformis*, *H. gottfriedi*, *H. inermis* and *H. monroi*. However, those in *H. adamaformis* are rhomboidal, in *H. gottfriedi* they have a clear external knob, in *H. inermis* they are all incurving and in *H. monroi* they have a subterminal internal spinule as opposed to the simple straight verticil spines (not the hook) in *H. helmata*. Records of *H. helmata* in species lists from the Eastern Mediterranean (e.g. Ben-Eliahu 1991: 518; Faulwetter *et al.* 2017: 109) are without descriptions and/or illustration, and still have to be corroborated.

Distribution: Mediterranean.

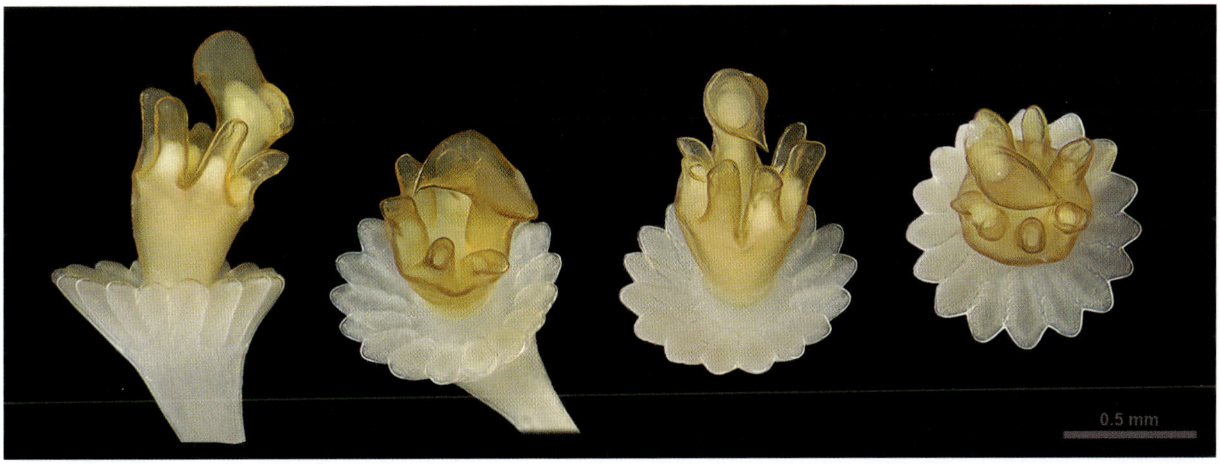

Fig. 58. *Hydroides helmata*, various views of the same operculum, San Marino, Island Rab, Croatia, AM W.43185 (photo E. Wong).

43. *Hydroides heterocera* (Grube, 1868)

Figure 59

Type information: Red Sea (not further specified, but collector von Frauenfeldt sampled crustaceans as far south as El Qoseir, Egypt, about 26°N in the northern Red Sea); deposition of material not given by Grube (1868: 639), however, it is not in Vienna, where von Frauenfeldt worked (Dr P. Frade, *pers. comm.*), nor in Berlin or Wroclaw where some of his material can be found.

Etymology: Not given by Grube (1868: 639), but he apparently Latinised the Greek ἑτερος = other, different, and κερας = horn, making *heterocerus* an adjectival form to be declined; named after the dimorphism in verticil spines.

Selected synonymy

Serpula (Eupomatus) heterocerus Grube, 1868: 639, pl. 7, fig. 8a–c (diagnosis [original]; Red Sea).

Serpula (Hydroides) uncinata **sensu** Gravier 1908: 114–115, figs 463–466, pl. 8, figs 286–287 (diagnosis; South Red Sea) [**non** Philippi, 1844, and synonymised by Zibrowius (1971a: 715)].

Hydroides heteroceros [*sic*] – Pixell 1913: 75, pl. 8, fig. 2a–c (Gulf of Suez, wall of Suez quay; Suakin Harbour, Sudan, Red Sea).

Hydroides heterocerus – Laubier 1966: 17 (name; first Mediterranean record as Lessepsian migrant, Lebanon); Ben-Eliahu & ten Hove 2011: 25–31, figs 5F–G, 8–10, presumably also fig. 11 (description of juveniles, synonymy, extensive discussion; Israel [Mediterranean], Suez Canal, Gulf of Suez, Suakin [Sudanese Red Sea], Somalia [Gulf of Aden]).

Hydroides heterocera – Zibrowius 1971a: 715, figs 38, 39 (redescription, synonymy; Beirut, Lebanon; Gulf of Suez; southern Red Sea; Djibouti, Gulf of Aden; Mozambique; Arabian Gulf); Sun *et al.* 2018, figs 1–3, 5, 6 (DNA, phylogeny; Al-Doha, Kuwait).

Diagnosis: Peduncle with well-defined, chitinised constriction, merging into base of funnel. The latter with 30 (24–38) chitinised T- to anchor-shaped radii. Verticil sitting or on short basis in funnel, bilaterally symmetrical, with 7 ((5) 6–9) slender, elongated, incurving spines; dorsal spine larger than the others and lacking any additional spinules, other spines similar in size and shape, each with pair of lateral spinules at 2/3 to 3/4 their length and a pointed tip; all with a sharply pointed basal spinule above small central disk, without central tooth.

Remarks: *Hydroides heterocera* is similar to *H. homoceros*, except for the absence of a dorsal hook and the position of the pair of lateral spinules in the verticil spines, at half-length in the latter. Ben-Eliahu & ten Hove (2011: 25–31, figs 8, 9) document the variability of the funnel radii, as mentioned in our diagnosis above. The latter (p. 28, table 7) give a depth distribution of (exceptionally 1) 8–33 m, and question the single record of 830 m (Ilan *et al.* 1994) as being an 'outlier record'. In phylogenetic analyses (Sun *et al.* 2018, figs 1–3, 5, 6, Fig. 13 here) generally the two taxa are found in the same main clade, though the COI barcoding shows a more variable picture. In the morphology of the verticil, large hook and slender spines with trifid tip (incurving tip + pair of lateral spinules) and basal spinule *H. heterocera* also shows similarity with *H. bifurcata*, but the position of the subdistal incurving tip and distal pairs of lateral spinules is completely different.

Distribution: Red Sea, Zanzibar, Madagascar, Arabian (Persian) Gulf, Sri Lanka, ?New Caledonia. Lessepsian migrant to the Levant Mediterranean: Israel, Lebanon, Cyprus, Turkey.

Fig. 59. *Hydroides heterocera*, various views of the same operculum, Gulf of Aden, Somalia, ZMA V.Pol. 3851 (photo E. Wong).

44. *Hydroides heterofurcata* Pillai, 1971

Figure 60

Type information: Talaimannar Pier, NW Sri Lanka; BMNH 1968.145 (holotype), BMNH 1968.146, 147 (paratypes).

Etymology: Named after the two types (Greek έτερος = other) of furcate verticil spines of the operculum (Pillai 1971: 114).

Selected synonymy

Hydroides heterofurcata Pillai, 1971: 113–114, figs 6F, 7A–C (description [original]; Sri Lanka); Sun 2017: 156, 169–170, fig. 4G (diagnosis; same); Sun *et al.* 2018, fig.1 (phylogeny after Pillai 1972).

Diagnosis: Peduncle with well-defined, chitinised constriction, merging into base of funnel. The latter with 26 (23–26) radii, seemingly bluntly conical, but in reality finely pointed. Verticil on short basis in funnel, bilaterally symmetrical, with 9–11 spines, including dorsal hook. The large dorsal spine split in two slightly geniculate hooks and with a very large upward basal spinule (trifurcate according to the original description); the rest bifurcate, V-shaped, latero-dorsally short and progressively increasing in size ventrally. Each prong of the fork long, slender, tapering, outcurving. Inner side of bifurcate spines smooth, without basal spinule. Small central disk.

Remarks: By its bifurcate verticil spines, the dorsal one as a twin hook – it is one of the most easily recognised species of the genus. Differences with the superficially similar *H. furcifera* are given in Remarks there.

Distribution: Sri Lanka.

Fig. 60. *Hydroides heterofurcata*, various views of the same operculum, Talaimannar Pier, Sri Lanka (photo E. Wong).

45. *Hydroides homoceros* Pixell, 1913

Figure 61

Type information: Not given by Pixell (1913), but her material came from the Maldive Islands (Miladumadulu Atoll and Minikoi), Suakin (Sudanese Red Sea) and from off Zanzibar, nine syntypes NHML 1924.6.13.147 (detailed in Ben-Eliahu & ten Hove 2011).

Etymology: Greek όμος = uniform and κερας = horn, after the opercular verticil spines.

Selected synonymy

Hydroides homoceros Pixell, 1913: 74–75, pl. 8, fig. 1 (description [original]; Red Sea; Maldives and Zanzibar, Indian Ocean); Ben-Eliahu & ten Hove 2011: 32–35, figs 5H–I, 12A–E (redescription, synonymy, discussion; Israel (Mediterranean, Lessepsian migrant), Suez Canal, Gulfs of Suez and Akaba, Suakin (Sudanese Red Sea), Seychelles, off Zanzibar, South Arabian Coast, Gulf of Oman, Arabian (Persian) Gulf); Satheesh & Wesley 2006: 324 (name; Kudankulam, S. India); Sun *et al.* 2018: 739–740, figs 1–3, 5, 6 (DNA, phylogeny; Al-Doha, Kuwait).

Diagnosis: Peduncle with well-defined, lightly chitinised constriction, merging into base of funnel. The latter has 20 (14–21) chitinised anchor-shaped to double-anchor-shaped radii (see Remarks). Verticil sitting or on short basis in funnel, radially symmetrical, with 7 (6–9) slender, elongated, incurving spines, all similar in shape and size; each with pair of lateral spinules at mid-length and a pointed tip; all with a sharply pointed basal spinule above small central disk, without central tooth.

Remarks: Ben-Eliahu & ten Hove (2011: 33, fig. 12A–E) draw the attention to the variability in shape of the funnel radii in *H. homoceros*, generally more complicated in structure than in the similar *H. heterocera*. The species further differs from the latter by the radial symmetry of the operculum, and the position of the pair of lateral spinules of verticil spines more proximal, at mid-length instead of at 2/3 to 3/4 of their length. The American *H. chilensis* and *H. crucigera* also show verticil spines with paired lateral spinules; however, these taxa have acutely pointed radii. A so-called *H. 'prisca'*-type operculum, developmental stage has been recorded (ten Hove & Ben-Eliahu 2005: 18). In the multigene phylogenetic analysis (Sun *et al.* 2018, fig. 2, Fig. 13 here, Clade A) it is recovered as a basal single clade with two (geographical) lineages next to the position of *H. heterocera*.

Distribution: Suez Canal, Red Sea, Arabian Gulf, Indian Ocean, Zanzibar, Seychelles (first reported here), Maldives, South India; established Lessepsian migrant to the Levant and once mentioned from ship's hull in Toulon, Mediterranean.

Fig. 61. *Hydroides homoceros*, various views of the same operculum, south-west La Digue Island, Seychelles, ZMA 4369 (photo E. Wong).

46. *Hydroides huanghaiensis* Sun & Yang, 2000

Figure 117A

Type information: Yellow Sea, 39°N, 122°70″E; holotype MBMCAS has been misplaced, most likely lost.

Etymology: Named after its type locality as 'Huanghai', which means 'Yellow Sea' in Chinese.

Selected synonymy

Hydroides huanghaiensis Sun & Yang, 2000: 134–135, figs 4A–F (diagnosis [original], discussion; Yellow Sea, 39°N, 122°70″E); 2014: 229–230, figs 122A–H, 134 (largely the same).

Diagnosis: Peduncle without, at best with very ill-defined constriction. Funnel with about 40 short, pointed radii. Verticil sitting in funnel, radially symmetrical, with 14 slender, smooth spines, all the same in shape and size and tapering into a long, pointed tip. Basal spinule absent. Central disk dome-like elevated, without central tooth.

Remarks: This species is very similar to (if not a synonym of) the sympatric *H. fusicola*, including the presence of a dome-like elevated central disk (Imajima 1976b, fig. 1a–b; see Remarks under *H. fusicola*). The only difference indicated, the absence of basal spinules in *H. huanghaiensis*, is somewhat arbitrary, since these spinules may be so minute as to be hardly visible. The taxon is also similar to *H. novaepommeraniae*, but often with a faint to distinct bilaterally symmetrical verticil; moreover, the latter has a tropical distribution, not cold-temperate.

Distribution: Yellow Sea.

47. *Hydroides humilis* (Bush, 1905)

Figure 62

Type information: Guaymas, Sonora, Gulf of California, Mexico, holotype unknown, not in YPM where Bush worked.

Etymology: Not given by Bush (1905: 235–236), apparently referring to the small size (Latin *humilis* = low, humble).

Selected synonymy

Eupomatus humilis Bush, 1905: 235–236, pl. 39, figs 39, 40, pl. 44, fig. 22 (diagnosis [original]; Gulf of California, Mexico).

Hydroides humilis – Bastida-Zavala & ten Hove 2003b: 92–93, figs 15A–F, 20 (redescription, synonymy, discussion; Bahia de los Angeles, Baja California to Panama, Pacific); Bastida-Zavala *et al.* 2016: 422–423, figs 5, 11J (diagnosis, synonymy, discussion; Oaxaca, Mexico); Sun 2017: 153, 170, fig. 2B (diagnosis; Baja California to Panama).

Diagnosis: Peduncle with ill-defined constriction. Funnel radii with 16 (12–20) pointed tips, lightly chitinised. Verticil sitting in funnel, either radially or bilaterally symmetrical, with 8 (7–10) yellow spines, curving outwards or straight. All spines similar in shape and size, with pointed or rounded tips, sometimes one simple dorsal hook present. Spines without any additional characters, though Bush (1905: 235) mentions a basal spinule. Verticil with small central disk, without tooth. Bayonet of collar chaetae with 2 teeth and accessory teeth or proximal rasp.

Remarks: At first sight the taxon resembles the juvenile stages mentioned for a range of (mostly Indo-West Pacific) taxa as '*Hydroides prisca*' (see '*Nomina inquirenda* and rejected names, occasionally still in use' later in this book), but the degree of chitinisation of both verticil and funnel in *H. humilis* is larger, making a juvenile state somewhat less likely. We agree with Bastida-Zavala *et al.* (2016: 422) that specific morphological and chaetal characters should be revised in juveniles of *Hydroides* spp. (from Baja California to Panama) to evaluate if the specific status of *H. humilis* is supported or if the nominal taxon represents a semaphoront (juvenile stage) of some, or several, *Hydroides* species.

Distribution: Gulf of California, Baja California to Panama.

Fig. 62. *Hydroides humilis*, various views of the same operculum, Paitilla Beach, Panama, ZMA V.Pol. 3632 (photo E. Wong).

48. *Hydroides ibarzabalae* nov. spec. Bastida-Zavala & Salazar-Vallejo

Figure 63

Type information: Cuba: Cayo Diego Pérez, Golfo de Batabanó, 15 m; holotype ANC 05.01.056).

Etymology: The species is dedicated to Diana Ibarzabal, Cuban researcher, who collected this beautiful specimen.

Synonymy

Hydroides sp. – Bastida-Zavala & Salazar-Vallejo 2000: 852, fig. 4a (diagnosis, discussion; Cuba).

Hydroides sp. 3 – Bastida-Zavala & ten Hove 2003a: 169–171, figs 38A–J, 40 (description, discussion; same specimen as above).

Diagnosis: Peduncle with ill-defined constriction. Funnel with 13 radii with unchitinised, rounded tip. Verticil sitting in funnel, bilateral symmetrical with 5 dark brown spines distally smoothly curving inwards. Most dorsal spine almost twice the length of the others, with acute tip strongly curving inwards. Remaining 4 spines almost conical, similar in shape and size, with pointed tip. All 5 spines with 2–3 pairs of short, acute lateral spinules (Bastida-Zavala & ten Hove 2003a, fig. 38A, B) and ditto basal inner spinule (Fig. 38C), the latter meeting each other in centre of verticil, without central tooth (Fig. 38C).

Description

Tube: Missing.

Colour and size: Body beige. Specimen incomplete.

Radiolar crown: With seven pairs of radioles, length 1.7 mm. Terminal filaments short.

Peduncle: Length including operculum 2.8 mm. Insertion left; constriction ill-defined (Bastida-Zavala & ten Hove 2003a, fig. 38a). Pseudoperculum present.

Operculum: Length 1.1 mm, diameter 0.4 mm. Funnel with 13 radii with un-chitinised, rounded tip (Bastida-Zavala & ten Hove 2003a fig. 38a). Interradial grooves 1/3 of funnel length. Verticil sitting in funnel, bilateral symmetrical with 5 dark brown spines, distally smoothly curving inwards (fig. 38a, b). The most dorsal spine almost twice the length of the others, with acute tip strongly curving inwards. The other 4 spines almost conical, similar in shape and size, with pointed tip. All 5 spines with 2–3 pairs of short, acute lateral spinules (fig. 38a, b) and ditto basal inner spinule (fig. 38c), the latter meeting each other in centre of verticil, without central tooth (fig. 38c).

Collar chaetae: Bayonet chaetae with 2 pointed, elongate teeth, distal blade smooth (Bastida-Zavala & ten Hove 2003a, fig. 38d–f); capillary chaetae present (fig. 38g).

Thorax: Length 1.0 mm, width 0.3 mm. Thoracic membranes well developed. Six chaetigers with limbate chaetae (Bastida-Zavala & ten Hove 2003a, fig. 38h) of two sizes, saw-shaped uncini with 8 teeth (fig. 38i).

Abdomen: Most abdominal chaetigers lost. Anterior abdominal chaetigers with flat-trumpet chaetae (Bastida-Zavala & ten Hove 2003a, fig. 38j). Anterior uncini saw-shaped.

Remarks: The holotype and only specimen of *Hydroides ibarzabalae* nov. spec. was originally labelled *H. parva* in the collections of the Instituto de Oceanología, Cuba. Nevertheless, with hindsight the shape of the operculum is unique in the genus, and very different from *H. parva*; it has radii with blunt tip, verticil spines smoothly curving inwards distally and with two or three pairs of lateral spinules (Fig. 63) and basal spinules meeting in the

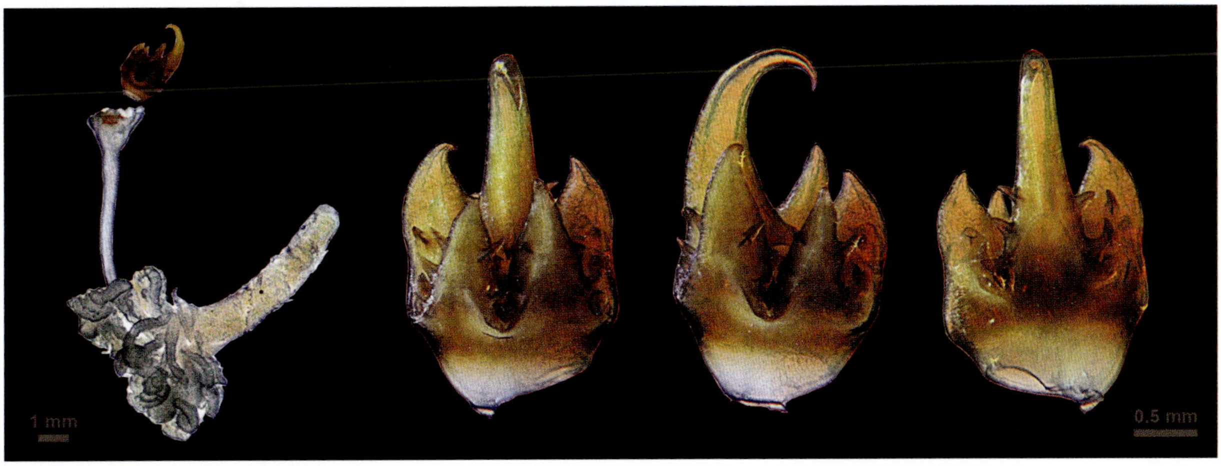

Fig. 63. *Hydroides ibarzabalae* nov. spec., holotype and various views of the opercular verticil, Cayo Diego Pérez, Golfo de Batabanó, Cuba, ANC 05.01.056 (photo S. Salazar).

centre; on the other hand *H. parva* has radii with chitinised, pointed tip; verticil spines abruptly curving inwards with an external knob on the 'knee' and only one pair of lateral spinules (Fig. 88), and basal spinules above small central disk. Moreover, *H. ibarzabalae* nov. spec. has bayonet chaetae with 2 teeth only as opposed to 2 teeth with accessory one in *H. parva.*

Hydroides ibarzabalae nov. spec. resembles *Hydroides caribensis* nov. spec. in having two or three pairs of lateral spinules. However, the former has a larger dorsal verticil spine (Fig. 63), the latter has all verticil spines of more or less the same shape and size (Fig. 35).

Distribution: Cuba.

49. *Hydroides inermis* Monro, 1933

Figure 64

Type information: James Bay, James Island, Galapagos, Ecuador, syntypes NHML 1933.7.10.426–430.

Etymology: Refers to the verticil spines without side teeth (Latin *inermis* = unarmed).

Selected synonymy

Hydroides inermis Monro, 1933: 1083–1085, fig. 27a–e (description [original]; Galapagos); Bastida-Zavala & ten Hove 2003b: 94–95, figs 16A–K, 20 (redescription, discussion; Galapagos, North Peru); Bastida-Zavala *et al.* 2016: 423, figs 5, 12A (diagnosis, synonymy, discussion; Oaxaca, Mexico); Sun *et al.* 2018, fig. 1 (phylogeny after Pillai 1972).

Diagnosis: Peduncle with well-defined constriction. Funnel radii with 31 (27–35) short, pointed, chitinised tips. Verticil sitting in funnel, bilaterally symmetrical, with 8 (7–9) broad spines, almost valvular, strongly incurving. The most dorsal spine decidedly bigger than the others, the rest of the spines similar in shape and size (although the latero-dorsal pair may be slightly larger), with blunt tip and partially obscuring the small central disk. No central tooth.

Remarks: With its very broad and blunt, almost valvular, verticil spines, *H. inermis* hardly can be mistaken for another species. With its large dorsal hook, the operculum superficially resembles *H. brachyacantha*, but the verticil spines in *H. inermis* have blunt tips and in *H. brachyacantha* they are pointed. Differences with *H. adamaformis*, *H. gottfriedi* and *H. monroi*, all with few (up to about 8), rather thick spines, including a large dorsal hook, are discussed under *H. helmata*.

Distribution: Oaxaca (Mexico), Galapagos, north Peru.

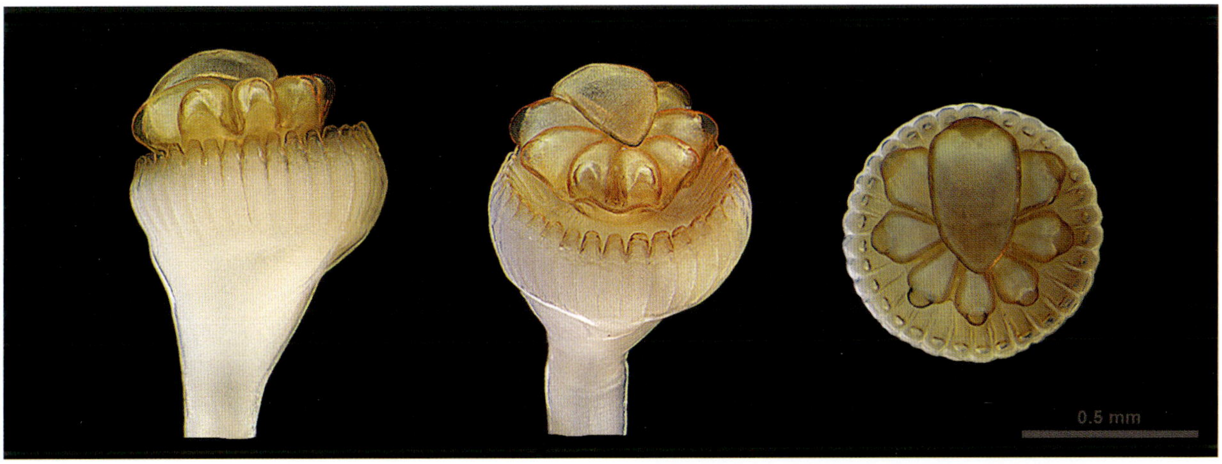

Fig. 64. *Hydroides inermis*, various views of the same operculum, Galapagos Islands, AM W.43198 (photo E. Wong).

50. *Hydroides inornata* Pillai, 1960

Figure 65

Type information: Negombo Lagoon, Colombo, Sri Lanka, holotype NMNH 1959.4.14.2.

Etymology: Not given by Pillai (1960: 14–15), but likely to refer to the fact that all verticil spines are without lateral spinules (Latin *inornatus* = unadorned).

Selected synonymy

Hydroides inornatus Pillai, 1960: 14–15, fig. 5F–I (diagnosis [original]; Sri Lanka; material studied); 1971: 104–107, fig. 5A–H (description, discussion, but not synonymy; Colombo, Sri Lanka); Huang & Mak 1982: 780 (Hong Kong); Mak 1982: 605 (short description, comparison with *H. exaltata* in remarks; same); Wang & Huang 1993: 1–12 (same); Sun & Yang 2014: 231–233, figs 123A–I, 119 (diagnosis, in Chinese; Dongshan, Fujian; Shantou, Guangzhou; Mirs Bay, Hong Kong).

Hydroides inoranta [*sic*] – Gaikwad & Ranade 1979: 50 (name; Ratnagiri, India, Arabian Sea).

Hydroides brachyacantha [**non** Rioja, 1941a] **sensu** Lakshmana Rao 1969: 67–68, pl. 4, figs a–g (diagnosis; Madras, Visakhapatnam and Bombay Harbours, east and west India).

Hydroides operculatus [**non** Treadwell, 1929] **sensu** Ishaq & Mustaquim, 1996: 56, fig. 3A–G (diagnosis; Pakistan)]; Kubal *et al.* 2012: 381–383, fig. 2A–J (det. ten Hove, diagnosis; Diu, Gujarat, India, Arabian Sea); Sun *et al.* 2012b: 17–18, fig. 9A description; Hong Kong); Sun & Yang 2014: 244–245, fig. 132 (diagnosis, in Chinese; Sharp Island, Tolo Harbour, Hong Kong); Tovar-Hernández *et al.* 2016: table 2, fig. 6 (name, DNA, phylogeny; Hong Kong, China, and synonymised by Sun *et al.* 2017b – see below).

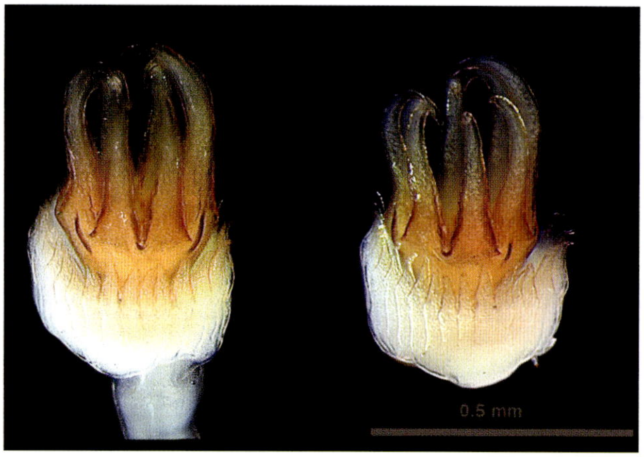

Fig. 65. *Hydroides inornata*, various views of the same operculum, Colombo, Sri Lanka, holotype NHML 159.4.14.2 (photo E. Kupriyanova).

Hydroides basispinosa – Sun & Yang 2014: 209–211, figs 108A–H, 117 (diagnosis, in Chinese; Dalian, Liaoning; Dongshan, Fujian; Mirs Bay, Hong Kong; referred herein to *H. inornata* conform Sun *et al.* 2017b).

Hydroides inornata – Sun *et al.* 2017b: 39–46, fig. 1B (discussion, figure of holotype and DNA supported resurrection from synonymy with *H. operculata*, where it was placed by Ishaq & Mustaquim 1996; Ratnagiri, Maharastra, India; Hong Kong, China); 2018: 741, figs 1–3, 5, 6 (DNA, phylogeny; same locality).

Not *Hydroides inornata* (?) [**non** Pillai, 1960] **sensu** Zibrowius 1968: 117, pl. 2, figs 27, 28 (Gulf de Marseille, Gulf de Gabès, Mediterranean; re-identified to *H. nigra* by Zibrowius 1971a: 711).

Not *Hydroides inornata* [**non** Pillai, 1960] **sensu** Amoureux *et al.* 1978: 147 (Elat, Israel, Red Sea; see *H. minax*, **fide** Ben-Eliahu 1991: 527).

Diagnosis: Peduncle without or with ill-defined constriction. Funnel with 35 (25–38) hardly chitinised, pointed radii. Verticil sitting in funnel (in two specimens on short basis), bilaterally symmetrical, with 7–9 long, slender incurving spines; with basal knob or short spinule over small central disk. The most dorsal of these spines (sometimes with the two adjacent ones) larger than the rest (equal among themselves).

Remarks: Species complex with the African/Arabian *H. operculata* **sensu stricto** and the tropical Australian *H. basispinosa* (see Sun *et al.* 2017b). As is evident from the synonymy, above, the taxon morphologically easily can be confused with a number of other taxa. For a full discussion see Remarks under *H. operculata*.

Distribution: Sri Lanka, India.

51. *Hydroides kimberleyensis* Pillai, 2009

Figure 66

Type information: Kimberley, Western Australia, east side of Fenelon Island, holotype: AM W.21391.

Etymology: 'named after the type locality, Kimberley, Western Australia'.

Selected synonymy

Hydroides kimberleyensis Pillai, 2009: 118–119, figs 12A–F, 13A–L, 14A–Q (description [original], discussion; Kimberley, WA); Sun *et al.* 2015: 44–46, fig. 13a–b, map 11 (description, discussion; Kimberley to Houtman Abrolhos, WA); Sun *et al.* 2018: 738, figs 2, 3, 5, 6 (DNA, phylogeny; Long Reef, Augustus Island, WA).

Diagnosis: Peduncle without or with ill-defined constriction. Funnel with 20 (14–24) slender, sharply tipped radii. Verticil sitting in funnel, with 5–6 uniform spines gently tapering to rounded tip; each spine triangular in cross-section, with 4–5 minute basal processes above small, flat central disk. Bayonet of collar chaetae with 2 teeth, sometimes with 1 accessory.

Remarks: *Hydroides kimberleyensis* is part of the *Hydroides tuberculata*-complex **sensu lato**. In the phylogenetic analyses (Sun *et al.* 2018 figs 2, 3, 5–7, Fig. 13 here) it was recovered together with *H. tuberculata* and *H. uniformis* in a single, well-supported clade (e.g. Fig. 2B, lower lineage): *Hydroides tuberculata*-complex **sensu stricto**. Apparently it is very similar to *H. uniformis*, in 2001 one of us (HAtH) identified the material of *H. kimberleyensis* described in Sun *et al.* (2015: 44–46) as *H. uniformis* (for a full discussion we refer to the last paper, and to the Remarks under *H. tuberculata*, below). It remains to be seen if the distant populations from the Kimberley coast and Houtman Abrolhos in a DNA-aided analysis belong to a single species; the tapering speeds of verticil spines and funnel radii show small differences.

Distribution: Western Australia, from Fenelon Island, Kimberley to Houtman Abrolhos.

Fig. 66. *Hydroides kimberleyensis*, various views of the same operculum, East Fenelon Island, Kimberley, Western Australia, AM W.21391 (photo E. Wong).

52. *Hydroides lambecki* Bastida-Zavala & ten Hove, 2003a

Figure 67

Type information: Curaçao, Netherlands Antilles, holotype ZMA V.Pol 3411.

Etymology: Named after Hugh J. P. Lambeck, a one-time assistant to ten Hove, who first recognised this to be different from *H. mongeslopezi*.

Selected synonymy

Hydroides lambecki Bastida-Zavala & ten Hove, 2003a: 139–140, figs 18A–M, 19A–D, 22, table 3 (description [original]; Quintana Roo (Mexico), Jamaica, Netherlands Antilles); Sun 2017: 154, 170, fig. 2M (diagnosis, figure of operculum).

Hydroides cf. *mongeslopezi* ten Hove & Nishi, 1996: 93.

Additional material, not mentioned in Bastida-Zavala & ten Hove (2003a):

Netherlands Antilles, Aruba, Pos Chikitu, reef, 13 August 1970. Reef with much sand and limestone; 23–26 m. From limestone and corals, legit H.A. ten Hove, stn 2025Ba (1 spec., ZMA V.Pol. 5083); Curaçao, Boca Santu Pretu, south-east, 1 May 1970. Rock, reef; 7–8 m. From corals and limestone, legit H.A. ten Hove, stn 2040A (4 spec., ZMA V.Pol. 5085, separated from V.Pol. 3412, and thus co-occurring with *H. mongeslopezi*); Bonaire, Playa Lechi (Paloe Lechi), S. entrance-channel, 13 July 1970. From muddy coral debris; 1.5–2 m, legit H.A. ten Hove, stn 2113Aa (2 spec., ZMA V.Pol. 5084, separated from V.Pol. 3410, and thus co-occurring with *H. mongeslopezi*). Barbados, 0.8 km off Holetown, shell debris on muddy sponge bottom, 100 m, legit P. Wagenaar Hummelinck stn 1442, 19 February 1964 (1 spec., ZMA V.Pol. 5082).

Diagnosis: Peduncle with ill-defined constriction. Funnel with 12 (11–14) radii, with rounded, not chitinised tips. Verticil sitting in funnel, bilaterally symmetrical, with 6 ((5)6–7) spines, incurving with sharp tip. Dorsal spine larger than ventral ones, but similar in shape; exceptionally all spines of same size; with up to 10 lateral to external spinules, grouped from base to halfway, 4–8 of them almost pair-wise. Spines with one internal spinule each, directed downwards and situated in almost basal position ventrally, but changing position to about halfway up in dorsal spine, where it is larger and directed upwards. Base of spines almost touching in centre, no central tooth.

Remarks: The taxon is rather similar to *Hydroides mongeslopezi*; in fact when one of us (HAtH) collected the material in the early 1970s *H. lambecki* was routinely identified as *H. mongeslopezi* (e.g. see ZMA V.Pol. 3410, above), later as *H.* cf. *mongeslopezi*. The main differences between the two species have been tabulated by Bastida-Zavala & ten Hove (2003a, table 3). *Hydroides lambecki* has 11–14 funnel radii; *H. mongeslopezi* has 17–28. The first has 6–8 verticil spines; *H. mongeslopezi* has 8–11. The internal spinules in *H. mongeslopezi* are all curving towards the bottom of the verticil, but in *H. lambecki* changing direction from downwards in ventral spines, upwards in dorsal spines.

Distribution: Quintana Roo (Mexico), Jamaica, Netherlands Antilles, Caribbean.

Fig. 67. *Hydroides lambecki*, various views of the same operculum, Discovery Bay, Jamaica, AM W.43191 (photo E. Wong).

53. *Hydroides lirs* Kupriyanova, Sun, ten Hove, Wong & Rouse, 2015

Figure 68

Type information: Lizard Island, Queensland, Australia, holotype AM W.43967, paratypes AM W.41749, W.42357, W.42358, ZMA V.Pol. 5035.

Etymology: 'The species name honours the Australian Museum's Lizard Island Research Station (LIRS)'.

Selected synonymy

Hydroides lirs Kupriyanova, Sun, ten Hove, Wong & Rouse, 2015 (description [original]); Sun *et al.* 2018, figs 2, 3, 5, 6 (phylogeny).

Diagnosis: Peduncle with long, well-defined, chitinised constriction. Funnel with 50 (34–62) chitinised long and narrow radii ending in pointed tips; radii each bearing a minute internal spinule basally. Verticil clearly stalked, with 11–12 spines, slender, curving outwards and a dorsal hook. Spines with pointed tips and sharp internal spinule at mid-length. Hook smoothly curving inward, with pair of subterminal, lateral spinules. Small, unadorned central disk.

Remarks: Phylogenetic analyses (Sun *et al.* 2018; for *H. lirs* DNA was analysed from holotype, Fig. 13 here) recovered *Hydroides lirs* as sister of a clade with *H. adamaformis* and *H. ancorispina*; it does not associate with one of the morphologically similar taxa, mentioned below. The species resembles *H. exaltata* in having a large incurving dorsal hook and outcurving verticil spines with an internal spinule each. It resembles *H. minax* in the presence of a pair of distal lateral spinules on the incurved dorsal hook, but internal spinules are missing here. Records under these names from before 2015 thus can and will contain material of *H. lirs*; for example, paratype ZMA V.Pol. 5035 was originally identified (HAtH in 1986) as *H. minax* and paratype AM W.41749 was originally identified as *H. exaltata*. The latter differs from *H. exaltata* in having a pair of lateral spinules on the dorsal hook, which is absent in *H. exaltata*. From *H. minax* it can be distinguished by the following characters: dorsal hook is smoothly curved in *H. lirs*, but strongly curved – almost angular – in *H. minax*; verticil spines in *H. minax* are short and lack internal spinules that are present in *H. lirs*. The basal teeth typical of the funnel radii in *H. lirs* were not observed either in *H. exaltata* or in *H. minax*. *Hydroides lirs* also resembles *H. pseudexaltata* Pillai, 2009, but can be distinguished from latter by the presence of lateral spinules in its dorsal hook, internal spinules in the verticil spines (as opposed to the absence of lateral and internal spinules in *H. pseudexaltata*) and its sharp-tipped funnel radii (as opposed to almost T-shaped tips in *H. pseudexaltata*).

Distribution: Queensland, Australia.

Fig. 68. *Hydroides lirs*, various views of the same operculum, Lizard Island, Queensland, Australia, AM W.41749 (photo E. Wong).

54. *Hydroides longispinosa* Imajima, 1976a

Figure 69

Type information: Konija, Amami-Oshima, Japan, holotype NSMT Pol. H120, paratype NSMT Pol. P123.

Etymology: Named after the long central verticil tooth, long in comparison with that in *Hydroides elegans*.

Selected synonymy

Hydroides longispinosa Imajima, 1976a: 240–246, fig. 5 (description [original] Ogasawara Islands, Southern Japan); 1987: 15 (distribution in Japan); Kupriyanova *et al.* 2015: 286–288, fig. 6 (diagnosis, synonymy; Lizard Island, Queensland, Australia); Bastida-Zavala *et al.* 2017: 36–37, figs 4H, 6 (diagnosis, extensive synonymy; Oahu, Hawaii); Sun *et al.* 2018: 738–740, 742, figs 2, 3, 5, 6 (DNA, phylogeny; Imperieuse and Long Reefs WA, Lizard Island Qld, Australia; Kavieng, Papua New Guinea).

Hydroides centrospina Wu & Chen, 1981: 354–355, 357, fig. 1a–i (description [original]; South China Sea).

Hydroides longispinosus – Fiege & Sun 1999: 116–118, fig 6A–B (description, synonymisation of *H. centrospina*; Hainan Island, South China Sea); Sun & Yang 2014: 233–234, figs 124A–H, 114 (largely the same as Sun & Yang 2000; Meizhou Island, Fujian; South China Sea; Sanya, Changlin, Hainan).

Hydroides centrospinus – Sun & Yang 2014: 211–212, figs 109A–I, 114 (diagnosis, in Chinese; Sanya, Hainan).

Diagnosis: Peduncle without constriction. Funnel not chitinised to very thinly so, with about 19 (15–22) blunt radii. Verticil sitting in funnel, radially symmetrical, with 14 (13–16) spines, lightly curving outwards. All verticil spines similar in shape and size, each with pointed tip, 7–9 irregularly spaced lateral spinules (at both sides), with up to 18 irregularly placed pointed internal spinules. Central disk flat to slightly elevated, central tooth rarely absent, generally longer than verticil spines, may be smooth, but mostly bearing numerous randomly placed small teeth distally. Bayonet of collar chaetae with 2 teeth and proximal rasp.

Remarks: As mentioned under *Hydroides azorica*, there is a group of morphologically similar taxa: *H. elegans*, *H. norvegica*, to a lesser extent the Eastern Pacific *H. ochotereana*, which can be distinguished from the Indo-West Pacific *H. longispinosa*, *H. multispinosa* and *H. nanhaiensis*. For the time being we have treated the latter three as being separate species. However, the ranges of numbers of verticil spines, lateral spinules of these, radii, and the absence/presence of a central tooth are overlapping between the nominal taxa and appear to be so between populations from different geographical areas: names as appearing in the literature, even if identified by one of us, should be treated warily. This caution has been intensified by the results of the genetic analysis (Sun *et al.* 2018, figs 2, 3, 5–7, Fig. 13 here) of assumedly *H. longispinosa* specimens, from Western Australia, Northern Australia and Papua New Guinea. In these, *H. longispinosa* is retrieved as a sister clade to both *H. elegans* and *H. norvegica*; however, the clade of *H. longispinosa* shows three lineages, in different widely distant geographical localities, suggesting polyphyly.

Sun & Yang (2014) mention both *H. centrospina* (pp. 211–212, figs 109A–I, 114; copy of Wu & Chen 1981)

Fig. 69. *Hydroides longispinosa*, various views of the same operculum, Lizard Island, Queensland, Australia, AM W.42333 (photo E. Wong).

and *H. longispinosa* (see above), and remark that both often can be found together, but that the former can be distinguished by its smooth central tooth instead of a spiny one. In our opinion this co-occurrence might support Fiege & Sun's (1999: 118) opinion that both forms are just due to infraspecific variability – they found a specimen with 2 opercula, one with smooth central tooth, the other adorned with lateral spinules.

Distribution: Reported suspiciously wide from the Indo-West Pacific, from WA and Qld, Australia, to South Japan and South China, Micronesia and Ponape Island, Tonga; Enewetak, Marshall Islands to Oahu, Hawaii.

55. *Hydroides longistylaris* Chen & Wu, 1980

Figure 70

Type information: Shellfish farms, Zhangpu (Zhangzhou), Fujian Province, China; holotype A-52, Weizhou Island, Guangxi Province, China; paratype A-51, both in IOCAS.

Etymology: Not given by Chen & Wu (1980: 249), but refers to the long pillar-like chitinised constriction of the opercular funnel (Latin *longus* = long, adjectival Latinisation of the Greek στυλος = pillar).

Selected synonymy

Hydroides longistylaris Chen & Wu, 1980: 248–249, figs 1–8 (diagnosis, in Chinese [original]; shellfish farms, Fujian and Guangxi Provinces, China); Fiege & Sun 1999: 118–119, fig. 8A–E (redescription; Hainan, China); Sun *et al.* 2012a: 15–17, fig. 8A–I (redescription, synonymy, discussion; Hong Kong, China); Sun & Yang 2014: 234–235, fig. 125 (redescription, in Chinese; Bailongwei, Guangxi Province; Xincun, Meixia, Hainan Province; South China Sea, China).

?*Hydroides* cf. *ezoensis* **sensu** Wang & Huang 1993: 7, 10 [**non** Okuda, 1934], from Hong Kong, probably see *H. longistylaris*, **fide** Sun *et al.* (2012a: 15).

?*Hydroides* cf. *fusicola* **sensu** Mak 1982: 603–604 [**non** Mörch, 1863], from Hong Kong, most probably see *H. longistylaris*, **fide** Sun *et al.* (2012a: 15, 17).

Diagnosis: Peduncle with very well-defined constriction: proximal part of funnel chitinised, narrow, extending for 1.5 mm or more, before abruptly widening into fleshy peduncle proper. Funnel with 24 (16–29) radii ending in pointed, lightly chitinised tips. Verticil on short stalk or sitting in funnel, radially symmetrical, with 7 (5–9) smooth spines, same in shape and size, curving outwards.

Remarks: *Hydroides longistylaris* is characterised by its long opercular peduncle, chitinised over almost half its length. Such a long, chitinised part of the peduncle (or funnel) occurs in some specimens of *H. recta* as well (see Fig. 70), but the latter taxon has 1 enlarged dorsal verticil spine.

The drawings by Chen & Wu (1980), Sun (2011) and Sun & Yang (2014) do not show basal spinules on verticil spines, nor a central disk. However, Fiege & Sun's (1999) SEM figure shows at least 3 of the verticil spines with a basal spinule, and a high magnification of Sun *et al.* (2012a) fig. 8A suggests 2 of such basal spinules.

As evidenced by our selected synonymy, above, the species shows some similarity with the taxa *H. ezoensis* and *H. fusicola* (certainly as figured under the name *H. uncinata* by Okuda 1937: 64, fig. 10b) essentially from temperate waters, but differs in having a smaller verticil, with spines shorter and generally in a lower number. It also resembles developmental opercular stages, with undifferentiated smooth radii as well as verticil spines, formerly known as *H. prisca* (see ten Hove & Ben-Eliahu 2005 and '*Nomina inquirenda* and rejected names, occasionally still in use' later in this book).

Under names as *H.* cf. *recta* and *H.* aff. *recta*, specimens have been recorded with opercular morphologies intermediate between those of *H. recta* and *H. longistylaris*; these may be juveniles or different taxa, a question which only can be solved with more material and preferably DNA analyses (see Sun *et al.* 2015: 75–76, fig. 24; Kupriyanova *et al.* 2015: 290, fig. 8).

Distribution: Fujian, Hong Kong, Hainan, Guangxi; southern East China Sea, South China Sea.

Fig. 70. *Hydroides longistylaris*, various views of the same operculum, Sharp Island, Hong Kong, AM W.40546 (photo E. Wong).

56. *Hydroides malleolaspina* Straughan, 1967a

Figure 71

Type information: Pialba, Queensland, Australia, holotype AM W.3999.

Etymology: Not given by Straughan (1967a: 222), but it is a compound noun from the Latin *malleolus* [= small hammer], referring to the dorsal hammer-shaped verticil spine, and *spina* [= thorn].

Selected synonymy

Hydroides malleolaspina Straughan, 1967a: 222–224, fig. 7a–e (diagnosis [original]; Queensland, Australia; material studied); Sun *et al.* 2015: 50–53, fig. 15, map 13 (diagnosis, synonymy, discussion; distribution in tropical Australia); 2018, fig. 1 (phylogeny, after Pillai 1972).

Pro parte *Hydroides ancorispina/us/malleolaspina/us*-complex – ten Hove 1994: 108 (discussion); Fiege & Sun 1999: 114–116, fig. 5A–C (description, differential diagnosis, discussion; Hainan, China).

Hydroides trihamulatus Pillai, 2009: 131–134, figs 24A–G, 25A–C, 26A–L (description, extensive discussion; Kimberley), **fide** Murray *et al.* (2010: 393–394).

Not *Hydroides malleolaspina* [**non** Straughan, 1967a] **sensu** Imajima 1982: 45, fig. 3k–s (Palau & Yap Islands; see *H. novaepommeraniae* **fide** Imajima & ten Hove 1984: 45–46).

Hydroides malleolaspinus – Sun & Yang 2014: 237–238, figs 126A–I, 106 (diagnosis, in Chinese; Gulei, Fujian; Hong Kong; Sanya, Hainan; South China Sea).

Diagnosis: Peduncle with indistinct to well-defined, thinly chitinised constriction. Funnel with 22 ((12)20–27) chitinised, T- to anchor-shaped radii. Verticil inserted on short stalk into oblique funnel, bilaterally symmetrical, with dorsal hook and 7–11 equal, outcurving spines. Dorsal hook large, somewhat hammer-like, with 2 lateral conical processes (bifid) distally and a subdistal stout beak-like or rounded medial process. Bases of spines almost meeting, central tooth absent. Bayonet of collar chaetae with 2 teeth, sometimes with 1 or 2 accessory ones.

Remarks: *Hydroides malleolaspina* was recorded as an Australian mainland species by Straughan (1967a), distribution supported and extended by Sun *et al.* (2015, map 13). Pillai (1971) described a new species *Hydroides ancorispina* from Ceylon, characterised by a dorsal verticil spine that is a smooth, flattened column instead of bifidly tipped with a large, beak-like medial process as in *H. malleolaspina*. Imajima & ten Hove (1984) synonymised *H. ancorispina* with *H. malleolaspina* since small specimens of *H. malleolaspina* may show a simple dorsal verticil spine similar to that of *H. ancorispina*. On the other hand, ten Hove (1994) pointed out that the fully grown specimens of what he identified as *H. ancorispina/malleolaspina*-complex from the Seychelles and the Malaysian Peninsula always have a rather straight, bulbous dorsal verticil spine similar to that in *H. ancorispina* from Ceylon, instead of being hammer-like as in *H. malleolaspina* from Australia – possibly two separate species are involved.

Pillai (2009) described a new species *Hydroides trihamulatus* (*sic*!) based on the comparison of his specimens with Straughan's (1967a) original description of *H. malleolaspina*. However, Straughan's description does not mention

Fig. 71. *Hydroides malleolaspina*, various views of the same operculum, Shoal Point, Queensland, Australia, AM W.3996 (photo E. Wong).

the beak-like medial extension present on the holotype. After comparison of the original material of *H. trihamulata* with the holotype of *H. malleolaspina*, Murray *et al.* (2010) synonymised *H. trihamulata*. For a further discussion see Remarks under *H. ancorispina*. In view of the reported opercular variability, convoluted synonymy and wide distribution we regard the taxon as a complex of a species, awaiting further analysis with help of DNA sequence data.

Distribution: Australia: NT, Qld, WA; South China Sea.

57. *Hydroides microtis* Mörch, 1863

Figure 72

Type information: Not specified beyond 'boreal America, on *Pecten irradians* Lamarck'. Mörch's original material cannot be found in the collections of AMNH, NHML, USNM, ZMH and ZMC. Read *et al.* (2017) guessed 44°N, 68°W as the mid-point of the coastal extent. In view of the large distribution, and observed variability in opercula (see below), a neotype was deemed necessary, for which we have selected a specimen from off Vero Beach, Florida, 27°33.2′N, 80°03.2′W, 33 m, deposited in the USNM 1143410.

Etymology: Not given by Mörch (1863: 380) but probably refers to small dilations at the end of the verticil spines (Greek μικρος = small; οτοσ = ear).

Selected synonymy

Hydroides (Eucarphus) microtis Mörch, 1863: 380, no figures (diagnosis [original]; boreal America).

Vermilia microtis – Quatrefages 1866: 536 (diagnosis after Mörch).

Eupomatus microtis – Pequegnat & Pequegnat 1968: 23, 44 (name only; fouling, off Panama City, Florida, USA).

Hydroides microtis – Robertson & Mau-Lastovicka 1979: 323 (name, parasitised; off Core Bank, North Carolina, USA); Bastida-Zavala & ten Hove 2003a: 132–135, figs 14A–N, 15 (redescription, synonymy; Gulf of Mexico, Eastern USA, Caribbean Sea and Guyana); Sun 2017: 38, 103, 152, 173, fig. 1I (diagnosis, discussion, figure of operculum).

Material additional to that mentioned in the most recent revision (by Bastida-Zavala & ten Hove 2003a). Though judging from the few published reports it seems to be a rare taxon (no new material records since 2003), in reality it is fairly common.

Georgia, off Sapelo Island, on oyster, 128–105 m, legit M.B. Gray stn 158, 1 May 1962 (2 spec., ZMC); 7.2 km at 340° off Sapelo Island, Whistle buoy, 20 m, legit M.B. Gray stn 198, 10 April 1963; Florida, off St Augustine to Fort Pierce, 10–64 m, Rock Shrimp Project, stn 1A, 3, 4, 5, 173, 194, 298 (100+ spec., FSBC I 17414–5, 32460, 39625–33, 39635; ZMA V.Pol. 3261, 3262, 3371, 3372, 3273, 3376); 27°10.8′N, 80°00.8′W, on *Oculina varicosa* Lesueur, 44 m, RV *Johnson*, stn J043/JSL II 187 (1 spec., HBOM 50: 0830); Harbor Branch Benthic Ecology Offshore Station, 27°33.2′N, 80°03.2′W, 33 m (neotype USNM 1143410 and 141 spec., FSBC I 54821–2, HBOM 50: 1560, 50: 2570–2581, MCZ, ZMA V.Pol. 3924, USNM 1143412); off Hutchinson Island, 27°23′–27°21′N, 80°12′–14′W, 8–11 m, stn 1–5 (more than 500 spec., USNM 55595, 1143413, 1143416, FSBC I 33355–63, 39475–94, 39583).

Eastern Gulf of Mexico, Florida, 29°48′N, 86°9.5′W, 45 m, BLM stn 2531, 1976 (1 spec., ZMA V.Pol. 3375); off Cape San Blas, 29°16.30′N, 85°32′W, from types of *Crucigera websteri*, 47 m, '*Albatross*' stn 2369 (1 spec., USNM 98317); 61 km south of Cape San Blas, 29°07′N, 85°20′W (2 spec., FSBC I 33354); Florida Middle Ground, 28°39.1′N, 84°16.3′W, 28.6 m (3 spec., FSBC I 19410, 44616, 45996); same, 28°35.0′N, 84°14.9′W, from coral, 31 m (10 spec., FSBC I 33351, ZMA V.Pol. 3674); off Cedar Key to St Petersburg, 28°02′–36′N, 83°14′–32′W, 20–22 m (4 spec., FSBC I 33352–4); 11 km west of Anna Maria Island, on shell of pagurid, 12 m, J.K. McNulty don. (1 spec., ZMA V.Pol. 3372); 27°37′–26°24′N, 82°28′–84°13′W, 18–73 m, RV *Hourglass* stn B, C, D, E, J, K, L, M (110 spec., ZMA V.Pol. 3675–7, FSBC I 33314–49); off southern

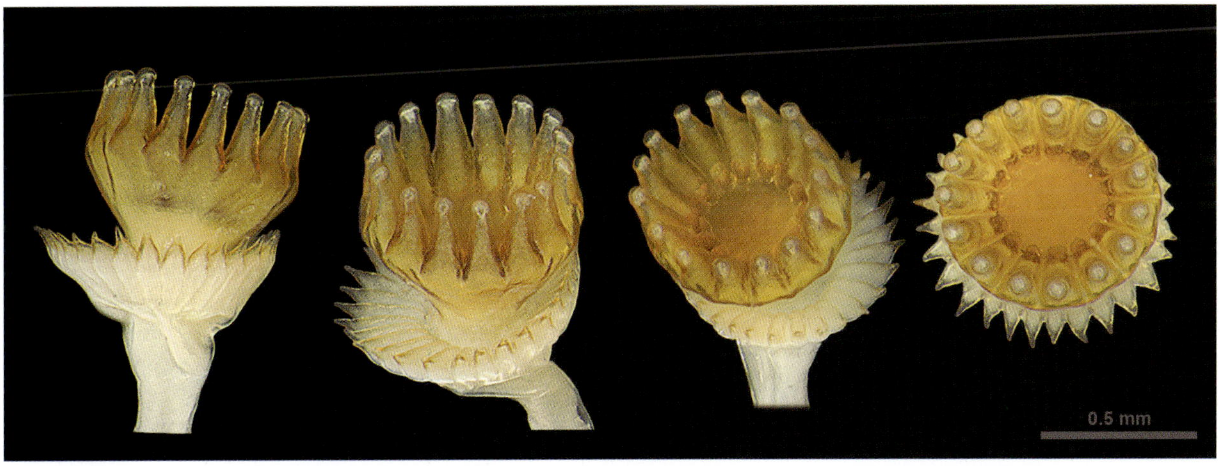

Fig. 72. *Hydroides microtis*, various views of the same operculum, Cape Canaveral, Florida, USA, AM W.43202 (photo E. Wong).

Florida, 25°17′48″N, 81°48′W, stn 52, 4 June 1983 (1 juvenile, USNM 130286); Alabama, 29°58.97′N, 88°04.30′W, hull of sunken ship *Liberty*, 25 m, February 1977 (1 juv., ZMA V.Pol. 3375).

Mexico, north of Yucatan Peninsula, 21°51′–22°03′N, 87°04′–06′W (2 spec., FSBC I 54823–4); Puerto Rico, Roosevelt Roads, 18°13′N, 67°37′W, U.S. Naval Oceanographic Office bio-fouling test panels, 7 m (1 spec., USNM 58577).

Netherlands Antilles, Curaçao, on Venezuelan destroyer in Schottegat, legit P. Wagenaar Hummelinck, stn 1218, 3 March 1955 (1 juvenile, ZMA V.Pol. 3631).

Venezuela, 40 km north of Margarita, on bivalves, 36 m, legit T. Blok, 9 December 1954 (ZMA Ver. P 1100, 1101 (microscopic slides)).

Surinam, shelf, 07°23′–06°29′N, 54°31′–56°43.8′W, 32–120 m, HNLMS *Snellius* and *Luymes* stn B24, G54, A68, K101B, K105, J113A (together 9 specimens, RMNH, VER. P. 10471 (microscopic slide), ZMA V.Pol. 3332–3).

Diagnosis: Peduncle without or with ill-defined constriction. Funnel with 26 (22–32) radii with pointed tip. Verticil on short stalk in funnel, radially symmetrical, with 11 equal spines (6–16), straight, each with truncated to globular dilated tip. Spines simple, smooth, with inner basal knob above well-developed central disk, without ornamentation. Bayonet of collar chaetae with 2 teeth (sometimes with two accessory teeth) and proximal rasp.

Description: Tube white, about 1.6 (up to 1.7) mm wide, with lumen of about 1.1 (up to 1.3) mm. Semicircular in cross-section, often with marked growth rings, covered by thin transparent layer. Three longitudinal ridges may be present. Damaged, older parts of tube may be closed off by tabulae (compare ten Hove 1973: 18–19); in this case the most recent tabula showed a fistula, three pairs of open slits and two or three of closed ones; the older was completely closed except for the fistula.

Radiolar lobes each with 8–12 radioles (x = 10.4, SD = 1.6, n = 9). Terminal filament of radioles as long as normal pinnule.

Peduncle smooth, subtriangular in cross-section, inserted just below first and second normal radiole; with indistinct constriction at the base of funnel. Pseudoperculum present.

Operculum with distal verticil inserted on short stalk into proximal, slightly oblique, almost radially symmetrical funnel. Verticil with 11–16 spines (x = 13.7, SD = 1.4, n = 18; however, see Remarks). Spines almost straight, slightly curved outward, each having blunt, wart-like basal tubercle; tubercle exceptionally missing. Spines with truncated to faintly globular tips, arranged around central disk. Central tooth absent. Funnel not chitinised, with 22–30 (x = 26.3, SD = 2.1, n = 18) acute chitinised radii. Grooves separating radii extending 1/5–1/3 of funnel length. Length of operculum about 1.3 (up to 1.5) mm, width 0.8 (1) mm.

Collar high, with entire edge; continuous with thoracic membranes, forming apron ventrally across anterior abdominal chaetigers. A triangular, almost finger-like protuberance of collar chaetiger present on each side just below junction of ventral and lateral collar lobes.

Thorax with collar chaetiger and 6 uncinigerous chaetigers. Collar chaetae stout bayonet with 2 large conical teeth, proximal rasp and distal striated limba, and narrowly limbate. Subsequent chaetae limbate, of two sizes. Uncini saw-shaped, with pointed fang and 5 curved teeth. Prostomial ocellar clusters not observed.

Abdomen with 49–66 (x = 56.8, SD = 5.6, n = 10) chaetigers. Uncini saw-shaped anteriorly, with fang and 4 teeth; rasp-shaped with 5–6 rows of teeth, fang and 10–12 teeth in profile view posteriorly. Chaetae flat trumpet-shaped with denticulate edge, each with slightly enlarged, blunt anterior tooth as wide as three normal denticles. 'Capillary' chaetae present in up to 20 posterior chaetigers. Pygidium bilobed.

Size: Length up to 22 mm, usually about 16 mm. Width of thorax 0.6–1.0 mm (x = 0.8, SD = 0.1, n 9). Radioles (including operculum) accounting for 1/5–1/3 of entire length.

Colour: No records.

Ecology: The season of attachment in the Northern Gulf of Mexico is June–December (Pequegnat & Pequegnat 1968: 23, 44). The known depth distribution is 7–128 m. See remark on parasitism under *H. dianthus* as well.

Remarks: Mörch's original material cannot be found in the collections of AMNH, NHML, USNM, ZMH and ZMC, although some of his other material has been located in these museums, and apparently is lost. However, as indicated by ten Hove & Wolf (1984: 55–25), Mörch's diagnosis is sufficient to identify the taxon with certainty. His remark 'vix auriculato' doubtlessly refers to, and is meant as opposed to, the other nominal species in his *Eucarphus* section (*dirampha*, *cuminghii* and *benzoni*, the latter two being synonyms of *H. dirampha*), which clearly have verticil spines with auriculate tips.

There is a marked dimorphism in the stoutness of the verticil spines, as noted also by ten Hove & Wolf (1984: 55–25). Scanning the material, it appeared that, in general, thin, elongate spines coincide with small size; stout, seemingly shorter spines occur in larger specimens. One specimen showed an older, fully chitinised, thin-spined operculum, and a soft, developing, much larger, stout-spined operculum on the opposite side.

Decidedly fewer verticil spines occur in the material from (tropical) Venezuela and Surinam (6–9, x = 7.5, SD = 0.9, n = 18, cf. Bastida-Zavala & ten Hove 2003a) than in that from Atlantic Florida (11–16, x = 13.7, SD = 1.4, n = 18, and the few specimens checked from the Gulf of Mexico. However, there is an almost complete overlap in these two groups of specimens in the number of funnel radii (20–29 (32), x = 23.8, SD = 2.6, n = 18). The taxon has a suspiciously large distribution. Further study, especially with DNA, is needed to evaluate if this is not a species complex.

In view of the lack of type material and the wide geographical range of the taxon, with differences between populations from 'N. America' (Mörch's original locality) and from South America, designation of a neotype appears to be necessary, for which a specimen from off Vero Beach, Florida, has been selected.

Differences with similar taxa *H. bandaensis*, *H. dirampha*, *H. sinensis* and *H. qiui* are enumerated in the Remarks with *H. qiui*. At the start of his career HAtH (and students) identified some material (e.g. from off Venezuela) as *H.* cf. *lunulifera*, indicative of superficial similarity with *H. dirampha*; later they noticed the difference in collar chaetae, with evident proximal rasp in this material, missing in *H. dirampha*. By its opercular morphology, rather undifferentiated verticil spines, *H. microtis* resembles the Western African *H. arnoldi*; however, the former has verticil spines globular, dilated tips around a well-developed central disk as opposed simply rounded tips and a small disk. *Hydroides microtis* has also verticil spines similar to *H. novaepommeraniae*; however, the operculum of the latter is bilaterally symmetrical. A summary of similar forms, even if just faintly so, is found in Tovar-Hernández *et al.* (2016 fig. 4).

Distribution: North-west Atlantic, warm temperate to tropical (from off Cape Hatteras, Gulf of Mexico to the shelf of Surinam).

58. *Hydroides minax* (Grube, 1878b)-complex

Figure 73

Type information: Philippines (not further specified), holotype unknown.

Etymology: Not given by Grube (1878b: 269–271) but no doubt refers to the enormous dorsal verticil spine (Latin *minax* = threatening sticking out).

Selected synonymy for this complex of species, arranged according to Provinces as given in Toonen *et al.* (2016), see Remarks.

Indo-Polynesian province, including Western Australia (WA) and North-Eastern Australia (NT, Qld)

Serpula minax Grube, 1878b: 269–271, pl. 15, fig. 5 (description [original]; Philippines).

Hydroides minax – Kupriyanova *et al.* 2008: 428, 430–431 (name, DNA data, phylogeny; Lizard Island, Qld); Pillai 2009: 120, figs 15A–G, 16A–F (description, discussion; Kimberley, WA); Sun & Yang 2014: 238–240, figs 127A–J, 128 (diagnosis, in Chinese; Weizhou Island, Guangxi; North part of Taiwan; Longya Bay, Hainan); Kupriyanova *et al.* 2015: 288–290, fig. 7 (redescription; Lizard Island, Qld, widely distributed in the Indo-West Pacific); Sun *et al.* 2015: 53–57, fig. 16, map 14 (diagnosis; WA, NT, Qld, Australia); Tovar-Hernández *et al.* 2016: table 2, fig. 6 (name, DNA, phylogeny; Qld, Australia); Sun *et al.* 2018: 733, 739, 742, figs 1–3, 5, 6 (DNA, phylogeny; Cassini Island, Long and Ashmore reefs, Eugene Medermott Shoal (WA), Gunyangara (NT), Lizard Island (Qld), Australia).

Not *Hydroides minax* **sensu** Gibbs 1971: 203 [**non** Grube, 1878b] from the Solomon Islands, see *H. albiceps*).

Red Sea

Serpula (*Hydroides*) *monoceros* Gravier, 1906: 110–111 (description; Bonhoure reef, Djibouti, Gulf of Aden); 1908: 115–117, figs 467–472, pl. 8, fig. 288 (description; same; **fide** Imajima 1976a).

Fig. 73. *Hydroides minax*-complex, various views of opercula. Top row: Kimberley, WA, Australia, AM W.21445; bottom row: Indonesia, AM W.28347 (photo E. Wong).

Hydroides monoceros – Pixell 1913: 76–77 (diagnosis; Suakin, Red Sea); Sun *et al.* 2018, fig. 1 (phylogeny, after Pillai 1972).

Hydroides nigra **sensu** Amoureux *et al.* 1978: 147 and Goren 1979: 24 [**non** Zibrowius, 1971a] from the Red Sea, reidentified to *H. minax* by Ben-Eliahu (1991: 527).

Western Indian Ocean

Hydroides monoceros – Pixell 1913: 76–77 (diagnosis; off Zanzibar).

Sino-Japanese

Hydroides minax – Ímajima 1976a: 129–130, fig. 5a–h (description, synonymy; Tanega-shima, south-west Japan); 1987: 15 (distribution in Japan); Fiege & Sun 1999: 119–120, fig. 9A–C (description; Hainan, China); Sun & Yang 2014: 238–240, figs 127A–J, 128 (diagnosis, in Chinese; Weizhou Island, Guangxi; north part of Taiwan; Longya Bay, Hainan).

Eastern Mediterranean

Hydroides minax – Zibrowius 1979: 133–134 (fouling aircraft carrier '*Foch*' on return to Toulon, Mediterranean, from a 7-month service in Western Indian Ocean); Zibrowius & Bitar 1981: 159–160 (name; established in Beyrouth, Lebanon); Ben-Eliahu & ten Hove 1992: 35, 37, 40–43, 45, 47, 52–53 (name only; recently became very abundant along Eastern Mediterranean Israeli coast); Çinar 2006: 223, 225, 228–229, fig. 5 (description; Mersin and Iskenderun Bays, South-Eastern Turkey).

Diagnosis: Peduncle with long, well-defined, chitinised constriction. Funnel with 22 (17–33) long, narrow, truncated chitinised radii. Verticil sitting in or on short basis in funnel, bilaterally symmetrical, with 8 ((6)7–11) spines, except for dorsal hook all outcurving and sharply tipped, lacking additional spinules. Dorsal hook stout, elongated, two to four times larger than remaining spines, somewhat swollen on its bend, strongly incurving with acute tip and two recurving lateral spinules, which may be of similar size as the terminal tip, but generally terminal tip larger. Verticil spines meeting in centre.

Remarks: Three genetic lineages were detected in *H. minax*, all in tropical Australia (material from other areas could not be checked yet). Two of the lineages included one specimen from Lizard Island (Qld) and one from Bremer Island (NT), forming a clade together with two of the five specimens from Augustus Island (WA), with the 11.5–29.4% genetic divergences in COI. The other three specimens of *H. minax* from Augustus Island also formed a monophyletic group, but were nested within the *H. albiceps* clade (Fig. 2). In the literature (e.g. Pillai 1960, 1971; Pixell 1913; Straughan 1967a; Vine & Bailey-Brock 1984), the nominal taxa *H. minax* and *H. monoceros* have been differentiated on the basis of three characters: (1) the funnel radii, with swollen tip, termed 'truncated' in this paper, in the former, with acute tips in the latter; (2) the implantation of the verticil centrally versus dorsally in the funnel; and (3) the angle of the dorsal hook, almost straight away from the worm to under an axis of almost 45° towards the remaining operculum. Although Gravier (1906) depicts the radii with clearly swollen tip in his *H. monoceros*, Pillai (1971, fig. 7D) shows them as acutely tipped. Unfortunately, often text and figures are not entirely congruent in various papers, not unexpected in view of the variability in these radii, even within a single specimen (e.g. see our figures of *H. recta*). Hartman (1959) synonymised *H. monoceros* with *H. minax*. We sorted all published figures known to us according to geographical provenance (as used in 'Selected synonymy', above), but failed to find a clear morphological pattern. Notwithstanding the fact that Vine & Bailey-Brock (1984, fig. 2B–F) show quite some variation from a Sudanese population, these three characters merit more systematic attention now that multiple genetic lineages have been demonstrated (Sun *et al.* 2018) in this complex.

Members of the *Hydroides minax*-complex might be confused with *H. exaltata*, *H. lirs* and *H. pseudexaltata*; a differential diagnosis is given in the remarks under *H. lirs*. A so-called *H. 'prisca'*-type operculum, developmental stage, has been recorded (ten Hove & Ben-Eliahu 2005: 18). Possible ecological differences (e.g. *H. minax* on coral rubble or rocks, and *H. lirs* embedded in living corals) have not yet been fully elaborated.

Distribution: This complex of species is widely distributed in the Indo-West Pacific, and established as Lessepsian migrant in the Levant Mediterranean; however, recent DNA-based work (Sun *et al.* 2018), mentioned above, makes a reconsideration necessary.

59. *Hydroides mongeslopezi* Rioja, 1958

Figure 74

Type information: Playa Norte and Isla Santiaguillo, Veracruz, Gulf of Mexico, type material lost (see Remarks). **Neotype**: Netherlands Antilles, Klein Curaçao, 0.9 km north-west of lighthouse, 16 November 1970, Reef, little sand, from corals, stn 2103A, legit H.A. ten Hove, 17–23 m (ZMA V.Pol. 3686.1).

Etymology: Named after I. R. Monges López.

Selected synonymy

Hydroides mongeslopezi Rioja, 1958: 257–259, fig. 4a–h (description [original]; Veracruz, Gulf of Mexico); Bastida-Zavala & ten Hove 2003a: 135–139, figs 16A–I, 17A–M, 19A–D, 22, table 3 (redescription, synonymy; Gulf of Mexico and Caribbean Sea); Sun 2017: 154, 173, fig. 2N (diagnosis; same); Sun *et al.* 2018, fig. 1 (phylogeny, after Pillai 1972).

Material additional to that mentioned in the most recent revision:

Florida, Fort Lauderdale, U.S. Naval Oceanographic Office bio-fouling test panels, 17 m, det. H. Zibrowius (1 spec., USNM 081921); 9 km off Fort Lauderdale, test-panels, 17–26 m, R. Turner, October 1965–October 1966 (1 spec., MCZ).

Eastern Gulf of Mexico, Florida, 26°24.6′–26°23.9′N, 80°47.0′–83°47.5′W, 73 m, '*Hernan Cortez*' 30 July 1980 (1 spec., FSBC I 33364); off Ten Thousand Islands, 24°32.5′N, 83°17.5–21.0′W, 60 m, '*Anton Bruun*' SEPBOP #19, stn 834, 18 October 1966 (1 spec., USNM 98410).

Bahama Islands, Tongue of the Ocean, 24°51′N, 77°49′W, U.S. Naval Oceanographic Office bio-fouling test panels (1 spec., ZMA V.Pol. 3530).

Jamaica, Discovery Bay, experimental asbestos caves, 40 m, J.B.C. Jackson 1974–5 (1 spec., ZMA V.Pol. 3226).

Puerto Rico, Vieques Sound, 18°19′N, 65°26′W, U.S. Naval Oceanographic Office bio-fouling test panels, 17 m (4 spec., USNM 081920, ZMA V.Pol.3529); 14.8 km off Isla Matei, reef, 31–36 m, legit H.A. ten Hove, stn 2136A (1 spec., ZMA V.Pol.3414).

Isles-des-Saintes, Ilet à Cabrit, 2–5 m, '*Président Théodore Tissier*' stn 167A (1 spec., det. H. Zibrowius, deposition unknown).

Netherlands Antilles, Aruba, common from tidal pools down to 25 m, legit H.A. ten Hove, stn 2003A, B, 2005B, 2008, 2023Ca, 2025Ba (15 spec., ZMA V.Pol.3683, AMNH, NSMT-Pol 98227, 98228, ZMH); Curaçao, common from tidal pools down to 41 m, legit P. Wagenaar Hummelinck stn 1028A, 1029A, 1334A, 1458, 1463, legit H.A. ten Hove, stn 2038B, 2041–2042, 2044, 2046, 2048, 2049, 2053–2056, 2063, 2069, 2070, stn 75–31; Carmabi buoy 1, 700 m north-west Piscadera Bay, from living corals, 10–20 m, legit Y.E. Steward-van Es (together 128 spec., RMNH 18146–52, ZMA V.Pol. 3411–3413, 3685, AHF, AMNH, NHML ZB.1981.105–118, MCZ, YPM, ZMB, ZMH, ZMC).

Klein Curaçao, down to 35 m, legit H.A. ten Hove, stn 2103A, B (4 spec., ZMA V.Pol. 3686.2); Bonaire, common from tidal pools down to 35 m, legit P. Wagenaar Hummelinck stn 1053, 1054, legit H.A. ten Hove, stn 2111–2117, 75–66, 75–72 (together 75 spec., RMNH 18145, 18153–7, FSBC I 27692, ZMA V.Pol. 3410, AM W.20156, 20158-9, 20161-2, NSMT); Klein Bonaire, common subtidally down to 35 m, legit H.A. ten Hove, stn 2105A, 2106, 2107A (16 spec., ZMA V.Pol. 3684, NHML).

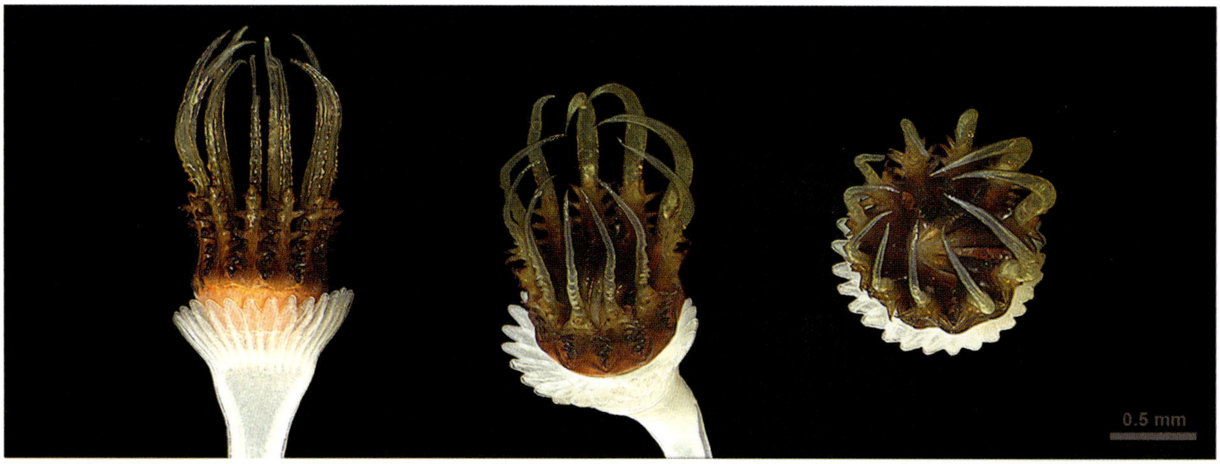

Fig. 74. *Hydroides mongeslopezi*, various views of the same operculum, Bonaire, Netherlands Antilles, AM W.20156 (photo E. Wong).

Diagnosis: Peduncle with ill- to well-defined constriction. Funnel with 23 ((12) 17–28) radii, with rounded, not chitinised tips. Verticil sitting in funnel, bilaterally symmetrical, but not always clearly so; 10 ((6)8–11) spines, incurving, with very darkly coloured (almost black) base. All verticil spines similarly shaped, though dorsal spines generally a little wider than ventral ones, with up to 15 darkly coloured lateral to external spinules, grouped from base to halfway, 8–10 of them almost pair-wise; spines ending in a long, sharp tip. Verticil spines with long basal spinule, almost in a semicircle dorsally, gradually changing to nearly straight in ventral spines. Base of spines almost touching in centre, no central tooth.

Remarks: Rioja apparently did not leave any collections in the Biological Institute of the National University of Mexico, and his material must be regarded as lost (M.E. Caso–Muñoz, *pers. comm.* with ten Hove). In retrospect *H. mongeslopezi* is easily distinguished from *H. lambecki*, but when one of us (HAtH) collected most of his Antillean material in the early seventies, and *H. lambecki* had not been described yet, both taxa were routinely identified as *H. mongeslopezi* (e.g. in ZMA V.Pol. 3410, now partly reidentified *H. lambecki* ZMA V.Pol. 5084), although from the mid-1970s these 'aberrant' specimens were named *H. mongeslopezi* atypical or *H.* cf. *mongeslopezi.* The main differences between the two obviously similar species have been tabulated by Bastida-Zavala & ten Hove (2003a, table 3). *Hydroides mongeslopezi* has 17–28 funnel radii, *H. lambecki* 11–14. The first has 8–11 verticil spines, *H. lambecki* has 6–8. The internal spinules in *H. mongeslopezi* are curving towards the bottom of the verticil, pointing distally upwards in *H. lambecki.* Our material of *H. mongeslopezi* **sensu stricto** differs in minor details from Rioja's figures. In view of this, and the early confusion between *H. mongeslopezi* with *H. lambecki*, a neotype for the former is advisable. Unfortunately, newly collected material, as for instance given in Bastida-Zavala & ten Hove (2003a) but also in the 'Material additional' (above) did not include specimens from the type locality, or even nearby. We selected an (in our view) 'typical' specimen to serve as neotype.

Distribution: Tropical West Atlantic.

60. *Hydroides monroi* Zibrowius, 1973a

Figure 75

Type information: Pointe Noire, Congo (Brazzaville), holotype and paratypes MRAC 34.952, 23.730.

Etymology: Named after C.C.A. Monro.

Selected synonymy

Hydroides monroi Zibrowius, 1973a: 29–31, fig. 3a–m (description [original], discussion; Congo, Zaire, Cabinda (Angola)); Sun 2017: 35, 156, 173–174, fig. 4J (diagnosis; same).

Hydroides uncinata **sensu** Monro, 1939: 223–224 (diagnosis; Congo, Zaire, Cabinda) [**non** Philippi, 1844, and renamed by Zibrowius (1973a, above)]; in part only, other parts of Monro's material belong to *H. arnoldi*, *H. augeneri* and *H. dipoma*.

Diagnosis: Peduncle without constriction. Funnel with 35 (31–39) chitinised, triangular radii. Verticil sitting in funnel, bilaterally symmetrical, with 7(–9) thickset spines. Dorsal hook smooth, without side spinules, bent or geniculate above the other spines; its development varying between the specimens, sometimes hardly larger than remaining spines. The latter straight, short, thickset, distally blunt, each with a subapical internal spinule of varying size, pointing more or less right-angled to the middle of the verticil. Subapical spinules may be missing in ventral spines. Occasionally 1 or 2 latero-dorsal spines intermediate in size and form between hook and remaining spines. All spines with blunt, basal spinule above smooth central disk.

Remarks: Differences with *H. adamaformis*, *H. gottfriedi* and *H. inermis*, all with few (up to about 8) rather thick spines, including a large dorsal hook, are discussed under *H. helmata*. A mistaken identification as *H. monroi* has been discussed under *H. brachyacantha*, indicative of the similarity between the species and *H. brachyacantha* **sensu auctores** (see Remarks there). However, the verticil spines of the latter complex (excluding the dorsal hook) all are geniculate, or have large incurving parts, and a sort of denticle or bump on the external curve. *H. monroi* on the other hand has thickset, blunt spines with a small subapical internal spinule.

Distribution: Congo, Cabinda, Zaire.

Fig. 75. *Hydroides monroi*, various views of opercula, types Pointe Noire, Congo (Brazzaville), MRAC 34.952 and 23.730 (photo J. Brecko © MRAC).

61. *Hydroides mucronata* Rioja, 1958

Figure 76

Type information: Isla de Sacrificios, Veracruz, Mexico; types lost.

Etymology: Refers to the pointed (Latin *mucronatus*) side spines of the verticil spines (Rioja 1958: 256 'que tienen forma de mucron').

Selected synonymy

Hydroides mucronata Rioja, 1958: 254–257, fig. 3a–h (description [original]; Veracruz, Gulf of Mexico); Sun *et al.* 2018 fig. 1 (phylogeny, after Pillai 1972).

Hydroides mucronatus – Bastida-Zavala & ten Hove 2003a: 141–142, figs 20A–J, 22 (redescription, synonymy, discussion; Campeche, Gulf of Mexico).

Fig. 76. *Hydroides mucronata*, Champotón, Campeche, Mexico, ECOSUR Serp-36 (photo H. Bahena Basave).

Diagnosis: Peduncle without or with ill-defined constriction. Funnel with 24 ((14) 20–26) unchitinised, bluntly triangular radii. Verticil sitting in funnel, radially symmetrical, with 7(–9) straight spines. All spines similar in shape and size, with attenuating, blunt tip. Spines all with a pair of lateral spinules about halfway, some with one internal spinule slightly more proximal; without basal or external spinules and/or wings. Central disk small, without central tooth.

Remarks: Rioja (1958) described *H. mucronatus* from Veracruz, the type material is regarded to be lost (Dr. M. E. Caso–Muñoz, *pers. comm.* to HAtH). Subsequent 'topotypical' collecting (Isla Verde, by Bastida-Zavala) was unsuccessful. The diagnosis above has been based on both Rioja (1958) and Bastida-Zavala & ten Hove (2003a). Although the single specimen from Campeche is smaller than those described by Rioja, it agrees in most other aspects, but for the presence of a middle internal spinule (Bastida-Zavala & ten Hove 2003a, fig. 20A; Fig. 76 here), which may have been overlooked by Rioja, the number of funnel radii (Veracruz 22–26, Campeche 14) and the shape of the latter (pointed *v* blunt), which may be a matter of appreciation. The straight, blunt verticil spines with a pair of lateral spinules about halfway makes the taxon unique (but see next species). All other species with one pair of lateral spinules have sharply pointed, incurving verticil spines.

Distribution: Veracruz to Campeche, Gulf of Mexico.

62. *Hydroides mucronatoides* nov. spec. ten Hove

Figure 77

Type information: Netherlands Antilles, Klein Bonaire, near jetty, 1 July 1970, barrier reef 20 m offshore, in surf, legit H.A. ten Hove, stn 2107; holotype ZMA V.Pol. 3339.1 (figured by Bastida-Zavala & ten Hove 2003a fig. 21A–B, two paratypes ZMA V.Pol. 3339.2.

Etymology: Named after *Hydroides mucronata*, the invariant Latin adjectival suffix *-oides* meaning 'similar to'.

Selected synonymy

Hydroides cf. *mucronatus* ten Hove 1984: 191, fig. 6a–c (diagnosis, possibly new species; Aruba, Curaçao, Puerto Rico); Bastida-Zavala & ten Hove 2003a: 142–143, figs 21A–O, 22 (description, synonymy, discussion; Gulf of Mexico, Bahamas and Caribbean Sea); Díaz-Castañeda & Almeda-Jauregui 1999: 202, 203, 207–209, 212, 215 (name, det. ten Hove, experimental fouling; Barbados).

Hydroides mucronatus – Sun 2017: 154, 174, fig. 3B (diagnosis, operculum; Veracruz, Gulf of Mexico).

Additional material (all identified by H.A. ten Hove and/or students):

Florida, 5 mi off Fort Lauderdale, legit R. Turner, October 1965–October 1966, det. H.A. ten Hove 1974 *Hydroides* cf. *mucronata* (1 spec.; MCZ).

Grand Bahama Island, off Freeport, Hydro-Lab, legit B.A. Vittor, 28 January 1974, on coral, 50 m, det. P.G. Johnson *Hydroides* near *parva*, redet. G. van Ee *Hydroides* cf. *mucronata* (2 spec., 1 loose operculum, deposition unknown); Bahamas, Tongue of the Ocean, U.S. Naval Oceanographic Office, Experimental fouling panels, J.R. de Palma ref. number 124 (1 spec., ZMA V.Pol 3532).

Puerto Rico, Joyuda, legit H.A. ten Hove, 4 October 1970, from rocks and boulders on sand, covered by a thin layer of muddy sand; 0.3–1 m, stn 2133 (4 spec., ZMA V.Pol 3335); La Parguera *Hydroides* cf. *mucronata* (1 spec., USNM 270934); Vieques Sound, U.S. Naval Oceanographic Office, Experimental fouling panels, J.R. de Palma ref. numbers 129, 12, det. H.A. ten Hove 1972 *Hydroides* nov. spec. (several, ZMA V.Pol. 3533).

Anguilla, Crocus Bay N., legit P. Wagenaar Hummelinck, 3 July 1973, limestone, igneous rock, 0–0.5 m, stn 1704 (1 spec., ZMA V.Pol 3336).

Netherlands Antilles, Saba, Port Bay, pier, legit P. Wagenaar Hummelinck, 7 July 1973, wooden piling, 0–1 m, stn 1705, det. S. Criens (2 spec., ZMA V.Pol 3341); St Eustatius, South of Tumble Down Dick Bay, legit P. Wagenaar Hummelinck, 10 July 1949, rocky andesite shore with magnetite sand, mid and low-tide zone, stn 1119 (1 spec., ZMA V.Pol 3340); Bonaire, Karpata, legit H.A. ten Hove, 20 February 1988, 15–18 m, on spur of reef (1 spec., USNM 1013610); Bonaire, Playa Lechi, North, inner bay, legit H.A. ten Hove, 13 July 1970, 0.3–0.5 m, from undersides of boulders, cemented to muddy rock, stn 2114 (6 spec., AM W.43204, Fig. 77); Bonaire, Punt Vierkant, legit H.A. ten Hove, 8 June 1970, tidal pool, sandy limestone, 0.3 m, stn 2116B (1 spec., ZMA V.Pol 3338); Bonaire, AISCO, Pekelmeer, 50 m south from main pumping station, legit H.A. ten Hove, 22 September 1975, boulders in limestone sludge on coral debris, 0.3–0.6 m, stn 75-61 (1 spec., ZMA V.Pol 5493); Curaçao, Playa Porto Marie (12°13′7″N, 69°5′12″W), legit H.A. ten Hove, 2 April 2012, from underside of floats bordering swimming area, 100 m offshore, 2 cm deep (1 spec.,

Fig. 77. *Hydroides mucronatoides* nov. spec., various views of the same operculum, Bonaire, Netherlands Antilles, Playa Lechi, AM W.43204 (photo E. Wong).

AM W.41463); Curaçao, Playa Hundu, legit P. Bouma & U. Frank, 20 March 1990, coral rubble, 0.2–1 m (1 spec., ZMA V.Pol 3759); Curaçao, 700 m NW of Piscadera Bay, legit Y.E. Steward–Van Es, 31 October 1977, Carmabi Buoy 1, depth 9 m (1 spec., ZMA V.Pol 4135); Aruba, Mangel Altu, mouth of Spaans Lagoen, near raised reef, legit H.A. ten Hove, 11 August 1970, coral debris, stn 2023Ca (4 spec., ZMA V.Pol 3337).

Panama, Caribbean coast, Punta de Toro, Fort Sherman, legit Jones, Dawson, Kaufman & Brown, NMNH-STRI, Panama Survey stn 125-2, sievings from *Thalassia* (1 spec., USNM 58581); Archipelago de San Blas, Program Carib I, RV *Alpha Helix*, stn 26, 30 June 1977, 09°32.8′N, 78°59.5′W, 0 m (2 spec., deposition USNM?); stn 31, 8 July 1977, 09°12.8′N, 78°08.7′W, 4 m, both samples det. D. Makhan (2 spec., deposition USNM?); Panama, Atlantic, U.S. Naval Oceanographic Office, experimental fouling panels, J. R. de Palma ref. number 102 (1 spec., ZMA V.Pol 3531).

Diagnosis: Peduncle with ill-defined constriction. Funnel not with 'ridge' of many radii, but replaced by 4 (2–7) globular processes. Verticil sitting on proximal bulbous part of operculum, radially symmetrical, with 12 (8–14) straight spines. All spines similar in shape and size, with attenuating, blunt tip. Spines with a pair of lateral spinules about halfway and a basal spinule above small central disk; without central tooth. Bayonet of collar chaetae with 3 or 4 teeth.

Remarks: This taxon differs from *Hydroides mucronata* in having a series of 2–7 globular processes instead of a funnel with bluntly triangular radii (ten Hove 1984, fig. 5a–c). In all material we have seen (hundreds of specimens, not all are included in the list above) only 3 or 4 of the opercula showed some similarity to a developing funnel with 7 rounded, fleshy teeth (one from Panama stn 31, another from Puerto Rico, Viequez Sound, and one from Grand Bahama Island); most frequently there are just 3 or 4 blunt knobs. At first sight, the small opercula might be mistaken for and have been identified as a juvenile *H. parva*. In specimens with 2–4 processes, the taxon shows a certain similarity with the genus *Crucigera*, especially *C. tricornis* Gravier, 1906 and *C. websteri* Benedict, 1887 (see ten Hove & Janssen-Jacobs 1984). A further difference is found in the verticil spines, with evident basal spinule in *H. mucronatoides* nov. spec., without in *H. mucronata*. Moreover, the bayonet collar chaetae are different, with 2 teeth in *H. mucronata*, 3 or 4 in *H. mucronatoides*. Compare taxonomic remarks in Bastida-Zavala & ten Hove (2003a) as well. In order to draw the attention to the occurrence of two similar taxa we decided to name the specimens without a true funnel *H. mucronatoides* nov. spec. More material and DNA data are needed ascertain the relation between the two.

Distribution: Bahamas, Florida, Gulf of Mexico and Caribbean Sea, Atlantic Panama.

63. *Hydroides multispinosa* Marenzeller, 1884

Figure 78

Type information: Eno-sima (Enoshima), Honshu, Japan, holotype NHMW 2026.

Etymology: Not given by Marenzeller (1884: 216) but evidently refers to multiple lateral spinules on the verticil spines.

Selected synonymy

Hydroides multispinosa Marenzeller, 1884: 216–217, pl. 4, fig. 2 (diagnosis [original]; Enoshima Island, Japan); Zibrowius 1972b: 443–444 (redescription of holotype, discussion); ?Sun *et al.* 2015: 58–59, fig. 17a–b, map 15 (diagnosis, discussion; distribution Australia; see remarks below).

Hydroides norvegica **sensu** Fauvel 1936: 63 (remark that *H. multispinosa* is the same as the 'cosmopolitan' *H. norvegica*; Enoshima, Japan).

Hydroides multispinosus – Imajima 1976b: 238–240, fig. 4a–k (description, synonymy, discussion; Tomioka Bay, Kagoshima Bay); 1987: 15 (distribution in Japan); ?Sun & Yang 2014: 211, 240–241, figs 129A–I, 114 (diagnosis, in Chinese; Dalian, Liaoning; Qingdao, Shandong; Sanya, Hainan; China).

Not *Hydroides multispinosa/-us* **sensu** Augener 1914: 139–142 (Port Jackson, NSW, Australia) [**non** Marenzeller, 1884]; Hartman 1954: 629; Reish 1968: 228 (both Marshall Islands); see *H. elegans*, **fide** Zibrowius 1971a: 721.

Diagnosis: Peduncle without constriction. Funnel not chitinised, with 25–27 blunt radii (holotype bioperculate). Verticil sitting in funnel, radially symmetrical, with 13 spines, straight to lightly curving outwards. All verticil spines similar in shape and size, each with pointed tip, 6–8 more or less pairs of lateral spinules and 2–4 pointed medial spinules. Central disk elevated, central tooth short. Bayonet of collar chaetae with 2–7 larger teeth merging into proximal rasp.

Remarks: Numbers in 'Diagnosis' based on holotype, in view of reasons sketched below. The type locality of *H. multispinosa*, Enoshima, Japan, is in temperate waters, so records from other areas very well may relate to other, similar taxa. Having been synonymised with *H. norvegica* (from cold-temperate Europe) by Imajima & Hartman (1964: 369–370), Zibrowius (1972b: 443–444) resurrected *H. multispinosa* after restudy of the holotype. Specimens of *H. multispinosa* reported from tropical Australia have fewer internal (0–4) and pairs of lateral spinules (4 or 5) on verticil spines and fewer funnel radii (19–21) when compared to the original description (Marenzeller 1884: internal spinules 2–4, pairs of lateral spinules (6–8), funnel radii (25–27). This and the localities are reasons to place a question mark with this record and that of the also tropical Hainan. For similar reasons, additional material and molecular studies are needed to determine whether *H. multispinosa* and *H. nanhaiensis* are separate species. *Hydroides multispinosa* can be distinguished from the harbour fouling *H. elegans* by the number of lateral spinules, which is 6–8 in *H. multispinosa*, but 2–4 in *H. elegans*. The most obvious differences with the somewhat similar *H. longispinosa* are the 13–15 verticil spines with many (up to 18) internal spinules in this taxon.

Distribution: Japan, ?South China Sea, ?Australia.

Fig. 78. *Hydroides multispinosa*, various views of the same operculum, Lord Howe Island, New South Wales, Australia, AM W.201331 (photo E. Wong).

64. *Hydroides nanhaiensis* Wu & Chen, 1981

Figure 79

Type information: Xi River estuary, Pearl River Delta, South China Sea, holotype MBMCAS.

Etymology: Named after its type locality as 'Nanhai' means 'South China Sea' in Chinese.

Selected synonymy

Hydroides nanhaiensis Wu & Chen, 1981: 355–357, fig. 2a–g (diagnosis [original]; South China Sea); Sun & Yang 2014: 241–243, figs 130A–G, 114 (diagnosis, in Chinese; Zhujiang River Estuary, Guangdong; South China Sea); Sun 2017: 49, 76, 89, 155, 174, fig. 3F (diagnosis; South China Sea).

Diagnosis: Peduncle without constriction. Funnel not chitinised, with 20–27 blunt radii (22 in holotype). Verticil sitting in funnel, radially symmetrical, with 9 or 10 spines, almost straight to lightly curving outwards. All verticil spines similar in shape and size, each with pointed tip, 6–8 pairs of lateral spinules and 5–8 internal spinules, including a distinct basal one. Central disk elevated, central tooth absent. Bayonet of collar chaetae with 2 teeth and proximal rasp.

Remarks: There is a discrepancy between text and figure of the funnel radii: the Chinese text gives a number of 22, the English 20, but visible in their fig. 2a already are 22, behind verticil another 5, thus in total 27. This species is very similar to *H. longispinosa* and *H. multispinosa*. According to Wu & Chen (1981) it can be separated from the other two species by the absence of a central tooth; however, Zibrowius (1972b: 443, fig. 3d) mentions opercula without central spinule in *H. multispinosa*. See Remarks under these taxa as well.

Distribution: South China Sea.

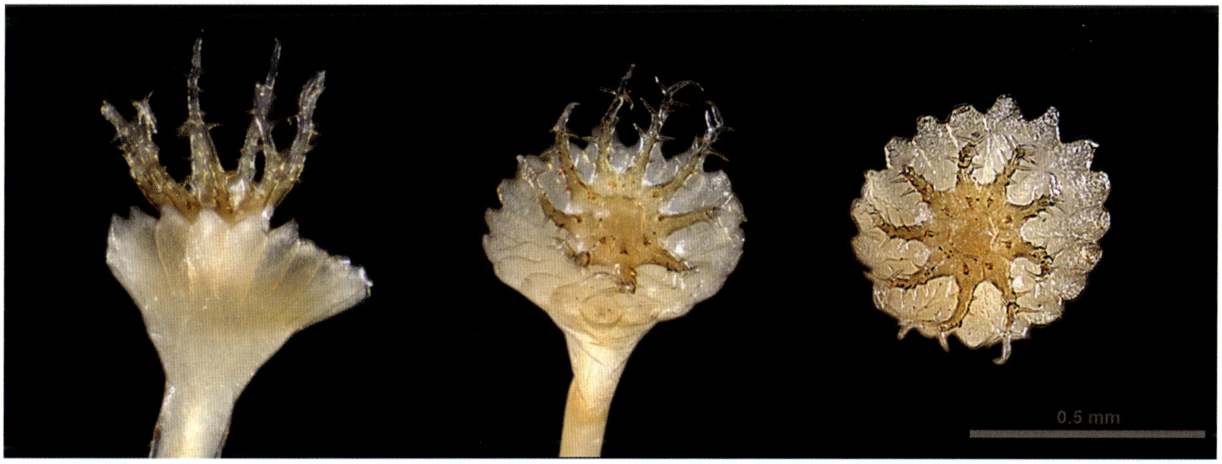

Fig. 79. *Hydroides nanhaiensis*, various views of the same operculum, Xi River estuary, Pearl River Delta, South China Sea, China, holotype MBMCAS (photo X. Wu).

65. *Hydroides nigra* Zibrowius, 1971a

Figure 80

Type information: Off Pointe Meloula, Tabarka, Tunisia, holotype and paratypes MHNH.

Etymology: Not given by Zibrowius (1971a: 711) but refers to the dark colour (Latin *niger, nigra, nigrum* = black, dark) of the operculum.

Selected synonymy

Hydroides inornata (?) [**non** Pillai, 1960] **sensu** Zibrowius 1968: 117, pl. 2, figs 27, 28 (Gulf de Marseille, Gulf de Gabès, Mediterranean; reidentified to *H. nigra* by Zibrowius 1971a: 711).

Hydroides nigra Zibrowius, 1971a: 711–713, figs 16–30 (description [original], discussion; off Tabarka, Tunisia); Bianchi 1981a: 67–68, fig. 23a–d (diagnosis; Mediterranean Sea); Kitsos *et al.* 2005: 258 (epibionthic on loggerhead turtles, Greece); Knight-Jones *et al.* 1991: 841, 854 (name; Turkey); Sun *et al.* 2017b: 40, 42–44, fig. 1 (DNA, phylogeny; Mallorca, Spain); 2018, figs 2, 3, 5, 6 (DNA, phylogeny; Spain); **pro parte** Grosse *et al.* 2021: 16–26, *not* fig. 3C (DNA, phylogeny; Mallorca – see Remarks).

Not *Hydroides nigra* **sensu** Amoureux *et al.* 1978: 147 and Goren 1979: 24 [**non** Zibrowius, 1971a] from the Red Sea, reidentified to *H. minax*-complex by Ben-Eliahu (1991: 527).

Diagnosis: Peduncle with well-defined, darkly chitinised constriction. Funnel with 23 (16–27) radii with long, sharply pointed, well-chitinised black tips. Verticil on short basis in funnel, bilaterally symmetrical, with 8 (6–10) black spines, differing in curvature, shape and size; all slender, incurving, sharply pointed, with basal spinule over small, smooth central disk. One dorsal hook distinctly larger than other spines, sometimes 2, or 1 large accompanied by a pair intermediate in size and shape. These hooks are either curving smoothly, or at most with a recognisable external knee. Other spines can be more abruptly curving, knee sometimes accentuated by clear external knob. This differentiation is increasing from dorsal to ventral spines.

Remarks: Similar to *Hydroides fusca* in dark colour of the operculum, but this taxon shows a radially symmetrical operculum. Recurved verticil spines occur too in *H. recurvispina* and *H. trompi*, but these turn clockwise. In phylogenetic analyses (Sun *et al.* 2018, figs 2, 3, Fig. 13), *H. nigra* forms a well-supported Mediterranean clade with *H. pseudouncinata*; both individual clades were equally well supported. Although phylogenetically an outgroup of the *Hydroides operculata*-complex, morphologically the operculum of *H. nigra* is not unlike some of those in that complex (Sun *et al.* 2017b: 44, fig. 1A–F). Grosse *et al.* (2021: 24, fig. 5) recovered two different taxa under the name *Hydroides nigra*; they once used cf. *nigra* to differentiate, presumably the cf. is their new material (their fig. 3C), which in our opinion morphologically (and ecologically, from fouling samples) more resembles *H. dianthus*, though clusters with one of the lineages of the polyphyletic *H. brachyacantha* from Pacific Mexico (see Remarks there as well). The other lineage presented under *H. nigra* is based on sequences by Sun *et al.* (2017b, fig. 1F; same as the left specimen in our present Fig. 80), and indeed what on morphological (and ecological, open coast) characters is regarded to be Zibrowius's (1971a) taxon.

Distribution: Mediterranean, from Spain to Turkey.

Fig. 80. *Hydroides nigra*, various views of opercula, Spain (left two) AM W.42073; Split, Croatia (right two) ZMA V.Pol. 4319 (photo E. Wong).

66. *Hydroides nikae* Sun, Wong, Tovar-Hernández, Williamson & Kupriyanova, 2016

Figure 81

Type information: Edithburgh Jetty, South Australia, holotype SAM E3530

Etymology: Named after Nika Mikhin, daughter of Kupriyanova.

Selected synonymy

Hydroides nikae Sun, Wong, Tovar-Hernández, Williamson & Kupriyanova, 2016: 55–56, figs 7, 8A–F (description [original]; SA); Sun *et al.* 2018: 742, figs 2, 5, 6 (DNA, phylogeny; Hallett Cove Beach, SA).

Diagnosis: Peduncle with ill-defined or without constriction. Funnel with 23 (18–26) chitinised, sharply tipped radii. Verticil inserted on short basis into funnel, bilaterally symmetrical, with 8 (6–9) spines, with pointed tip, strongly curved inwards, gradually becoming smaller from dorsal to ventral side. Number of larger spines (hooks) variable, from 1–4, long enough to cover central disk, either externally smooth or with small external knob; smaller spines all with small, round knob on their bend. All verticil spines with short basal spinule. Central tooth absent.

Remarks: The species shows a variable opercular morphology. The dorsal hook may or may not show an external knob similar to the situation in *H. amri*; the numbers of larger hooks are also similar in both taxa, 1–4 in *H. nikae*, 1–5 in *H. amri* (see remarks there as well). Molecular characters are a more effective method to diagnose these two species (Sun *et al.* 2016; 2018); however, though these analyses consistently place *H. nikae* rather distant (at least seven steps) from the morphological similar *H. brachyacantha*-complex, and associated with e.g. *H. tuberculata*-complex (in both analyses), its occurrence in a clade (Sun *et al.* 2016, fig. 2) with *H. externispina* and *H. tambalagamensis* is quite unexpected. At first glance the operculum of *H. nikae* is not unlike that of *H. tuberculata*-complex, but the latter has only a single hook with 4 (or 5) other spines, all alike.

Distribution: South Australia.

Fig. 81. *Hydroides nikae*, various views of the same operculum, Edithburg, South Australia, Australia, holotype SAM E3530 (photo E. Wong).

67. *Hydroides nodosa* Straughan, 1967a

Figure 82

Type information: Tannum Sands, Gladstone, Qld, Australia, holotype AM W.4013; paratype QM G3895.

Etymology: Not given by Straughan (1967a: 227), but likely to refer to the internal 'rounded projection' at the base of each verticil spine (Latin *nodosus -a -um* = 'knotty').

Selected synonymy

Hydroides nodosa Straughan, 1967a: 227, fig. 10a–h (diagnosis [original], material studied; Gladstone, Qld); Zibrowius 1971a: 725–726 (extended diagnosis, discussion; Gladstone (type material), Townsville, Qld, Australia); Sun *et al.* 2015: 59–63, fig. 18a, b, map 16 (diagnosis, discussion; Northern Territory, Queensland, Australia); 2018, fig. 1 (phylogeny, after Pillai 1972).

Diagnosis: Peduncle without constriction. Funnel with 14 (9–23) not chitinised, rounded radii. Verticil sitting in funnel, with 8 (6–10) slightly incurving spines, sharply pointed; each with a pair of lateral spinules at 1/3 of their length, and one to three of them may have a second lateral spinule on one side; each with a sharply pointed basal spinule above small central disk; short central tooth. Bayonet of collar chaetae with 2 teeth and proximal rasp.

Remarks: Zibrowius (1971a: 725–726), reviewing the (sparse) type material, was the first to point out that Straughan (1967a) had failed to notice the minute lateral spinules of the verticil spines, as well as a basal spinule and central tooth in the small opercula of her new species. He compared the small, in his opinion questionably juvenile, specimens with *H. elegans*, and pointed out that they consistently differed in the fact that verticil spines all are incurving in *H. nodosa*, as opposed to the straight spines in *H. elegans*. Though the bayonet collar chaetae of the latter show 2–5 teeth and proximal rasp, as compared with 2 teeth and proximal rasp in *H. nodosa*, the collar chaetae in both species are similar (**fide** Zibrowius 1971a: 726). It should be noted that Straughan's fig. 10c, d of the collar chaetae is rather sketchy; re-examined by one of us (HAtH 1983) they clearly show 2 large teeth and a proximal rasp. Moreover, *H. nodosa* has 1, exceptionally 2 spinules on each side of the verticil spines, while *H. elegans* has 2–4. *Hydroides nodosa* might be confused with *H. bispinosa*, also with incurving verticil spines with one pair of lateral spinules. However, the operculum of *H. nodosa* shows a central tooth (absent in *H. bispinosa*) and sometimes a second lateral spinule in 1–3 verticil spines. In view of the developmental ranges as discussed (e.g. in the paragraph on '*H. prisca*', below), and in particular more intermediate stages as mentioned in Ben-Eliahu & ten Hove (2011: 30, fig. 10A, B), for *H. heterocera*, we reviewed the species co-occurring with *H. nodosa* in tropical Australia (compilation in Sun *et al.* 2015). There is no taxon present with the combination of incurving verticil spines with lateral and basal spinules and a central tooth, of which *H. nodosa* might be a juvenile. Moreover, the regular occurrence of material, 16 samples in seven spread-out localities give us confidence that *H. nodosa* indeed is a species by itself. Unfortunately, we do not have DNA data. In collections we have noticed (old, preliminary) misidentifications as *H. norvegica/elegans*, for what we reidentified as *H. nodosa*.

Distribution: Australia: NT, Qld.

Fig. 82. *Hydroides nodosa*, various views of opercula, Cairns Inlet, Queensland, Australia (left two) W.29095; Tannum Sands, Queensland, Australia (right two) AM W.4013 (photo E. Wong).

68. *Hydroides norvegica* Gunnerus, 1768

Figure 83

Type information: Norway; the existence of type material is not mentioned by Moen (2006) or Bakken *et al.* (2011) and thus unlikely.

Etymology: Named after its type locality, Norway.

Selected synonymy

Hydroides norvegica Gunnerus, 1768: 51–53, pl. 2, figs 11–13 (Norwegian description [original; Trøndelag region, Norway) translated and commented upon by Moen (2006: 118–119); Zibrowius 1971a: 717–721, figs 48–55 (extensive description, discussion, synonymy; Atlantic Ocean from Norway to Portugal, Mediterranean Sea from French-Spanish border to Corfu Greece); Kitsos *et al.* 2005: 258 (epibionthic on loggerhead turtles, Greece); Rzhavsky *et al.* 2014: 46–48, fig. 8A–I, pl. 1B (description, discussion, synonymy; Arctic Seas); Sun *et al.* 2016: table 1, fig. 2 (DNA; outgroup for *H. brachyacantha*-complex); 2018: 733, 738, 742, figs 1–3, 5, 6 (DNA, phylogeny; Tromsø and Trondhjem, Norway; off Manfredonia, Italy, Adriatic Sea).

Hydroides norvegicus – Tovar-Hernández *et al.* 2016: table 2, fig. 6 (name, DNA, phylogeny; Norway).

Diagnosis: Peduncle without or with ill-defined constriction. Funnel not chitinised, with 30 (23–39) blunt radii. Verticil sitting in funnel, with 15 (11–20) spines, straight or slightly curving outwards, radially symmetrical. Spines each with 3–5 irregularly spaced lateral spinules (at both sides), as well as a basal spinule; 1 or 2 pointed internal spinules may be present. Central disk almost flat to slightly elevated, without tooth. Bayonet of collar chaetae with 2 teeth (exceptionally 1–3, some accessories and poorly developed proximal rasp) (ten Hove 1974).

Remarks: For several decades the nominal taxon *Hydroides norvegica* was thought to include all forms with multiple lateral spinules to their verticil spines. Zibrowius (1971a: 717–725) judged the distribution of *H. norvegica* **sensu auctores**, in the Mediterranean from harbour fouling down to a depth of almost 400 m, to be ecologically unlikely and demonstrated that the harbour-fouling material was morphologically different from that occurring in open waters between 40 and 400 m depth, which he identified as *H. norvegica* **sensu stricto**. For the widespread harbour fouling taxon he reintroduced the name *H. elegans*. In a further paper Zibrowius (1972b) described two new species, *H. azorica* and *H. capensis*, previously mistakenly identified as *H. norvegica*, and after restudy of the holotype (Zibrowius 1972b: 443–444) resurrected *H. multispinosa*, synonymised with *H. norvegica* by Imajima & Hartman (1964: 369–370). Furthermore, it shows similarity with *H. longispinosa*, *H. nanhaiensis* and, to a lesser extent, *H. ochotereana* (see Remarks under *H. azorica*). The bayonet collar chaetae in *H. norvegica* have 2 conical teeth at the base of the blade, while they have 2–5 pointed short teeth at the base the blade and a large proximal rasp in *H. elegans* (e.g. ten Hove 1974, figs 1–9; also see Fig. 5D).

In the phylogenetic analyses by Sun *et al.* (2018, figs 2, 5), *H. norvegica* formed a reasonably well-defined clade with *H. elegans* and *H. longispinosa*, though the latter taxon appears to be polyphyletic.

Distribution: Arctic Europe seas to Mediterranean (deeper water, 30–400 m (Zibrowius 1971a: 721; Ben-Eliahu & Fiege 1996: 7).

Fig. 83. *Hydroides norvegica*, various views of the same operculum, Irish Sea, Great Britain, AM W.43201 (photo E. Wong).

69. *Hydroides novaepommeraniae* Augener, 1925b-complex

Figure 84

Type information: Neu Pommern 'Hanam-Hafen', most probably a misspelling of Hannan (or Garua) Harbour, Willaumez Peninsula, north coast of New Britain, Bismarck Archipelago, Papua New Guinea; holotype ZMH V.9749.

Etymology: Not given by Augener (1925b), but evidently a Latinised derivation of the island of collection, New Britain, in Germany once named Neu-Pommern, after the coastal lands along the Pommersche Bucht (Pomeranian Bay), Baltic).

Selected synonymy

Hydroides (*Eupomatus*) *Novae-Pommeraniae* Augener, 1925b: 69–70, fig. 5–5a (description [original], material studied; Bismarck Archipelago, Papua New Guinea); Okuda 1937: 308–309, figs 54, 55 (description; Truk, Micronesia).

Hydroides malleolaspina **sensu** Imajima, 1982: 45, fig. 3k–s (description; Palau Islands) [**non** Straughan, 1967a] reidentified by Imajima & ten Hove (1984: 47).

Hydroides novaepommeraniae – Imajima & ten Hove 1984: 45–47, fig. 4a–d (redescription, partly based on holotype, synonymy; New Britain, Palau Islands, Truk Islands, Philippines).

Not *Hydroides novaepommeraniae* **sensu** Dew 1959: 26–27, fig. 5; Straughan 1967a: 221–222 [**non** Augener, 1925b] (both see *H. recta* **fide** Imajima & ten Hove 1984: 47).

Not *Hydroides novaepommeraniae* **sensu** Ben-Eliahu 1991: 527 (name; Levantine coasts of Israel) [**non** Augener, 1925b], recognised to be secondary opercular chimaera's of a couple of different *Hydroides* species by Ben-Eliahu & ten Hove (1992: 42); see discussion on the rejected name *Hydroides prisca* below as well.

Not *Hydroides novaepommeraniae* **sensu** Kupriyanova *et al.* 2008: 428–430 (Lizard Island, Qld; DNA data) [**non** Augener, 1925b], reidentified to *H.* cf. *recta* by Kupriyanova *et al.* 2015: 290); Tovar-Hernández *et al.* 2016: table 2, fig. 6 (name, DNA, phylogeny; Lizard Island, Qld, Australia); Sun *et al.* 2018: 733, figs 2, 3, 5, 6 (DNA, phylogeny; same). This specimen shows characters intermediate between *H. recta* (see remarks under this taxon as well) and *H. longistylaris*, and cannot be attributed to one of these taxa with certainty.

Diagnosis based on Augener and Okuda only (but see Remarks, below): Peduncle without or at most with ill-defined constriction. Funnel with 21–23 radii, long, tapering to a not really acute point. Verticil sitting in funnel, almost radially symmetrical, with 7–10 similarly sized outcurving spines (see Remarks). Spines simple, smooth, without basal spinule, arranged around small, flat, central disk. Up to 4 dorsal spines may be slightly larger than remaining ones, with a finely truncated tip (only visible in apical view).

Remarks: First described from a single 26 mm (in a later measurement 30) long specimen with 10 verticil spines and 23 radii (opercular formula 10/23), by Augener (1925b) from New Britain (Papua New Guinea), its operculum figured anew by ten Hove (Imajima & ten Hove 1984, fig. 4c, d). Two slightly shorter specimens (17 mm), opercular formula 7/21, were recorded from nearby Micronesia (1000 km north) by Okuda (1937). Both show a rather undifferentiated operculum, with just spines and

Fig. 84. *Hydroides novaepommeraniae*-complex, various views of the same operculum, Bay of Sanggar, Indonesia, AM W.43183 (photo E. Wong).

radii tapering to a not really acute point, and have been regarded to belong to the same taxon by Okuda, as well as by us. However, Imajima & ten Hove (1984: 45–47) documented the variability of opercula in a series of samples from Micronesia (e.g. Truk), the Palau Islands and the Philippines (ship fouling), and found that tips of verticil spines (as well as radii) within single opercula might range from rather acute (ventrally) to truncated or club-shaped (mostly dorsally only, but sometimes all around, approaching the form of *H. bandaensis* and *H.* cf. *bandaensis*); sometimes spines are even acute in side view, but definitely truncated in apical view (p. 46). Exceptionally one dorsal spine even may be distinctly larger than the others (Imajima 1982, fig. 3K, L, as *H. malleolaspina*; Imajima & ten Hove 1984, fig. 4a, b). For a full discussion see Imajima & ten Hove (1984). Based on this variability they placed the nominal taxon *Hydroides grubei* Pillai, 1965 (pp. 166–168, fig. 22A–G), with one larger dorsal verticil spine, into the synonymy of *H. novaepommeraniae.* On the other hand, Ben-Eliahu & ten Hove (1992: 42) and ten Hove & Ben-Eliahu (2005: 8) argued that the small specimens (not yet fully developed?) described from Sri Lanka as *Hydroides grubei* by Pillai (1965, 5.5 mm long), and mentioned as fouling in the Mediterranean by Zibrowius (1979: 132–133) and Zibrowius & Bitar (1981: 159–160), should be regarded to be undeterminable juveniles, like the rejected name *H. prisca*. Later, Sun & Yang (2000: 123, fig. 3J–K; 2014: 228–229, figs 121A–I, 106) reported *H.* cf. *grube(i)* from a number of localities in the South China Sea, and regarded it to be different from *H.* cf. *novaepommeraniae* on the basis of the lower number of verticil spines (5–8 *v* 8–11), moreover with one larger dorsal spine in *H.* cf. *grubei* as opposed to all similarly sized in *H.* cf. *novaepommeraniae* (**sensu** Sun & Yang 2014: 243–244, figs 131A–E, 117). By the use of cf., they clearly indicated doubts to their identifications.

Finally, the taxon is also similar to *H. huanghaiensis*, but always with a radially symmetrical verticil; moreover, the latter has a cold-temperate distribution, not tropical. In view of the reported opercular variability, convoluted synonymy and wide distribution we regard the nominal taxon (for the time being thus including forms identified as *H.* cf. *grubei* and possibly even *H.* cf. *bandaensis*) as a species-complex, awaiting further analysis with the help of DNA. The specimen figured by us seems not the most typical, with its (almost) radially symmetrical operculum, as opposed to the clearly bilaterally symmetrical one with 4 dorsal truncated spines of the holotype figured by Imajima & ten Hove (1984, fig. 4c–d); however, upon close inspection of our Fig. 84, the 5 dorsal verticil spines have truncated tips as compared to the remaining narrowly tipped spines, as seen most clearly in the middle photograph where these 5 are directed to the bottom of the plate.

Distribution: New Britain, Palau and Truk Island, ?Philippines.

70. *Hydroides ochotereana* Rioja, 1941a

Figure 85

Type information: Acapulco, Guerrero, Mexico, type specimens lost.

Etymology: Named after D. Isaac Ochoterena. The incorrect Latinisation to *ochotereana* led to subsequent variant spellings, but should have been regarded to be an (unchanging) noun in apposition. For a full explanation see Read *et al.* (2017: 30).

Selected synonymy

Hydroides ochotereana Rioja, 1941a: 164–167, pl. 2, figs 1–12, pl. 3, fig. 1, pl. 4, fig. 16 (description [original]; Western Mexico); Bastida-Zavala *et al.* 2016: 423–424, figs 5, 12B–C (diagnosis, synonymy; several localities Western Mexico); Sun 2017: 76, 153, 155, 176, 288, 296, 316, fig. 3E (diagnosis, DNA; partly as *H. ochoterena*; Oaxaca and Guerrero, Western Mexico); Sun *et al.* 2018: 738, 742, figs 2, 3, 5, 6 (DNA, phylogeny; Oaxaca and Guerrero, Western Mexico).

Hydroides ochoterenai – Bastida-Zavala & de León-González 2002: 389 (name only; Western Mexico).

Hydroides ochoterena – Bastida-Zavala & ten Hove 2003b: 95–97, fig. 17A–M (redescription, synonymy; Acapulco, Western Mexico to Northern Colombia.

Diagnosis: Peduncle without or with ill-defined constriction. Funnel with 29 ((15)27–32) chitinised, pointed radii. Verticil sitting in funnel, with 18 ((11)16–21) spines, straight or slightly curving outwards, radially symmetrical. Spines each with (8)12–18 lateral spinules, pointed or split tip; the basal spinules may be as long as their verticil spines. Central disk not observed. Bayonet of collar chaetae with 2 teeth and questionably with 2 accessory ones.

Remarks: The operculum of *Hydroides ochotereana* shows some resemblance to the opercula of *H. azorica* (Atlantic), *H. longispinosa* (including *H. centrospina*), *H. multispinosa* and *H. nanhaiensis* (the latter three Indo-West Pacific). It immediately can be differentiated from these taxa by the high number of long lateral spinules on the verticil spines, (8)12–18, almost twice that of the next highest in this aspect, *H. azorica*, with 6–10. The Indo-Pacific taxa have collar chaetae with proximal rasp, absent in *H. ochotereana* (and *H. azorica*). In both Bayesian inference trees, the position of *H. ochotereana* is not well-defined (Sun *et al.* 2018, figs 2, 3, Fig. 13 here), but outside the well-defined clade of *H. elegans*, *H. longispinosa* and *H. norvegica*.

Neither Rioja, in his original description, nor Bastida-Zavala & ten Hove (2003b) mentioned basal spinules, as long as or even somewhat longer than the verticil spines. This remarkable feature is rather casually mentioned in Sun (2017: 176, fig. 3H) 'with or without one long basal spinule extended above verticil tips and straight pointed upwards'. The character is so unexpected (and maybe hidden in the bunch of verticil spines) that it was only noticed for the first time when one of us (HAtH in 2012) selected material partly studied by Bastida-Zavala & ten Hove (2003b: 95, Galapagos: ZMA V.Pol. 3498) for a loan to Sun, and found that four of the seven specimens had such long internal basal spinules, as it were a second verticil within the usual one. It would be interesting to know more about the distribution, percentual as well as geographical, of this phenomenon.

Distribution: Acapulco, Western Mexico to Equador (Galapagos Islands).

Fig. 85. *Hydroides ochotereana*, various views of the same operculum, Galápagos Islands, ZMA V.Pol. 3498 (photo E. Wong).

71. *Hydroides operculata* (Treadwell, 1929)

Figure 86

Type information: Gulf of Aden, Somaliland, holotype AMNH 1994.

Etymology: Not given by Treadwell (1929: 12), but probably refers to the fact that his specimen was well endowed with 2 opercula.

Selected synonymy

Eupomatus operculata Treadwell, 1929: 12 (diagnosis [original]; material studied; Berbera, Somaliland, Gulf of Aden).

Hydroides uncinata **sensu** Day 1967: 805–806, fig. 38.4h, i [**non** Philippi, 1844] (diagnosis; Morrumbene, Mozambique, **fide** Zibrowius 1971a: 695).

Hydroides operculata – Sun *et al.* 2015: 63, 65 (in part, remarks on type material only); 2017b: 39–48, fig. 1A (DNA-based revision of complex; Al-Doha, Kuwait); 2018: 733, 738–741, figs 1–3, 5, 6 (DNA, phylogeny; Al-Doha, Kuwait).

Diagnosis: Peduncle at most with ill-defined constriction. Funnel with 38–39 hardly chitinised, pointed radii. Verticil sitting in funnel, radially or somewhat bilaterally symmetrical, with 6 or 7 long, slender incurving spines; with short basal spinule over small central disk, 1 or 2 dorsal spines larger than the rest, less frequently all spines of the same size. Collar chaetae not mentioned by Treadwell, but with 2 teeth (HAtH, from type). (Diagnosis based on types and specimen from Mauritius only).

Remarks: *Hydroides operculata* was originally diagnosed from a single specimen collected in the Gulf of Aden, Somalia. Based on very similar morphologies, Zibrowius (1971a: 695) synonymised the nominal species *H. basispinosa* (including *H. gradata*) from Queensland, Australia with *H. operculata*, and later Ishaq & Mustaquim (1996: 166–168) referred *H. inornata* from Sri Lanka and Pakistan to *H. operculata*. After these synonimisations, the distribution of the morphospecies *H. operculata* extended from Queensland to Hong Kong, via the Indian subcontinent to East Africa; moreover, it was supposed to have successfully colonised the (Eastern) Mediterranean (see below).

However, Sun *et al.* (2017b) examined the taxonomic status of *H. operculata* **sensu lato** using standard molecular techniques, which revealed three well supported clades, suggesting a phylogenetic pattern as might be expected from a biogeographical viewpoint. They argued that the taxa *H. inornata* (India and Hong Kong) and *H. basispinosa* (Australia, and including *H. gradata*) should be reinstated to full species, and that the name *H. operculata* should be restricted to populations from the Arabian (Persian) Gulf to Eastern Africa.

The nominal species *H. operculata* has frequently been mentioned from the Mediterranean, for example by Zibrowius (1979a: 133–134; arrived in Toulon, France, fouling ship returned from African side of Indian Ocean), Zibrowius & Bitar (1981: 159–160; established as Lessepsian migrant in Lebanon), Ben-Eliahu & ten Hove (2011: 107; Indian Ocean, but not yet recorded from the Red Sea or the Gulfs of Aqaba and Suez). Ben-Eliahu (1976: 106–107, fig. 1) and Çinar (2006, fig. 6) are the only papers figuring an operculum. With its two large dorsal spines, Ben-Eliahu's figure is reminiscent of the holotype of *H. operculata*. On the other hand, Çinar's figure is most reminiscent of those given for *H. inornata* from Sri Lanka (Pillai 1960), Pakistan (Ishaq & Mustaquim 1996, fig. 3) or Hong Kong (Sun *et al.* 2012a, fig. 9). However,

Fig. 86. *Hydroides operculata*, various views of the same operculum, Umm Rasas, Oman, ZMA V.Pol. 4043 (photo E. Wong).

opercular variability as we have seen in material classified now as *H. basispinosa*, respectively *H. inornata*, is larger than captured in these figures, and in our present state of knowledge morphology alone cannot be trusted for identifications to species level.

In absence of DNA-based analyses of African and of Mediterranean samples, our reappraisal of the *H. operculata*-complex can only be tentative. The status of *Hydroides* cf. *operculatus* (Bastida-Zavala & ten Hove 2003a: 104, 108, 115, 141, 172–174, figs 39D–O, 40), from the Caribbean, remains unresolved too.

Distribution: Somalia, Gulf of Aden; Oman and Arabian Gulf; Mozambique.

72. *Hydroides panamensis* Bastida-Zavala & ten Hove, 2003b

Figure 87

Type information: Paitilla Beach, Western Panama, holotype USNM 1013581, paratypes USNM 1013582, ZMA V.Pol 5078, ECOSUR 0038.

Etymology: Named for its distribution.

Selected synonymy

Hydroides panamensis Bastida-Zavala & ten Hove, 2003b: 97–99, figs 18A–N, 24 (description [original], discussion; Western Panama to Ecuador); Bastida-Zavala *et al.* 2016: 424–425, figs 6, 12D (diagnosis, synonymy, discussion; several localities in Western Mexico); Tovar-Hernández *et al.* 2016: 595, 696, 700, table 2, figs 4H, 5B, D, F (discussion, DNA, phylogeny; Paitilla Beach, Panama; Costa Rica); Sun *et al.* 2018: 733, 739, 742, figs 1–3, 5, 6 (DNA, phylogeny; Cuajiniquil, Guanacaste, Costa Rica and Panama Bay, Panama).

Diagnosis: Peduncle with ill-defined constriction. Funnel with 27 (25–29) radii with pointed tips. Verticil bilaterally symmetrical, sitting in or inserted on short basis into funnel; with 11(–13) spines, similar in shape and size, with short, acute basal spinule above small central disk. The dorsal spines may be slightly larger, all are strongly curving inwards with a pointed tip, especially the ventral spines. Dorsal spines with one pair of lateral spinules on the bend, at 2/3 of spine length; these spinules may gradually change from a pair in lateral position in dorsal spines to one external spinule in ventral spines, thus change from trifid tip to bifid tip (Bastida-Zavala & ten Hove, 2003b, fig. 18c). Verticil without central tooth.

Remarks: *Hydroides panamensis* might be confused with the Panamian *H. crucigera* and *H. chilensis*, or with the Caribbean *H. bispinosa* and *H. parva*; all have incurving verticil spines with one pair of lateral spinules. In the four mentioned taxa those spinules are always in lateral position, while they shift to a more external position in the ventral spines of *H. panamensis*, even to the extent that the pair is merging into a single spine. The main differences between *H. panamensis* and the very similar *H. dolabrus* are verticil spines mostly with trifid tip *v* mostly bifid; basal spinules long *v* short; and funnel radii pointed *v* rounded (Tovar-Hernández *et al.* 2016: 700–701, fig. 5). The latter authors depict opercula of a few more or less 'similar' forms in their fig. 4. In phylogenetic analyses *H. panamensis* forms a clade with *H. dolabrus* (see Tovar-Hernández *et al.* 2016, fig. 6; Sun *et al.* 2018, Fig. 13 here), with which taxon it easily can be confused (Tovar-Hernández *et al.* 2016: 200).

Distribution: Pacific coasts, Costa Rica to Ecuador.

Fig. 87. *Hydroides panamensis*, various views of the same operculum, Paitilla Beach, Panama, AM W.43184 (photo E. Wong).

73. *Hydroides parva* (Treadwell, 1902)

Figure 88

Type information: Boqueron Bay, Puerto Rico, two syntypes USNM 016173.

Etymology: Not given by Treadwell (1902: 210), but he regarded his specimens to be 'very small' (Latin *parvus* = small).

Selected synonymy

Eupomatus parvus Treadwell, 1902: 210, figs 79, 80 (diagnosis [original]; Puerto Rico).

Hydroides parvus – Bastida-Zavala & ten Hove 2003a: 127–129, figs 12A–K, 15 (redescription, synonymy, discussion; Campeche, Gulf of Mexico, Caribbean to Recife, Brazil).

Hydroides parva – Bastida-Zavala *et al.* 2017: 37–38, figs 4I, 6 (diagnosis, synonymy; Biscayne Bay, Florida); Sun *et al.* 2018, figs 1–3, 5, 6 (DNA, phylogeny; Veracruz, Mexico and Colon, Panama).

Diagnosis: Peduncle with ill- to well-defined constriction. Funnel with 22 radii ((17)19–26) with chitinised, pointed tip. Verticil sitting or inserted on very short basis in funnel, radial to bilateral symmetrical with 8 (6–10) spines, abruptly curving inwards with a sharp pointed tip and an external knob on the 'knee'. Spines with pair of lateral spinules at mid-length and a sharply pointed inner basal spinule above small central disk; sometimes one or two dorsal spine(s) slightly bigger than the others. Bayonet of collar chaetae with 2 teeth and 1 accessory.

Remarks: The opercular verticil of *Hydroides parva* is very similar to that of *H. bispinosa*, and has been confused with that taxon in routine identifications. See Remarks under the latter species.

Distribution: Gulf of Mexico, South Florida to Recife, Brazil.

Fig. 88. *Hydroides parva*, various views of the same operculum, Aruba, Netherlands Antilles, AM W.43200 (photo E. Wong).

74. *Hydroides perezi* Fauvel, 1918

Figure 89

Type information: 28 km from the coast of Oman, Arabian (Persian) Gulf (24°55′–25°10′N, 54°40′–55°10′E), syntypes MNHN.

Etymology: Named after the French biologist Ch. Perez (1873–1952).

Selected synonymy

Eupomatus perezi – ten Hove 1970: 58–59, figs 9–16, 26 (diagnosis, synonymy (in part); Arabian Gulf).

Hydroides perezi Fauvel, 1918: 342–344, figs 2a–j (description [original]; Arabian (Persian) Gulf); Vine & Bailey-Brock 1984: 142, fig. 2I–J (diagnosis; Port Sudan); Hassan 1998: 51 (diagnosis, figure of operculum and tube; Aqaba, Jordan); Sun 2017: 130, 170, 177, fig. 5F (diagnosis; figure of operculum); Sun *et al.* 2018, fig. 1 (phylogeny, after Pillai 1972).

Not *Hydroides perezi* **sensu** Straughan 1967a: 219–220, fig. 6o (Havannah Island) [**non** Fauvel, 1918] and reidentified by Sun 2017: 126–127 as *Hydroides tuberculata*.

Not *Hydroides perezi* **sensu** Ben-Eliahu 1976: 106, 110, fig. 2 (Gulf of Elat, Israel; restudied by HAtH 1999, and reidentified herein to *H. dafnii*) [**non** Fauvel, 1918].

Not *Hydroides perezi* **sensu** Bailey-Brock & Magalhães 2013: 4–6, fig. 4 (description, discussion; Guam, Mariana Islands) [**non** Fauvel, 1918], reidentified herein to *H. tuberculata*-complex **sensu stricto** – see Remarks, below.

Diagnosis: Peduncle without or with ill-defined constriction. Funnel with 18 (15–26) chitinised, sharp-tipped radii. Verticil sitting in funnel, bilaterally symmetrical, with dorsal hook and 4 (3–7) bulbous spines with broadly rounded tip, the outward side of each more or less flat, the inward side rounded; dorsal spine with a sharp point horizontally directed inward. No lateral, internal, external or basal spinules though a minute, almost terminal external knob may be present. Bases of verticil spines almost meeting in centre. No central tooth. Tube may be bluish. Bayonet of collar chaetae with 2 large teeth only.

Remarks: The broadly rounded verticil spines of *H. perezi* are unlike the tapering tips in *H. dafnii* (taxon resurrected above, but long regarded to be a synonym of *H. perezi*), and also different from the more triangular spines in the *H. tuberculata*-complex (with which it has been confused). However, Bailey-Brock & Magalhães (2013) recorded *H. perezi* from Guam (Mariana Islands), a very unlikely record on biogeographical grounds; moreover, they depict collar chaetae with 2 teeth and some accessories for their material, typical for *H. tuberculata*-complex **sensu stricto**, and to which this record from Guam should be referred. Morphologically, *H. perezi* can be placed in the *H. tuberculata*-complex **sensu lato**. Field notes (of HAtH) mention faintly bluish tubes for a specimen (identified as *H. perezi*) from Gordon Reef, Sharm-El-Sheikh, Northern Red Sea, Egypt, a colour often found in *H. tuberculata* **sensu stricto**. Additional studies are needed to resolve the complex of taxa including *H. dafnii*, *H. perezi*, *H. tuberculata* and *H. uniformis*. It is remarkable that the Red Sea, as well as tropical Australia and neighbouring areas, shows a dimorphy in the opercula of morphological 'twins', with tapering *v* broad spines: *H. dafnii* and *H. perezi*, respectively *H. uniformis* and *H. tuberculata*-complex **sensu stricto**. Between the latter pair there is a genetical difference (Sun *et al.* 2018, fig. 2, Fig. 13 here). See remarks under *H. tuberculata* as well.

Distribution: Iranian (Arabian) Gulf, Red Sea.

Fig. 89. *Hydroides perezi*, various views of the same operculum, Bahrain, ZMA V.Pol. 3001 (photo E. Wong).

75. *Hydroides plateni* (Kinberg, 1867)

Figure 90

Type information: La Plata, Argentina, type material apparently lost (Zibrowius 1972c: 154).

Etymology: Named after its type locality, though not mentioned by Kinberg (1867: 351).

Selected synonymy

Eupomatus Plateni Kinberg, 1867: 351–352 (diagnosis [original]; Argentina).

Hydroides plateni – Bastida-Zavala & ten Hove 2003a: 151–153, figs 26A–R, 28 (redescription, synonymy, discussion; Uruguay, Argentina); Ferrero *et al.* 2005: 259–260, 262, 265, 267–268, fig. 6e (comparison with very similar Mid-Holocene 'reefs'; Rio de la Plata estuary, Mar del Plata Harbour, Argentina; material studied HAtH); Sun 2017: 151, 177, fig. 1B (diagnosis; Argentina); Albano & Obenat 2019: 4, 8, 10, 12 (ecology, fouling; Mar del Plata harbour).

Diagnosis: Peduncle without constriction. Funnel with 41 (24–53) radii with rounded to triangular tips. Verticil sitting in funnel, radially symmetrical, small, generally not or hardly surpassing the funnel perimeter; 11 (8–14) short spines, slightly incurving or almost straight, rarely more elongated. All spines similar in shape and size, with acute tip; rarely the dorsal spines larger. Spines with or without basal spinule; without external and lateral spinules and/or wings. Central disk of variable size, without central tooth. Bayonet of collar chaetae usually with 2 teeth (occasionally with 2 accessories).

Remarks: *Hydroides plateni* shares a simple verticil and a large funnel with *H. microtis* and *H. protulicola*; moreover, in both *H. protulicola* and *H. plateni* the verticil is small in relation to the funnel. However, verticil spines are pointed in *H. plateni*, distally expanded in *H. microtis*; the bayonet chaetae of *H. plateni* have two teeth, sometimes with 1 or 2 accessory ones, in *H. protulicola* they show a proximal rasp; verticil spines are straight or incurving in *H. plateni*, always outcurving in *H. protulicola*.

Distribution: South Brazil to Patagonia.

Fig. 90. *Hydroides plateni*, various views of the same operculum, Buenos Aires, Argentina, AM W.43199 (photo E. Wong).

76. *Hydroides protulicola* Benedict, 1887

Figure 91

Type information: Off Cape Hatteras, North Carolina, USA, 17 syntypes USNM 972, ZMA V.Pol. 3211, ZMA VER.P.1089 (microscopic slide).

Etymology: Evidently named after its habitat (on the genus *Protula*; Latin *colo* = to live).

Selected synonymy

Hydroides protulicola Benedict, 1887: 550, pl. 20, fig. 17; pl. 21, figs 18–23 (diagnosis [original], material studied; off Cape Hatteras, USA); ten Hove & Wolf 1984: 55.21–55.23, figs 15–17, 15–18a–m (diagnosis, synonymy; Louisiana to Florida, northern Gulf of Mexico); Bastida-Zavala & ten Hove 2003a: 153–154, figs 27A–T, 28 (redescription, synonymy, discussion; North Carolina to Texas, USA); Sun 2017: 153, 177, fig. 2A (diagnosis).

Eupomatus protulicola – Rioja 1946: 199, figs 10–13; Pequegnat & Pequegnat 1968: 23, 43, 44 (name, below 10 m, season of attachment April–November, growth rate; off Panama City, Florida, USA).

Diagnosis: Peduncle without or at most with ill-defined constriction. Funnel with 31 ((16)25–38) chitinised radii, with acute tips. Verticil sitting or on short basis in funnel, radially symmetrical, small, generally not or hardly surpassing the funnel perimeter; with 12 (8–16) outcurving spines, their acute tips straight or sometimes lightly twisted clockwise. All spines similar in shape and size, with small knob-like basal spinule; without external and lateral spinules and/or wings. Central disk of variable size, without central tooth. Bayonet of collar chaetae with 2 teeth and proximal rasp.

Remarks: *Hydroides protulicola* shares a relatively unadorned verticil and a large funnel with *H. microtis* and *H. plateni*; with the latter it also shares outcurving verticil spines, small in relation to the large funnel. *Hydroides microtis* has verticil spines with distally expanded tips as opposed to the other two taxa, with acute tips. Both *H. protulicola* and *H. microtis* have bayonet chaetae with proximal rasp, missing in *H. plateni*. Tips of verticil spines clockwise twisted are more clearly expressed in *H. recurvispina* and *H. trompi*.

Small specimens of *Hydroides protulicola* may be mistaken for *H. sceptrifera* nov. spec.; see Remarks on the latter species. One might confuse juvenile *H. protulicola* with *H. microtis* as well (in fact some routine identifications of material from Florida were a mix of both taxa upon second inspection), also because both species have similar collar chaetae. Typically, however, verticil spines are pointed and curved outward in *H. protulicola*, blunt and almost straight in *H. microtis*.

Distribution: Eastern USA from Cape Hatteras southwards and northern Gulf of Mexico.

Fig. 91. *Hydroides protulicola*, various views of the same operculum, off Cape Hatteras, North Carolina, USA, ZMA V.Pol. 3388 (photo E. Wong).

77. *Hydroides pseudexaltata* Pillai, 2009

Figure 92

Type information: off North East Heywood Island, Kimberley, Western Australia, holotype AM W.202930.

Etymology: Pillai (2009: 122) named after the superficial similarity of its operculum to that of *H. exaltata*.

Selected synonymy

Hydroides pseudexaltatus Pillai, 2009: 120–122, fig. 17A–K (description [original], discussion; Kimberley, WA).

Hydroides pseudexaltata – Sun *et al.* 2015: 65–67, fig. 20, map 18 (diagnosis, discussion; WA).

Diagnosis: Peduncle with well-defined, lightly chitinised constriction. Funnel with 21 T-shaped radii. Verticil on short basis in funnel, bilaterally symmetrical, with 9 spines, all lacking additional spinules; dorsal spine large hook, somewhat swollen on its knee-like bend, strongly incurving with acute tip; remaining verticil spines much shorter, ending in simple outcurving acute tips. Spines almost meeting each other basally, small central disk. Bayonet of collar chaetae with 2 teeth.

Remarks: The operculum of the only known specimen of *H. pseudexaltata* is similar to that of *H. exaltata* (Marenzeller, 1884), but for its T-shaped radii, which are acute in *H. exaltata*. Moreover, all verticil spines of *H. exaltata* have a short basal spinule as opposed to the completely smooth spines in *H. pseudexaltata*. Bayonet of collar chaetae may show 2 teeth with up to 3 accessory ones in *H. exaltata*. There is also some similarity with the operculum of *H. minax*, but this has long, narrow truncated radii, a longer dorsal hook with a pair of side spinules and a longer chitinised constriction.

Distribution: Kimberley, Western Australia.

Fig. 92. *Hydroides pseudexaltata*, various views of the same operculum, Kimberley, Western Australia, AM W.202930 (photo E. Wong).

78. *Hydroides pseudouncinata* Zibrowius, 1968 nominotypical subspecies *pseudouncinata*

Figure 93

Type information: Marseille, France, holotype MNHN.

Etymology: Named because of its similarity to *Hydroides uncinata*.

Selected synonymy

Hydroides pseudouncinata Zibrowius, 1968: 112–114, pl. 2, figs 8–17 (description [original], synonymy, in part, only the European records; Marseille, France); Amoureux 1978: 30 (diagnosis, see Remarks below; Tanger, Morocco); Bick 2006: 87 (name, epizoic on hermit crab dwelling; Ibiza, Spain); Sun *et al.* 2017b: 42–44, fig. 2 (DNA, phylogeny; Mallorca, Spain; Sicily, Italy; Greece); 2018: 733, figs 1–3, 5, 6 (DNA, phylogeny; same as Sun 2017).

Hydroides pseudouncinata pseudouncinata – Zibrowius, 1971a: 707–709, figs 10–12 (diagnosis of the new subspecies, synonymy; specification of the holotype, from Gaby Island, Marseille; further records from entire Mediterranean, from Turkey to Spain; Bianchi 1981a: 65–66, figs 7c, 22a–c (diagnosis; Mediterranean Sea, Atlantic coast of Portugal);

Hydroides pseudouncinatus – Lehrke *et al.* 2007: 198–199, figs 2, 3 (name, DNA, phylogeny; Banyuls, Mediterranean France); Kupriyanova *et al.* 2008: 428–430, figs 2–4 (name, DNA, phylogeny; Marseille, Banyuls, Mediterranean France); Tovar-Hernández *et al.* 2016: table 2, fig. 6 (name, DNA, phylogeny; same).

Diagnosis: Peduncle well-defined, darkly chitinised constriction. Funnel with 30 (18–40) radii, elongated, acute, all with a subterminal internal spinule. Verticil sitting or at most on short basis in funnel, radially to lightly bilaterally symmetrical, with 7 ((5)6–10) subequal spines smoothly curving inwards. All spines similar in shape and size, with slender, pointed tip; dorsal spines maybe slightly larger. No lateral, inner or outer spinules. Spines meeting in centre of verticil, no central tooth.

Remarks: *Hydroides pseudouncinata* is most similar to *Hydroides gracilis*, but the former species is distinct in having funnel with elongated pointed spinule bearing an additional internal spinule each. In a phylogenetic analysis (Sun *et al.* 2017b, fig. 7) it was recovered as sister clade to *Hydroides nigra*; however, in the latter taxon the verticil spines are abruptly curving in a knee, sometimes accentuated by clear external knob. It should be noted that Amoureux *et al.* (1978: 30) explicitly mention that of his 35 specimens from Tanger, 23 are typical, nine to be attributed to the subspecies *H. pseudouncinata africana*, and three are intermediate in morphology of the operculum.

Distribution: Mediterranean Sea, Atlantic coast of Portugal.

Fig. 93. *Hydroides pseudouncinata pseudouncinata*, various views of the same operculum, Spain, AM W.42072 (photo E. Wong).

79. *Hydroides pseudouncinata africana* Zibrowius, 1971a

Figure 94

Type information: Holotype and 13 paratypes (MNHN); off Rio de Oro, Mauritania, Atlantic coast of Africa.

Etymology: named after its continent of collection, Africa.

Selected synonymy

Hydroides pseudouncinata africana Zibrowius, 1971a: 709–711, figs 13–15 (diagnosis [original], synonymy, discussion; Morocco to Guinea, Atlantic coasts of Africa); Amoureux 1974: 460 (short characteristic; Madagascar; questionable, see Remarks); 1978: 30 (diagnosis; Tanger, Morocco; see Remarks above).

Diagnosis: Peduncle with well-defined, chitinised constriction. Funnel with 41 (29–48) radii, elongated, acute, all smooth. Verticil sitting or at most on short basis in funnel, radially to lightly bilaterally symmetrical, with 8 (7–10) subequal spines smoothly curving inwards. All spines similar in shape and size, with slender, pointed tip; dorsal spines maybe slightly larger. No lateral, inner or outer spinules. Spines meeting in centre of verticil, no central tooth.

Remarks: The only difference with *Hydroides pseudouncinata pseudouncinata* is the absence of a subterminal internal spinule in its funnel radii. Judging by biogeographical and morphological evidence, it might be a full species; however, in absence of DNA support, treating it as a subspecies, in our opinion, is more informative. The record by Amoureux (1974: 460) from Madagascar is far outside the biogeographical range recorded by Zibrowius and should be checked; it rather might be regarded as a member of the *H. operculata*-complex.

Distribution: Morocco, Western Sahara, Mauritania, Sénégal, Gambia, Guinea; Cape Verde Islands (this paper); ?Madagascar.

Fig. 94. *Hydroides pseudouncinata africana*, various views of the same operculum, Cape Verde Islands, West Africa, ZMA V.Pol. 4057 (photo E. Wong).

80. *Hydroides qiui* Sun, Wong, ten Hove, Hutchings, Williamson & Kupriyanova, 2015

Figure 95

Type information: Northern Territory, Darwin Harbour, East Arm Port, 12°29′30″S, 130°52′59″E, holotype MAGNT W25505.

Etymology: Named *qiui* after Dr Jian-Wen Qiu.

Selected synonymy

Hydroides qiui Sun, Wong, ten Hove, Hutchings, Williamson & Kupriyanova, 2015: 67–71, figs 21, 22a–g, map 19 (description [original]; Darwin, NT, Australia); Sun *et al.* 2018, figs 2, 5, 6 (DNA, phylogeny; Darwin, NT, Australia).

Diagnosis: Peduncle with well-defined, lightly chitinised constriction. Funnel with 21 (17–25) chitinised, T-shaped radii. Verticil inserted with short basis into funnel, radially symmetrical, with (8–)9 equal spines, almost straight to slightly outcurving, with T- to anchor-shaped tips. Spines simple, smooth, almost joining at centre of verticil, small central disk without spine. Bayonet of collar chaetae with 2 elongated teeth.

Remarks: Among all *Hydroides* species described, only four species have all verticil spines similar in size and shape, with truncated, T-shaped to anchor-shaped tips: *Hydroides bandaensis*, *H. dirampha*, *H. qiui* and *H. sinensis*. *Hydroides sinensis* can be immediately distinguished by its large subapical internal spinules. The most obvious differences between *H. dirampha* and the remaining two taxa are its sharply pointed funnel radii, as opposed to truncated or T-shaped, and the basal spinules in the verticil spines of *H. dirampha*. A fifth taxon with all verticil spines equal in size and shape, with dilated tips, is *H. microtis*; however, here these tips are globular and rounded. The collar chaetae of the four mentioned taxa may be distinguishing as well: with 2 teeth and proximal rasp in *H. microtis*, 2 rather short teeth and 1 or 2 accessories in *H. bandaensis*, and even the 2 teeth only in *H. dirampha*, *H. sinensis and H. qiui* seem to be different, longer and more slender in the last. Juveniles of the nominal taxa *H. ancorispina*, *H. malleolaspina* and *H. novaepommeraniae* may also show verticil spines with dilated tips, but the dorsal-most spine is larger, bulbous or even hammer-shaped. In the phylogenies generated by Sun *et al.* (2018), *H. qiui* is not associated with its morphological look-a-likes. It is recovered in a morphologically rather diversified clade with *H. adamaformis*, *H. ancorispina* and *H. lirs*, as sister clade to the morphologically most different *H. lirs*, with all acute radii and verticil spines. Obviously, more research is needed.

Distribution: NT, Australia.

Fig. 95. *Hydroides qiui*, various views of the same operculum, Darwin, Northern Territory, Australia, holotype MAGNT W25505 (photo E. Wong).

81. *Hydroides ralumiana* Augener, 1927

Figure 119B

Type information: Ralum, New Britain in the Bismarck Archipelago of Papua New Guinea, holotype ZMB Verm. 6574, lost (Hartwich 1993: 132).

Etymology: Named after its type locality.

Selected synonymy

Hydroides (*Eupomatus*) *ralumianus* Augener, 1927a: 150, fig. 5 (description [original]; Ralum, Papua New Guinea; type lost in World War II **fide** Hartwich 1993: 132).

Not *Hydroides* (*Eupomatus*) *ralumianus* **sensu** Fauvel 1930: 502, 506, 558–559 [**non** Augener, 1927]; Île de Pins, New Caledonia; referred herein to *H. externispina*; see Remarks.

Not *Hydroides ralumianus* **sensu** Day 1957: 118 [**non** Augener, 1927]; from Moçambique. According to the short diagnosis and the figure in Day (1967: 806–807, fig. 38.4m, m') this is more likely to be a member of the *H. brachyacantha*-complex in its wide sense (see Remarks as well).

Not *Eupomatus ralumianus* **sensu** Poore *et al.* 1975: 30 [**non** Augener, 1927] (name only; Port Phillip Bay, Victoria. Material studied by ten Hove in 1986, probably juvenile of *H. amri* (**fide** Sun *et al.* 2015: 14)).

Diagnosis: Peduncle with well-defined constriction. Funnel with 28–30 pointed, chitinised radii. Verticil on short basis in funnel, bilaterally symmetrical, with 9–10 spines ending in sharply pointed tips, curved inwards. Two dorsal spines large, smooth, without lateral spinules, and hovering over the other spines. Remaining verticil spines all with a pair of outcurving lateral spinules at mid-length, a smaller pair near their base, as well as 1 outcurving external spinule on the bend at 2/3 of their length (subterminal). All verticil spines almost meeting in the centre, no basal spinules or central tooth.

Remarks: Apart from the original description of *Hydroides ralumiana*, there is only one more record which, with a short diagnosis, has been attributed to this species, by Fauvel (1930, repeated in 1947), from the 2000 km more southern New Caledonia, still within the same Indo-Polynesian Province. The original figure of Augener at first sight is indeed very similar to *H. externispina*, but for the fact that the latter has only one pair of lateral spinules to the verticil spines as opposed to two pairs, as well as a pair of small lateral spinules to the two dorsal hooks, explicitly mentioned as smooth for *H. ralumiana*. However, Ben-Eliahu (unpublished) restudied Fauvel's specimen (in Paris), and reidentified it to *H. externispina* since she only observed verticil spines with only one pair of lateral spinules, as well as a basal spinule, not mentioned by Fauvel. With low resolution these basal spinules might be mistaken for proximal lateral spinules. Moreover, *H. externispina* – not yet described in Fauvel's time – in the meantime has been found in New Caledonia (Zibrowius, ten Hove det., unpublished, ZMA V.Pol. 2558). New material from the type locality of *H. ralumiana* (Papua New Guinea) is necessary to solve the question if *H. externispina* is a junior synonym of *H. ralumiana*.

In the superficially similar *H. glasbyi*, with a similar number of verticil spines including two hooks, lateral spinules are completely absent while the other spines show a subterminal external spinule. At first sight *H. ralumiana* might be confused with *H. tambalagamensis* due to the similar dark curved lateral spinules on the verticil spines of both species. However, all spines are alike in *H. tambalagamensis*, while *H. ralumiana* has two large hooks.

Day's (1957, and figure in Day 1967) misidentification may have been only based on the presence of two large dorsal hooks and subterminal external spinules in the remaining spines (characters indeed present in members of the *H. brachyacantha*-complex in its wide sense), the most prominent features in Augener's (German) description of *H. ralumiana*. No doubt Poore's identification has been based on Day.

Distribution: Bismarck Archipelago (Papua New Guinea) to New Caledonia.

82. *Hydroides recta* Straughan, 1967a

Figure 96

Type information: Pretty Beach, Cairns, Qld, Australia, holotype AM W.3998; four paratypes AM W.4006, one paratype ZMA V.Pol.3463, ZMA VER.P.1090 (microscopic slide of collar chaeta).

Etymology: Not given by Straughan (1967a: 224–225), but the Latin *rectus* = straight, perpendicular, probably refers to the enlarged dorsal verticil spine with a 'pointed process perpendicular to it'.

Selected synonymy

Hydroides recta Straughan, 1967a: 224–225, fig. 8a–j (diagnosis [original], material studied; Cairns, Townsville, Mackay, Qld, Australia); Sun *et al.* 2015: 71–74, fig. 23a–b, map 20 (redescription, discussion; Australian distribution); 2018, figs 1–3, 5 (DNA, phylogeny; Darwin NT, Mission Beach, Qld, Australia).

Hydroides novaepommeraniae **sensu** Dew 1959: 26–27 (Thursday Island, Townsville, Qld) [**non** Augener, 1925] **fide** Imajima & ten Hove (1984: 47).

Diagnosis: Peduncle with ill-defined constriction. Funnel with 19 ((14)17–22) truncated to T-shaped radii, usually thinly chitinised. Verticil smaller than funnel, on short stalk, bilaterally symmetrical, with 8 (5–10) short, slender spines; dorsal spine slightly larger, ending in a blunt tip, with a small, internal, sub-terminal spinule, sharply pointed. Other spines with pointed tips, curving outwards. Central disk small to absent. Bayonet of collar chaetae with 2 teeth, exceptionally 3.

Remarks: The main difference between *Hydroides recta* and the nominal taxon *H. longistylaris* is the extended, narrow, chitinised base of the funnel, ending in the constriction separating it from the peduncle in the latter. However, though this chitinised base is short in the holotype of *H. recta*, it may be a lot longer in specimens from other localities (Ela Beach, Papua, QM G5260, G5262; ten Hove, unpublished) approaching the situation as figured for *H. longistylaris*. The funnel radii in *H. recta* are blunt, rectangular, while they are pointed in *H. longistylaris*.

Fiege & Sun (1999: 120–121, fig. 10A–E) describe and discuss *Hydroides* cf. *recta* from Hainan. However, judging by the 'budding' radii in their fig. 10A, not fully developed and thus not showing 'diagnostic' pointed or rectangular tips, this is a regenerating specimen, and indeed not to be attributed to a known species with certainty. *Hydroides* cf. *recta* **sensu** Sun & Yang (2014: 247–249, figs 135A–L, 128, several localities in Yellow and South China Seas) morphologically rather approaches *H. albiceps*; the identity of these forms should be ascertained with help of DNA analysis. An ostensibly different *Hydroides* cf. *recta* has been diagnosed by Sun *et al.* (2015: 74–76, fig. 74; Kupriyanova *et al.* 2015: 290, fig. 8) from Lizard Island, Australia. Morphologically this material shows characters between *H. recta* and *H. longistylaris*, and too cannot be attributed to one of these taxa with certainty. This material had been mentioned as *H. novaepommeraniae* by Kupriyanova *et al.* 2008: 428–430 [**non** Augener, 1925], from Lizard Island, Qld, and its DNA data used under this name in phylogenetic analyses by Tovar-Hernández *et al.* (2016: table 2, fig. 6; name, DNA) and by Sun *et al.* (2018: 733, figs 2, 3, 5, 6; same).

Fig. 96. *Hydroides recta*, various views of opercula, Cairns, Queensland, Australia (left three) AM W.3998; Weipa, Queensland, Australia (right two) AM W.30417 (photo E. Wong).

Imajima & ten Hove (1984) re-examined the records of *H. novaepommeraniae* by Dew (1959) and Straughan (1967a) and re-identified them to *H. recta*. In the type of *H. novaepommeraniae*, the verticil is larger and wider than the funnel, while in *H. recta* the funnel is wider than the verticil. Moreover, neither Dew nor Straughan mentioned the special dorsal verticil spine with a sharp, subterminal spinule, which is present in some of their (minutely sized) specimens.

Distribution: Australia: NT, Qld; Papua New Guinea.

83. *Hydroides recurvispina* Rioja, 1941a

Figure 97

Type information: Acapulco, Guerrero, Mexico, type specimens lost.

Etymology: Not given by Rioja (1941a: 167–169), but no doubt refers to the verticil spines which are sharply curving backwards on themselves (Latin *recurvus* = backward curved, *spina* = thorn).

Selected synonymy

Hydroides recurvispina Rioja, 1941a: 167–169, pl. 1, figs 11–15, fig. la–c (description [original]; Acapulco, Mexico); Bastida-Zavala & ten Hove 2003b: 99–101, figs 19A–10, 24 (redescription, synonymy, discussion; Baja California, Mexico to Panama); Sun *et al.* 2018: 733, figs 1–3, 5, 6 (DNA, phylogeny; Guanacaste, Costa Rica; Pacific Panama).

?**Not** *Eupomatus recurvispina* **sensu** Kudenov 1975: 228 (Puerto Penasco, Gulf of California, Mexico) [**non** Rioja, 1941a], most probably see *H. trompi* **fide** Bastida-Zavala & ten Hove, 2003b: 105.

Not *Hydroides recurvispina* **sensu** Tovar-Hernández *et al.* 2009: 13–15, figs 3m, 8d–f (Mazatlán, Mexico) [**non** Rioja, 1941a], see *H. dolabrus* **fide** Tovar-Hernández *et al.* 2016: 696.

Diagnosis: Peduncle with well-defined constriction. Funnel with 24 ((19)20–29) pointed, chitinised radii. Verticil sitting in funnel, radially symmetrical, with 9 (8–10) sharply incurving spines, with sharp tips twisted clockwise. All spines similar, simple, smooth, without lateral spinules, but with an external distal knob on the bend, as well as a knob- (to wart-) like basal spinule. Small central disk without tooth.

Remarks: *Hydroides recurvispina* resembles *H. trompi* in the clockwise twist of the verticil spine tips; however, verticil spines of *H. recurvispina* have a pronounced knob on their bend, whereas in *H. trompi* they are smoothly curving without external knob. Furthermore, *H. recurvispina* has been mistaken for *H. dolabrus* (see 'Selected synonymy' above), mainly on the occurrence of a dorsal knob on the bend of the verticil spines. However, upon closer inspection these are all rounded in *H. recurvispina*, while in *H. dolabrus* they are more triangular, and the most dorsal spine has a trifid tip (not always easy to see). Remarkably *H. recurvispina* from the Pacific sides of Costa Rica and Panama is recovered in phylogenetic analyses as an isolated lineage, separate from a clade with, for example, *H. dolabrus* and *H. panamensis* (Sun *et al.* 2018 fig. 2, 3, 5, Fig. 13 here).

Distribution: Acapulco, West Mexico, Gulf of California to Panama.

Fig. 97. *Hydroides recurvispina*, various views of the same operculum, Guanacaste, Costa Rica, AM W.46941 (photo T. Villalobos).

84. *Hydroides rhombobula* Chen & Wu, 1980

Figure 98

Type information: Dongshan, Fujian Province, China, holotype MBMCAS.

Etymology: Not given by Chen & Wu (1980: 247–248) but apparently refers to the shape of the verticil spines, being a combination of the Greek ρομβος = rhombus, a parallelogram with oblique angles, and the diminutive female Latin suffix *-ula*.

Selected synonymy

Hydroides rhombobulus Chen & Wu, 1980: 247–248, figs 1–9 (diagnosis, in Chinese [original]; Dongshan, China); Sun *et al.* 2012a: 19–20, fig. 3D–J (redescription, synonymy, discussion; Hong Kong, China); Sun & Yang 2014: 249–252, figs 136A–I (copy of Chen & Wu 1980), 137A–G, 138 (diagnosis, in Chinese; Dongshan and Zhangpu, Fujian; Mirs Bay, Sharp Island, Hong Kong; Weizhou Island, Guangxi).

Hydroides rhombobula – Sun 2017: 71, 130, 152, 179, fig. 1F (diagnosis; same).

Diagnosis: Peduncle with ill-defined constriction. Funnel with 16–20 rather triangular, sharply tipped radii. Verticil sitting or on short basis in funnel, with 5 or 6 large rhomboid valves, smooth, no lateral, internal, external or basal spinules; sometimes with a hint of a small, subdistal external knob. Spines equal in size, though the dorsal one may be slightly larger, all thick, almost triangular in cross-section, leaving only a small central disk.

Remarks: *Hydroides rhombobula* resembles *H. xishaensis* in shape of the operculum. However, *H. rhombobula* has rhomboid valves in its verticil and bayonet chaetae with 2 teeth only, while *H. xishaensis* has triangular valves, and bayonet chaetae with 2 teeth and up to 3 accessory teeth. Sun & Yang (2014: 250) mention co-occurrence of *H. rhombobula* with the nominal taxon '*H. priscus*', but it is not unlikely that the latter specimens are just juveniles of *H. rhombobula*, secondary ontogenetic opercular stages, like reported for more Indo-West Pacific taxa (see ten Hove & Ben-Eliahu 2005, notably paragraph 4d). For a more general comparison see Remarks for *H. tuberculata*-complex **sensu lato**.

Distribution: Fujian, Hong Kong, China.

Fig. 98. *Hydroides rhombobula*, various views of the same operculum, Sharp Island, Hong Kong, AM W.41407 (photo E. Wong).

85. *Hydroides rzhavskyi* nov. spec. ten Hove

Figure 99

Type information: Netherlands Antilles. Holotype and one paratype, Bonaire: ZMA V.Pol. 5004, Kralendijk, Flamingo Beach Hotel, sloping reef, sand flats, 33 m, 27 July 1970, legit H.A. ten Hove, stn 2115B; one paratype, Curaçao: ZMA V.Pol.5003, Playa Kalki, reef with small sandy areas, 13–17 m, 28 May 1970, legit H.A. ten Hove, stn 2037B.

Etymology: Named after the late Dr Alexander V. Rzhavsky, colleague polychaetologist.

Synonymy

Hydroides sp. 4 Bastida-Zavala & ten Hove 2003a: 171, figs 39A, C, 40 (description, discussion; same specimens as above).

Diagnosis: Peduncle with ill-defined constriction. Funnel with 26–31 chitinised, blunt tipped radii. Verticil inserted on short basis in funnel, radially symmetrical, with 7–8 spines abruptly and strongly curving inwards with sharp tip, all similar, without lateral spinules, with basal spinule above small central disk.

Description: Tube: white, about 1.2 mm wide, with lumen of about 0.8 (up to 0.9) mm; two tubes lacking peristomes and transversal ridges; one specimen with two longitudinal ridges, while the other tubes with three.

Radiolar lobes: With 7–9 radioles each. Radiolar length up to 5.5 mm, terminal filament very long in all specimens.

Peduncle: Smooth, almost circular in cross-section; length including operculum 7.5 mm ($n = 2$). Insertion left (1 spec.) or right (2); constriction ill-defined (Bastida-Zavala & ten Hove 2003a, fig. 39A). Pseudoperculum present in all specimens.

Operculum: With distal verticil inserted on short stalk into proximal slightly oblique funnel. Funnel with 29 radii (26–31) with chitinised blunt tip. Interradial grooves 1/2–2/3 of funnel length. Verticil inserted on short basis in funnel, with seven (1 spec.) or eight (2) spines, curving inwards strongly and abruptly, colour yellow to light brown. All spines similar in shape and size, with pointed tip. Spines with one basal internal spinule (split in one; Bastida-Zavala & ten Hove 2003a, fig. 39C) curving over small central disk without central tooth; without external and lateral spinules and/or wings. Length of operculum about 1.4 mm, diameter 0.7 mm.

Collar and thoracic membranes: Unknown, specimens were dried out and very fragile.

Thorax: With collar chaetiger and 6 uncinigerous chaetigers. Collar chaetae bayonet with 2 blunt-elongate teeth, distal blade smooth; capillary chaetae present. Subsequent chaetigers with limbate chaetae of two sizes, saw-shaped uncini. Length of thorax 1.5–2.2 mm.

Abdomen: With 66–97 chaetigers. Anterior and middle abdominal chaetigers with flat trumpet-shaped chaetae. Posterior chaetigers with 'capillary' chaetae. Anterior uncini saw-shaped, posterior uncini rasp-shaped.

Size: Length up to 20.0 mm (only entire specimen), width of thorax about 0.8 mm.

Colour: Body yellow.

Remarks: When collected in 1970, the material was set aside as *Hydroides* nov. spec. At the time three specimens only were deemed to be insufficient to describe a new species on. In 2002–03 that too was the reason that

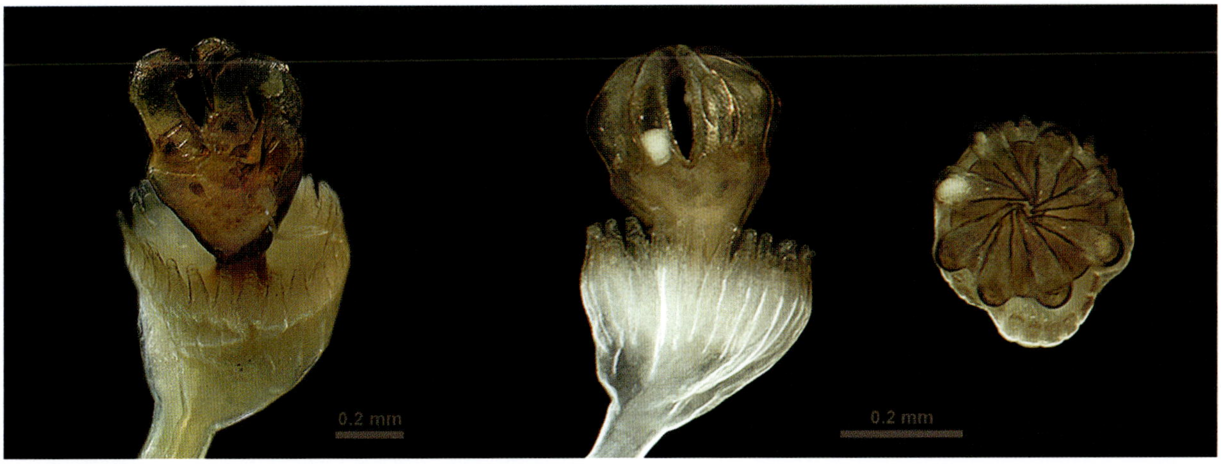

Fig. 99. *Hydroides rzhavskyi* nov. spec., various views of opercula, Netherlands Antilles, paratype ZMA V.Pol.5003 (left), holotype ZMA V.Pol. 5004 (middle and right) (photo H.A. ten Hove).

Bastida-Zavala & ten Hove (2003a) refrained from naming this species. Having reviewed over 100 species in this genus now, I (HAtH) am confident to name these three specimens, from two different islands. Superficially *H. rzhavskyi* nov. spec. resembles *H. bispinosa* in overall compactness of the operculum, but it differs in its long funnel radii and absence of lateral spinules in the verticil spines.

Distribution: Bonaire and Curaçao, Netherlands Antilles.

86. *Hydroides salazarvallejoi* Bastida-Zavala & ten Hove, 2003a

Figure 100

Type information: Cabo de la Aguja, Santa Marta, Colombia (Atlantic side), holotype and paratype ZMA V.Pol.3776.

Etymology: Named after Dr Salazar-Vallejo.

Selected synonymy

Hydroides salazarvallejoi Bastida-Zavala & ten Hove, 2003a: 158–162, figs 32A–P, 33, table 5 (description [original], discussion; Puerto Rico to Martinique, Antilles; Atlantic Colombia to Panama; Pacific Costa Rica to Colombia); 2003b: 101, fig. 12N (marginal differences with Atlantic material; Miraflores Locks, Panama (Pacific); Sun 2017: 157, 179–180, fig. 5C (diagnosis; Santa Marta, Atlantic Colombia).

Diagnosis: Peduncle with well-defined lightly chitinised constriction. Funnel with 18 ((12)14–25) well-chitinised radii with pointed tips, outcurving. Verticil on short basis in funnel, bilaterally symmetrical, with 6 (5–8) spines, including dorsal hook. The latter much larger than the others, with a strongly incurving tip, almost in position of an internal spinule, with a long distal knob on the bend. Shape of dorsal hook approaching that of a strongly bent pickaxe, remaining 4–7 verticil spines with blunt to club-shaped tip; smooth, with basal spinule but without further appendages. Insertion of laterodorsal spines encroaching base of dorsal spine. No central disk.

Remarks: See Remarks for *Hydroides glandifera*; Bastida-Zavala & ten Hove (2003b) mentioned marginal differences between Pacific and Atlantic populations.

Distribution: Antilles, Caribbean; Atlantic Colombia to Panama; Pacific Costa Rica to Ecuador.

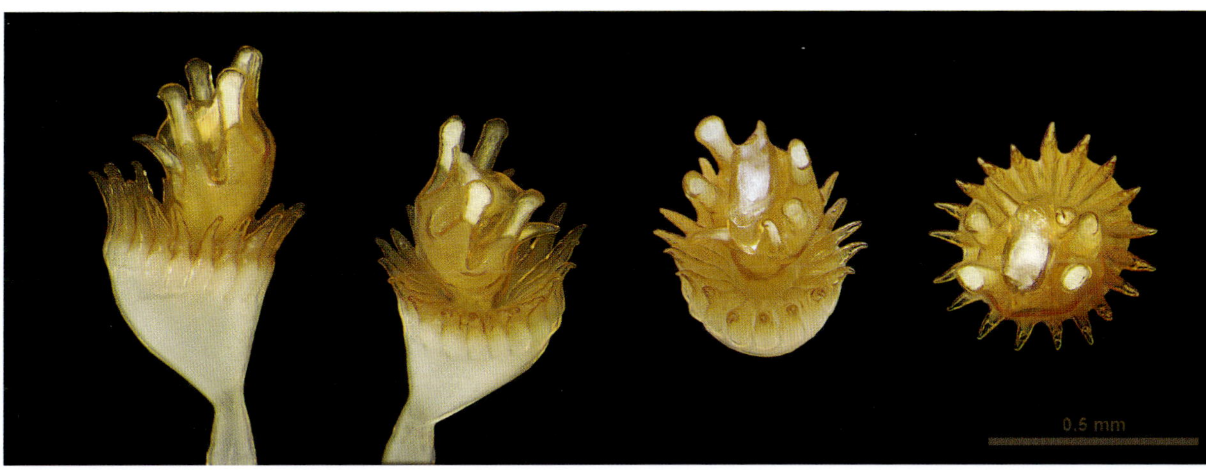

Fig. 100. *Hydroides salazarvallejoi*, various views of the same operculum, La Parguera, Puerto Rico, ZMA V.Pol.5072 (photo E. Wong).

87. *Hydroides sanctaecrucis* Krøyer in Mörch, 1863

Figure 101

Type information: Saint Croix, Lesser Antilles, five syntypes in ZMC, labelled: 'Kr. St. Croix, legit Oerstedt'.

Etymology: Named after its type locality.

Selected synonymy

Hydroides (*Eucarphus*) *sanctae-crucis* Krøyer in Mörch, 1863: 378–379, pl. 11, fig. 12 (diagnosis in Latin [original], figure of verticil spines; St Croix, Lesser Antilles).

Eupomatus sanctaecrucis – Fauvel 1919: 478–479, fig. 2e–h (description; French Guyana).

?*Hydroides vizagensis* Lakshmana Rao, 1969: 9, pl. 6a–h (diagnosis [original]; Visakhapatnam Harbour, India, Gulf of Bengal; in our opinion most probably a synonym of *H. sanctaecrucis*, see remarks under *H. vizagensis*, below).

Hydroides sanctaecrucis – Bastida-Zavala & ten Hove, 2003a: 148–151, figs 25A–R, 28 (redescription, synonymy, discussion; Florida, US Virgin Islands, Puerto Rico; Atlantic Panama to French Guyana); 2003b: 101–102, figs 19P, Q, 24 (name; first records from Pacific Mexico to Panama); Lewis *et al.* 2006: 665–671, figs 1, 2 (diagnosis, discussion; first record for Darwin, NT, Cairns, Qld, Australia); Sun & Yang 2014: 252–253, fig. 139A–B (diagnosis in Chinese; Tolo Harbour, Victoria Harbour, Hong Kong, China); Sun *et al.* 2012a: 20–21, fig. 10A–G (description; Hong Kong, China); 2015: 76–79, fig. 25, map 22 (diagnosis, synonymy; ship-transported to the Indo-West Pacific); 2017a: 1, 3, 6 (name, DNA, phylogeny; Mazatlán, Pacific Mexico; St Petersburg, Florida, USA; Hong Kong, China; Darwin, NT, Australia); 2018: 733, 739, figs 1–3, 5, 6 (DNA, phylogeny; Colon (Atlantic) and Panama (Pacific); Maharashtra, India; Darwin, NT, Australia); Tovar-Hernández *et al.* 2016: table 2, fig. 6 (name, DNA, phylogeny; Florida, USA; Mazatlán, Mexico); Bastida-Zavala *et al.* 2017: 38–40, figs 4J, 6 (diagnosis, extensive synonymy; Biscayne Bay, Tampa Bay, Florida); Ferrario & Minchin 2017: 1–6, figs 1–3 (diagnosis, discussion; Taiwan); Phuttapreecha *et al.* 2018: 147, 150 (name, fouling; Thailand); Keppel *et al.* 2019: 469, fig. 7 (diagnosis, fouling; Galapagos).

Hydroides (*Eupomatus*) *dianthoides* Augener, 1922: 49–50 (diagnosis, in part *H. dianthus*, in part *H. sanctaecrucis*; Haiti, material studied by HAtH).

Hydroides santaecrusis (in errore) – Dueñas 1981: 99–100, 135 (name; Cartagena, Colombia).

Not: *Hydroides sanctaecrucis* **sensu** Díaz Díaz & Liñero Arana 2001: 13–14 [**non** Krøyer in Mörch, 1863] from Cariaco Gulf, Venezuela and referred to *H. bispinosa* by Bastida-Zavala & ten Hove (2003a: 125).

Not: *Hydroides sanctaecrucis* **sensu** Rech *et al.* 2018: 10, 13–15 [**non** Krøyer in Mörch, 1863] from Venezia, Italy, probably a misidentification of *H. dianthus* according to Langeneck *et al.* 2020: 261, table 1.

Diagnosis: Peduncle with ill- to well-defined constriction. Funnel with 24 (17–34) chitinised, pointed radii. Verticil bilaterally symmetrical, sitting or on short basis in funnel, with 12 (10–14) slender spines, all pointing in ventral direction and with wings extending for more than half of spine length. Three to 5 dorsal spines curving inwards, slightly larger than others, and with external knee rather than spinule at 2/3 of their length, gradually changing into spines pointing ventrally and thus outwards, with distinct external spinules. Verticil spines

Fig. 101. *Hydroides sanctaecrucis*, various views of opercula, Cairns, Queensland, Australia (left three) AM W.33432; Aruba, Netherlands Antilles (right two) AM W.37768 (photo E. Wong).

with or without a small basal spinule. Central disk without central tooth.

Remarks: This species shares a verticil with all spines curving ventrally with *H. dianthus*, but the latter does not have external spinules. For an in-depth comparison see Zibrowius (1971a: 697–705). *Hydroides sanctaecrucis* is common in the Caribbean (Bastida-Zavala & ten Hove 2003a, fig. 28). As a probably ship-transported species, it is a dominant fouling serpulid in Pacific Panama and Oaxaca, Mexico (Bastida-Zavala & ten Hove 2003b: 102; Bastida-Zavala 2008: 29). Long (1974: 28) listed the species as fouling from Hawaii, but this record was doubtful as some other taxa listed by Long, and never corroborated in later studies, nor present in the collections of the Bishop Museum, Hawaii. In the meantime, *H. sanctaecrucis* has been recorded as an invasive species in Northern Australia and Singapore (Lewis *et al.* 2006; Hutchings 2008). In Hong Kong, *H. sanctaecrucis* was found in low densities on the pier and the fish farm, where *H. elegans* dominated the fouling community (Sun *et al.* 2012a: 21). These Indo-Polynesian records make Long's old record less improbable, but a confirmation still is desirable.

There is a very small genetic distance (only 0.53%) between Caribbean and Indo-Polynesian populations of *H. sanctaecrucis* (Sun *et al.* 2018, Fig. 13 here), making a relative recent anthropogenic distribution likely.

Distribution: St Croix, Caribbean, from Florida to French Guyana, both sides of Mexico and Panama; ship-transported to Singapore, Thailand, Hong Kong, Taiwan and tropical Australia.

88. *Hydroides sceptrifera* nov. spec. ten Hove

Figure 102

Type information: Netherlands Antilles, Bonaire, 250 m north of Witte Pan, sand flat below reef, 46 m, legit H.A. ten Hove, stn 2117B, 3 July 1970; holotype ZMA V.Pol. 3913.01, and Ver. P1279, 1280 (microscopic slides), seven paratypes ZMA V.Pol. 3913.02 (one photographed in Fig. 102); three paratypes USNM 1143383, 3 paratypes NHML 1992.292–294.

Additional material: Netherlands Antilles, Bonaire, as above, ZMA V.Pol. 3913.03 (about 60 fragments of specimens, 11 operculate, and many tubes from type locality, not studied in detail, nor included in type series).

Eastern Gulf of Mexico, 26°24′N, 83°22′W, 55 m, RV *Hourglass* stn L (1 spec., FSBC I 33512).

Panama, Caribbean Sea, 08°54.42′N, 077°41′W, 30 m, RV *Alpha Helix* stn ND-19, 29 June 1977 (1 spec., USNM 098421).

Etymology: The specific name is derived from the Greek *skeptron* = sceptre, and *pherein* = to carry.

Synonymy

Hydroides sp. 2 Bastida-Zavala & ten Hove 2003a: 169, figs 37P–S, 40 (description; Netherlands Antilles, Klein Bonaire, N, 0.8 km east of Westpunt, sand flat below reef, scarce corals, 49 m, legit H.A. ten Hove, stn 2105C, 1 July 1970, 1 spec., ZMA V.Pol. 3914).

Diagnosis: Peduncle with distinct constriction. Funnel with 18 (16–21) unchitinised, rounded radii. Verticil sitting or on short basis in funnel, radially symmetrical, with 7–10 equal, outcurving spines tapering to blunt tip. Lateral spinules, medial teeth, basal tubercle absent. Flat central disk without tooth. Bayonet of collar chaetae with 2 large and upcurving teeth.

Description

Tube: White, about 0.6 (up to 1.0) mm wide, with lumen of about 0.5 (up to 0.8) mm. Walls thin (about 0.05 mm), almost transparent. Tube circular in cross-section, generally smooth, without distinctive marks. Rarely collected erect parts may show thin collar-like rings. Granular layer not observed.

Radiolar lobes: With up to 9 radioles each, generally 7 as in type series. All specimens preserved in tubes and details on pinnules difficult to observe, even after extraction. Terminal filament of radioles simple, as long as longest pinnules.

Peduncle: Smooth, subcircular in cross-section, inserted just below first and second normal radiole; with distinct constriction at base of funnel.

Operculum: With distal verticil inserted on short stalk into proximal radially symmetrical funnel. Verticil with 7–10 equal, yellowish spines arranged around flat central disk without central tooth. Spines all curving outward, tapering to blunt tip. Lateral spinules, medial teeth and basal tubercle absent. Funnel not chitinous, with 16–21 ($x = 17.6$, $SD = 1.5$, $n = 13$) rounded, marginal radii. Grooves separating radii extend 1/3 of funnel length. Length of operculum about 0.8 (up to 1.0) mm, width about 0.4 (up to 0.6) mm.

Collar and thoracic membranes: Collar low, with entire edge; continuous with thoracic membranes, which form apron ventrally across 1–6 anterior abdominal chaetigers. A triangular protuberance of collar chaetiger on each side just below junction of ventral and lateral collar parts.

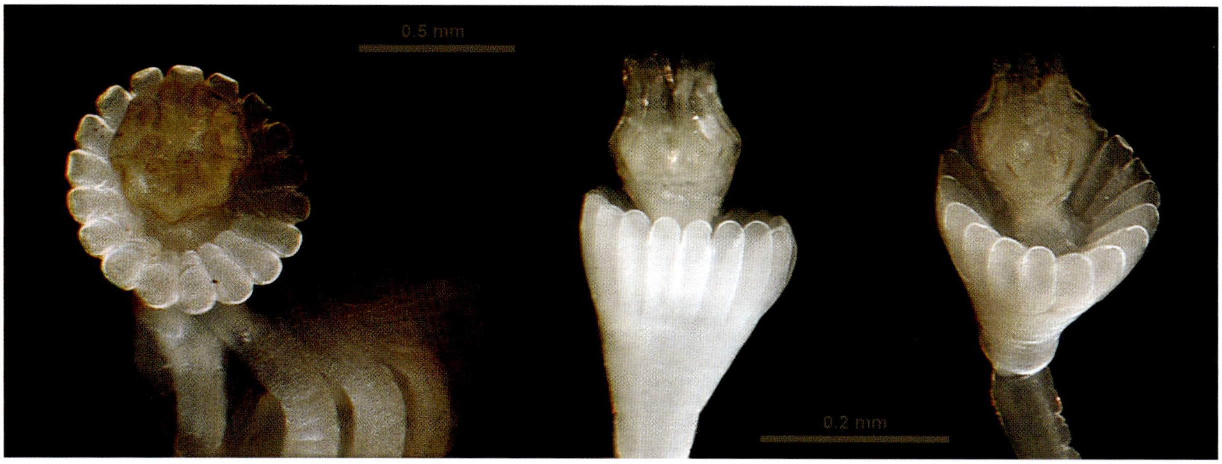

Fig. 102. *Hydroides sceptrifera* nov. spec., various views of operculum, Bonaire, Netherlands Antilles, ZMA V.Pol. 3913.02 (photo H.A. ten Hove).

Thorax: With collar chaetiger and 6–9 uncinigerous chaetigers. Collar chaetae stout bayonet and narrowly limbate. Bayonet chaetae with 2 very characteristically bulging teeth, but without proximal rasp or distal striated 'limba'. Subsequent chaetae limbate, of two sizes. Uncini saw-shaped, with fang and 5 teeth. Pair of simple ocellar clusters deeply embedded in prostomium.
Abdomen: Usually with about 70 (up to 88) chaetigers. Uncini saw-shaped anteriorly, with fang and 5 teeth; rasp-shaped with four rows of teeth, fang and 9 teeth in profile view posteriorly. Chaetae flat trumpet-shaped with denticulate edge, each with anterior non-denticulate zone approximately covering the width of 10 denticles. 'Capillary' chaetae in up to 13 posterior chaetigers. Pygidium with two small rounded lobes.
Size: Length up to 15 mm, usually about 12 mm. Width of thorax 0.35–0.60 mm ($x = 0.44$, $SD = 0.09$, $n = 12$). Radioles (including operculum) account for 1/4 of entire length in small specimens, 1/6 of length in large specimens.
Colour: No records.
Remarks: Specimens of this species easily might be (in fact have been) mistaken for small specimens of *Hydroides protulicola* and *H. microtis*. However, collar chaetae of the latter two species show a proximal rasp and a distal striated limba, both absent in *H. sceptrifera* nov. spec. Not of special size or form in *H. protulicola* and *H. microtis*, the 2 teeth of collar chaetae in *H. sceptrifera* nov. spec. are fairly characteristic in being large and upcurving (Bastida-Zavala & ten Hove 2003a, fig. 37R). Moreover, funnel radii of *H. sceptrifera* are rounded, whereas they are sharply pointed in the other two species, even in juvenile specimens. Differences in opercular structures are shown in our plates.
The two specimens of *H. sceptrifera* nov. spec. from Florida and Panama have 7 chaetigers. The range was 8–10 in the material from Bonaire and Klein Bonaire; in one specimen the left side of the thorax had 8, the right side 10 chaetigers. Half of the specimens from the Netherlands Antilles had different numbers of thoracic parapodia on each side. The specimen from Klein Bonaire, diagnosed in Bastida-Zavala & ten Hove (2003a) showed two opercula.
Ecology: The material from the Netherlands Antilles was collected from boulders and other coral debris, partly buried in the sand flat at the base of the fringing reef. The species was never found on the coral reef proper, although that habitat was sampled far more extensively by one of us (HAtH). The known depth distribution is 30–55 m. The type series consists of an aggregate of tubes, suggesting that the species may be gregarious.
Distribution: Tropical West Atlantic.

89. *Hydroides similis* (Treadwell, 1929)

Figure 103

Type information: Baja California, Western Mexico, holotype AMNH 1993.

Etymology: Not given by Treadwell (1929: 11–12), but the Latin adjective *similis* = similar to, and Treadwell thought it in many characters similar to another species he described at the same time (1929: 12), *H. californicus*, synonymised in the meantime with *H. crucigera*.

Selected synonymy

Eupomatus similis Treadwell, 1929: 11–12, fig. 31 (diagnosis [original]; Baja California, Mexico).

Hydroides brachyacantha **sensu** Bastida-Zavala 1993: 35 (Bahîa de La Paz, Gulf of California) [**non** Rioja, 1941a].

Hydroides similis – Bastida-Zavala & ten Hove 2003b: 102–104, figs 21A–U, 24 (redescription, synonymy, discussion; Baja California, Pacific Mexico to Panama); Tovar-Hernández *et al.* 2009: 12 (comparison with similar taxa in remarks); Bastida-Zavala *et al.* 2016: 427, figs 6, 12F (very common in the Gulf coast of Baja California Peninsula); Sun 2017: 35, 156, fig. 4E (diagnosis, DNA; Guaymas, Sonora, Marina La Paz, Baja California Sur); Sun *et al.* 2018, figs 2, 3, 5, 6 (DNA, phylogeny; Guaymas and La Paz, Gulf of California).

Diagnosis: Peduncle with ill-defined constriction. Funnel with 32 (25–39) chitinised radii with pointed tips. Verticil sitting in funnel, bilaterally symmetrical, with 14 (12–16) spines, strongly curving inwards. Three to 4 dorsal spines larger than the others; the latter are similar in shape and size, with small to pronounced external knob on the bend, with pointed tip. Spines with basal spinule of varying length; without external and lateral spinules and/or wings. Small central disk without tooth.

Remarks: The original 'similarity' with *Hydroides crucigera* (as *H. californicus*) as perceived by Treadwell (1929) is, in our present understanding of the genus, immediately refuted by the side spinules of the verticil spines in *H. crucigera*. *Hydroides similis* shows some resemblance to *H. brachyacantha*, both have strongly incurving verticil spines, ventral ones with a knob; in the past it was even misidentified as such (see Bastida-Zavala & ten Hove 2003b: 102, 104). However, *H. similis* has 2–4 larger dorsal spines, while there is only one covering the central disk in *H. brachyacantha*. In a phylogenetic analysis (Sun *et al.* 2018, Fig. 13 here) *Hydroides similis* is recovered as sister clade of *H. deleoni*, to which it shows a basically morphological similarity in the bilaterally symmetrical verticil with one dorsal hook (respectively three or four) accompanied by a ventral set of rather smooth incurving spines with a small external knob.

Distribution: Baja California to Panama.

Fig. 103. *Hydroides similis*, various views of the same operculum, Baja California Sur, Mexico, ZMA V.Pol. 4988 (photo E. Wong).

90. *Hydroides similoides* Bastida-Zavala & ten Hove, 2003a

Figure 104

Type information: La Parguera, Isla Magueyes, Puerto Rico, holotype and seven paratypes ZMA V.Pol. 5038, 5039.

Etymology: Named for its resemblance to *H. similis*.

Selected synonymy

Hydroides similoides Bastida-Zavala & ten Hove, 2003a: 155–158, figs 30A–O, 31A–D, 33, table 4 (description [original], discussion; US Virgin Islands, Puerto Rico; Atlantic Panama to Venezuela, ?Veracruz, Mexico); De Assis *et al.* 2009: 1–3, fig. 1A–I (diagnosis; Paraíba, Brazil); Sun 2017: 156, 180–181, fig. 4D (diagnosis; Puerto Rico).

Hydroides cf. *brachyacanthus* Stock 1995: 3 (St Thomas, parasitised).

Diagnosis: Peduncle with ill- to well-defined constriction. Funnel with 17 (12–22) very lightly chitinised radii with blunt tip. Verticil sitting in funnel, at most on very short basis, bilaterally symmetrical, with 11 (8–13), strongly curving inwards. Two or 3 dorsal spines smoothly incurving and larger than the others; the latter similar in shape but decreasing in size ventrally, curving abruptly, geniculate, their bends accentuated by a prominent knob, all with pointed tip. Spines with basal spinule; without external and lateral spinules and/or wings. Small central disk without central tooth.

Remarks: The operculum of this tropical Western Atlantic taxon is reminiscent of that of the tropical Pacific American *H. similis* (Treadwell, 1929), by its 2 or 3 dorsal verticil spines larger than the others. It differs from *H. similis* in its lower number of radii (12–22 *v* 26–39), moreover with blunt instead of pointed tips. It also resembles *H. brachyacantha* (in fact was routinely identified as *H.* cf. *brachyacantha* by HAtH and students between 1965 and 1985) in its bilateral symmetry and geniculate ventral verticil spines. However, *H. brachyacantha* has a single dorsal hook and pointed radii, while there are 2 or 3 hooks and rounded radii in *H. similoides* (see Bastida-Zavala & ten Hove 2003a: 157, 159). For a comparison with the Caribbean similar taxa *H. deleoni* and *H. zibrowii* nov. spec., see Remarks under the latter.

Distribution: Puerto Rico, Caribbean to Paraiba, Brazil.

Fig. 104. *Hydroides similoides*, various views of the same operculum, Joyuda, Puerto Rico, AM W.43196 (photo E. Wong).

91. *Hydroides simplidentata* Pillai, 2009

Figure 105

Type information: Reef north-west of Buffon Island, Kimberley, Western Australia, holotype AM W.21415.

Etymology: Refers to the simple unmodified spines at the base of the enlarged coronal (verticil) spine (Pillai 2009: 125).

Selected synonymy

Hydroides simplidentatus Pillai, 2009: 122–125, fig. 18A–J (description [original], discussion; Kimberley, WA).

Hydroides simplidentata – Sun *et al.* 2015: 79–81, fig. 26a, b; map 23 (redescription, discussion; WA, Komodo, Indonesia).

Diagnosis: Peduncle without or with ill-defined constriction. Funnel with 16 chitinised, pointed radii. Verticil bilaterally symmetrical, sitting in funnel, with 6 spines and a dorsal hook. The latter vesicular, shaped like a swollen beak with rounded tip, curving towards centre of verticil; spines pressed against base of hook, seemingly simple and straight, but if squeezed open they show a subapical tip pointing inwards and an extended, conical knob on their bend. No central disk. Bayonet of collar chaetae with 2 teeth and 2 accessory ones.

Remarks: Although Pillai (2009) compared his *H. simplidentata* with *H. brachyacantha*, the species is more similar to what he described in 1971 (pp. 107–108) as *H. rostrata*, a homonym renamed *H. gottfriedi* Read, ten Hove, Sun & Kupriyanova, 2017 (pp. 35–36). In fact, a field identification by ten Hove (1984) from an Indonesian specimen of what now has been recognised as *H. simplidentata* was *H. rostrata*; see discussion in Sun *et al.* 2015: 124). The main difference is in the shape of smaller verticil spines, with a large, conical knob in *H. simplidentata*, small and pointed in *H. gottfriedi*. By its large, vesicular dorsal hook *H. simplidentata* is resembling *H. glandifera*, but the remaining spines in the former are not fused to the hook like in *H. glandifera*, but form a funnel including it.

Distribution: Kimberley, Western Australia; Komodo, Indonesia.

Fig. 105. *Hydroides simplidentata*, various views of the same operculum, Kimberley, Western Australia, holotype AM W.21415 (photo E. Wong).

92. *Hydroides sinensis* Zibrowius, 1972a

Figure 106

Type information: Qingdao, China, Yellow Sea, holotype ZINRAS AN 1/10 1930.

Etymology: Named after its distribution in Chinese seas.

Selected synonymy

Hydroides sinensis Zibrowius, 1972a: 89–92, fig. 1a–h (description [original], synonymy; Qingdao, Yellow Sea); Sun & Yang 2000: 128, fig. 7A–E (diagnosis, in Chinese, not the synonymy; South China Sea); 2014: 253–254, figs 140A–H, 112 (largely the same as 2000); Sun 2017: 103, 154, 181, 288, 297, 317–318, fig. 2L (diagnosis, DNA, phylogeny; Qingdao, China); Sun *et al.* 2018: 738, 742, figs 2, 3, 5, 6 (DNA, phylogeny; Qingdao, China).

Hydroides lunulifera **sensu** Uschakov & Wu 1963: 158 [**non** Claparède, 1870] **fide** Zibrowius (1972a: 89).

Diagnosis: Peduncle without or with ill-defined constriction. Funnel with 38–42 radii, short, cone-shaped with lightly chitinised, more or less rounded or truncated tips. Verticil inserted with short basis into funnel, almost radially symmetrical, with 8–10 uniform spines, thicker dorsally than ventrally. Spines almost straight to slightly curving outwards, with bluntly dilated, truncated tips; at mid-length all with a well-developed internal spinule, shaped like the head of a pickaxe and directed to the centre of the verticil. Proximal of these there is a row of 3 or 4 radial spinules, the most basal almost touching were spines meet. No central disk.

Remarks: *Hydroides sinensis* shows a superficial resemblance to *H. dirampha* and *H. qiui* in the dilated verticil spine tips. However, these have crescent-shaped tips, while they are bluntly dilated, truncated in *H. sinensis*. Funnel radii are rounded truncated in *H. sinensis*, acute in *H. dirampha*, anchor-shaped in *H. qiui*. *Hydroides sinensis* shares pickaxe-shaped internal spinules with the West African *H. arnoldi*, but the verticil spines of the latter have smoothly rounded tips, not dilated like in *H. sinensis*. A summary of similar forms, even if just faintly so, is found in Tovar-Hernández *et al.* (2016, fig. 4).

Distribution: Yellow Sea.

Fig. 106. *Hydroides sinensis*, various views of the same operculum, Qingdao, China, AM W.45634 (photo E. Wong).

93. *Hydroides spongicola* Benedict, 1887

Figure 107

Type information: off Venice, Florida, USA, 27°04′N, 83°21′15″W, nine syntypes USNM 975, one syntype ZMA V.Pol. 3209, ZMA VER.P.1087 (microscopic slide).

Etymology: Evidently after its habitat (sponges) and the Latin *colo* = to live.

Selected synonymy

Hydroides spongicola Benedict, 1887: 549–550, pl. 20, figs 11, 12; pl. 21, figs 13–16 (diagnosis [original], material studied; off south-west Florida, USA); Bornhold & Milliman 1973: 370 (tube mineralogy, material studied; Curaçao, Netherlands Antilles); Bastida-Zavala & ten Hove 2003a: 121–125, figs 8A–O, 9A–F, 10 (redescription, synonymy, discussion; Florida, USA, and Curaçao, Netherlands Antilles); Humann *et al.* 2013: 82 (colour photograph of worms in sponge; Florida, Caribbean); Sun 2017: 153, 181, fig. 2D (diagnosis, DNA; Florida, USA).

Eupomatus spongicola – Jones 1962: 207 (differential diagnosis and extensive discussion).

Diagnosis: Peduncular constriction very ill-defined, if any. Funnel with 51 (45–61) chitinised, pointed radii, wider than verticil, which is sitting in funnel or on short basis, radially symmetrical, with 15 (13–18) slender incurving spines. All spines essentially similar in shape; proximally they may or may not be flattened into lateral wings for up to more than half length. Spines with long basal spinule, over but not crossing central disk, spinules without or rarely with knob in the middle. Without central tooth. Some populations show up to 50% specimens without opercula, but with two pseudopercula only. Exclusively associated with sponges (*Neofibularia nolitangere* (Duchassaing & Michelotti, 1864)).

Remarks: This species is very similar to *H. alatalateralis*, *H. elegantula* and *H. floridana* in having wings along its slender, incurving verticil spines. It differs from these species by its distinctly higher number of funnel radii, with longer tips. For a full discussion, see remarks for *H. alatalateralis*.

Distribution: Gulf of Mexico, Bahamas, Caribbean.

Ecological remarks. Unique is that all material in Bastida-Zavala & ten Hove (2003a: 122) has been mentioned as associated with sponges; in the actual material (often dredged) the direct relation could not always be ascertained owing to degradation of the sponge by fixation in formalin, especially in the scores of samples of the RV *Hourglass* (of which a few mentioned in Bastida-Zavala & ten Hove). In fieldwork on Curaçao and Klein Bonaire, ten Hove only found this taxon in *Neofibularia*, very commonly. The symbionts of the sponges *Agelas clathrodes* (Schmidt, 1870), *Ircinia strobilina* (Lamarck, 1816) and *Neofibularia nolitangere* in Curaçao were studied by Peeze Binkhorst (unpublished). *Hydroides spongicola* was only found in *Neofibularia*, and in 19 out of 20 specimens of this sponge; the only specimen in which *H. spongicola* was not found was a few centimetres small. Up to 1000 specimens of *H. spongicola* may be encountered in a large sponge. Regeneration of damaged *Neofibularia* placed in a transparent plastic tunnel (10 m long, 2 m wide, 1.2 m high) on a horizontal sandy plain, 9 m deep, was similar to that in the open water, but in a black tunnel the sponges bleached and after 83 days had lost all their *H. spongicola* symbionts. The serpulid may benefit

Fig. 107. *Hydroides spongicola*, various views of the same operculum, Curaçao, AM W.43518 (photo E. Wong).

from the water currents generated by the sponge, which may provide a constant supply of food. The fairly aggressive sponge – aptly named 'do-not-touch-me-sponge' – may moreover provide protection to the worms, as suggested by the frequent loss of the protecting operculum by the worms, a phenomenon not known in any other species of the genus. Possible benefits for the sponge and why it disappeared under dark conditions are not clear. The known depth distribution is 3–73 m.

One previous record (Curaçao, stn 2073Aa; ten Hove 1989: 136) of the occurrence of *H. spongicola* in the sponge *Mycale (Carmia) microsigmatosa* Arndt, 1927 (det. R.W.M. van Soest 1989) may be inaccurate (e.g. caused by a mix-up of material in collecting or sorting).

94. *Hydroides steinitzi* Ben-Eliahu, 1972

Figure 108

Type information: Egypt, Suez Canal, Little Bitter Lake (opposite Kabrit (or Kabret)); holotype and two paratypes HUJ SLC 85/P116.

Etymology: Named after the marine biologist Professor Heinz Steinitz.

Selected synonymy

Hydroides steinitzi Ben-Eliahu, 1972: 78–80, figs 1, 2 (diagnosis [original]; Little Bitter Lake and Gulf of Suez, Egypt); Ben-Eliahu & ten Hove 2011: 35–39, figs 13–15 (name, synonymy, extensive discussion; Suez Canal; South Red Sea: Gulfs of Suez and Aqaba, Dahlak Archipelago, South Red Sea; Philippines); Sun 2017: 157, 181, fig. 4S (diagnosis, figure of operculum).

Diagnosis: Peduncle with shallow to well-defined constriction. Funnel with 10–12 chitinised, rounded to bluntly pointed radii. Verticil radially symmetrical, sitting on funnel and composed of a single bulbous spine with (5–)6 bulging out-pockets; in apical view somewhat resembling a peeled orange. Without central disk.

Remarks: Its verticil is unusual in being composed of a single element, somewhat resembling a peeled orange when seen in apical view. This bulbous spine shows some similarity to that of *H. bulbosa*, but accompanied by 7–11 'normal' verticil spines. The variation in length of bulbous spine as shown by the two right-hand photographs in Fig. 108 is bridged by a second specimen from the Philippines, which is very similar to Ben-Eliahu (1972: paratype, fig. 2). Magnification of the right apical view of the bulb (Fig. 94) shows five very shallow longitudinal furrows delimitating its out-pockets. The rather disjunct distribution, Red Sea and Philippines, necessitates a closer look; unfortunately, no DNA data are available. For a full discussion, see under *H. bulbosa*.

Distribution: Suez Canal, Red Sea, Philippines, found once on a ship's hull in Toulon, Mediterranean, upon return from the Suez Canal.

Fig. 108. *Hydroides steinitzi*, various views of opercula, Suez Canal, Egypt (left three) ZMA V.Pol. 3513; Banacon Island, Phillipines (right two) AM W.38610 (photo E. Wong).

95. *Hydroides stoichadon* Zibrowius, 1971a

Figure 109

Type information: Cap du Merlan, Parc Nacional de Port Cros, France, Mediterranean, holotype and two paratypes MNHN.

Etymology: Not given by Zibrowius (1971a: 716–717), but a personal communication by him explains that Στοιχαδας is the old Greek name for the Îles d'Hyères near Stoechades Canyon, Mediterranean France.

Selected synonymy

Hydroides stoichadon Zibrowius, 1971a: 716–717, figs 40–47 (description [original], extensive discussion; coasts of the Provence, France); Bianchi 1981a: 59–60, fig. 19a–e (diagnosis; France, Yugoslavia); Ben-Eliahu 1991: 518, 521, 523 (name; Cyprus); Kitsos *et al.* 2005: 258 (epibionthic on loggerhead turtles, Greece); Boaventura *et al.* 2006: 342 (name in list; concrete artificial reef of Ancão, Faro, South Portugal); Vasconcelos *et al.* 2007: 167 (name in list; epibiontic on gastropod; South Portugal); Sun 2017: 154, 182, fig. 2S (diagnosis, figure of operculum).

Diagnosis: Peduncle without constriction. Funnel with 18 ((13)15–23) not chitinised, rounded radii. Verticil inserted on very short basis in funnel, radially symmetrical, with 10–11 identical short, lightly incurving spines, distally blunt to pointed, sometimes flattened externally and each with a pair of lateral denticles at mid-length. Distinct central disk without spine.

Remarks: By its (almost) radially symmetrical verticil with spines showing a pair of lateral spinules, *Hydroides stoichadon* resembles *H. bispinosa*, *H. chilensis*, *H. crucigera*, *H. homoceros*, *H. mucronata*, *H. mucronatoides* nov. spec. and *H. parva*. The first, *H. bispinosa*, shows funnel radii with rounded tips, just as *H. stoichadon*, but its verticil spines have (with the exception of the most dorsal one) a basal spinule, missing in *H. stoichadon*. Funnel radii of *H. homoceros* are anchor-shaped to double-anchor-shaped; *H. mucronatoides* nov. spec. does not have funnel radii, but 2–6 globular processes; the remaining four species all have pointed, acute radii. For more differences (e.g. curvature and thickness of spines, collar chaetae) one should consult the respective diagnoses and figures of mentioned resembling taxa.

Distribution: Atlantic Algarve, South Portugal to Cyprus in the Eastern Mediterranean.

Fig. 109. *Hydroides stoichadon*, various views of the same operculum, Famagusta, Cyprus, SMF 5169 (photo E. Wong).

96. *Hydroides tambalagamensis* Pillai, 1961

Figure 110

Type information: Tambalagam Lake, Sri Lanka, holotype BMNH 1960.3.13.30.

Etymology: Named after its type locality.

Selected synonymy

Hydroides tambalagamensis Pillai, 1961: 36–38, fig. 12A–G (description [original]; Tambalagam Lake, Sri Lanka); Imajima 1976a: 123–126, fig. 2a–j (redescription; Tanega-Shima, Japan); 1987: 16 (distribution in Japan); Imajima & ten Hove 1984: 49 (name, synonymy; Majuro, Marshall Islands, Lizard Island, Qld, Australia); Sun & Yang 2000: 128–130, fig. 6K–S (diagnosis, in Chinese, synonymy; South China Sea); 2014: 253–254, figs 140A–H, 112 (largely the same as 2000; Dongshan, Fujian; Mirs Bay, Hong Kong; South China Sea); Sun *et al.* 2012a: 21–23, fig. 11A–J (description; Hong Kong); 2015a: 82–85, fig. 27, map 24 (diagnosis, synonymy; WA, NT, Qld, Australia); 2018, figs 1–3, 5, 6 (DNA, phylogeny; Ashmore Reef, WA; Lizard Island, Dunk Island, Heron Island, Qld, Australia); Kupriyanova *et al.* 2015: 290–293, fig. 9 (diagnosis; Lizard Island, Qld, Australia).

Hydroides spiculitubus Pillai, 2009: 125–128 (Kimberley, WA) **fide** Sun *et al.* (2015); Bailey-Brock *et al.* 2012: 973, fig. 3F–I (description; Enewetak Atoll, Marshall Islands).

Diagnosis: Peduncle with well-defined constriction. Funnel with 31 (23–39) sharply pointed dark-brown chitinised radii. Verticil inserted on a short stalk in funnel, with 7(–8) sub-equal spines, curving outwards and ending in pointed tip. All spines bearing a pair of outcurving lateral spinules at about half of their length, an inwardly curved radial spinule at the same level or slightly more distal, and a small, pointed basal spinule above small central disk.

Remarks: Pillai (2009) described specimens with a *Hydroides tambalagamensis*-type operculum and a detachable inner tube as *H. spiculitubus*. However, Sun *et al.* (2015) synonymised *H. spiculitubus* with *H. tambalagamensis* based on the identical 18S sequences of specimens with and without inner tubes. While presence or absence of inner tubes is rarely mentioned in descriptions of species, indeed it is not a character to separate these two nominal species. Nevertheless, on biogeographical grounds it would be desirable to check if the material from Sri Lanka genetically is the same as the Australian population.

At first sight *H. tambalagamensis* might be confused with *H. externispina* due to the similar dark curved lateral spinules on the verticil spines of both species. However, all spines are alike in *H. tambalagamensis*, while *H. externispina* has two large hooks. Vine & Bailey-Brock (1984: 140, fig. 2G–H) depict *Hydroides* aff. *tambalagamensis* from Sudan. Unfortunately, there is no explanatory text, and the figures show insufficient detail. However, the figures are not like any taxon we presently know, though by the 2 hooks they figure there is some similarity with *H. externispina*. In 1982 one of us (HAtH) had identified this specimen, the 2 dorsal spines indeed are slightly larger than the other 4, but they are just like the others with a pair of outcurving lateral spinules at about half of their length and an inwardly curved radial spinule at the same level or slightly more

Fig. 110. *Hydroides tambalagamensis*, various views of opercula, Sharp Island, Hong Kong (left three) W.40552; Heron Island, Queensland, Australia (right two) AM W.4108 (photo E. Wong).

distal; however, basal spinules (as present in *H. tambalagamensis*) are missing. This, and the lower numbers of both verticil spines and funnel radii led to the identification as **affinis** *tambalagamensis*. It very well might be a separate taxon, but for this decision more material will be necessary.

In their phylogenetic analyses (Sun *et al.* 2018, figs 2, 3, 5) the position of *H. tambalagamensis*, split over different clades, is rather difficult to interpret. Unfortunately, it has not been possible to obtain COI from the WA material, more material will be needed to explain the cladograms.

Distribution: Widely distributed in the Indian Ocean and Indo-West Pacific: Sri Lanka; Dongshan, Fujian, Mirs Bay, Hong Kong, South China Sea; Tanega-Shima, Japan; Majuro, Marshall Islands; Indonesia, Banda Sea (reported herein, material collected and identified by ten Hove in 1984); WA, NT, Qld, Australia.

97. *Hydroides tenhovei* Bastida-Zavala & Leon–Gonzalez, 2002

Figure 111

Type information: Cabo San Lazaro, western coast of Baja California Sur, Mexico, 24°50′N, 112°15′W, holotype and four paratypes ZMA V.Pol. 4994.

Etymology: Named after Harry ten Hove, who first recognised this to be an undescribed species.

Selected synonymy

Hydroides tenhovei Bastida-Zavala & De León-Gonzâlez, 2002: 389–394, figs 2a–m, 3a–h (description [original], discussion; Cabo San Lazaro and Punta San Juanico, Baja California, western Mexico); Bastida-Zavala & ten Hove 2003b: 104, fig. 12O, 24 (diagnosis, distribution; western coast of Baja California Sur, Mexico); Sun 2017: 157, 182, fig. 5D (diagnosis; Baja California).

Diagnosis: Peduncle with well-defined chitinised constriction. Funnel with 33 (20–42) chitinised, pointed radii. Verticil bilaterally symmetrical, sitting in funnel, with 12 (9–14) spines, including 3 dorsal incurving hooks. These shaped like a pickaxe with a globular head, and have partly fused bases; the middle one larger than the other 2. Remaining spines sharply pointed, directed outwards, with rounded basal spinules (absent in dorsal hooks) above small central disk.

Remarks: *Hydroides tenhovei* is quite unique in its opercular verticil, with three dorsal incurving pickaxe-like hooks accompanied by a number of outcurving other spines. It is somewhat similar to *H. salazarvallejoi*, with one pickaxe-like hook, fused at half-length with two larger spines, but with only up to 7 other verticil spines, straight to outcurving but blunt instead of pointed. Other taxa with 1–3 dorsal hooks all have incurving verticil spines in addition. Differences with some other superficially similar taxa such as *H. brachyacantha*, *H. glandifera*, *H. malleospina* and *H. similis* have been given in the original description.

Distribution: Western coast of Baja California, Mexico.

Fig. 111. *Hydroides tenhovei*, various views of the same operculum, Baja California Sur, Mexico, AM W.43189 (photo E. Wong).

98. *Hydroides trilobula* Chen & Wu, 1978

Figure 112

Type information: Xisha Islands, Hainan, China, holotype MBMCAS.

Etymology: not given by Chen & Wu (1978: 145), but evidently refers to the three vesicular verticil spines [three lobes], small ones [the diminutive Latin suffix -*ulus*].

Selected synonymy

Hydroides trilobulus Chen & Wu, 1978: 143–145, pl. 2 figs 1–9 (diagnosis, in Chinese [original], shorter in the English abstract; Xisha Islands, China); Sun 2017: 22, 155, 182, fig. 3M (diagnosis; Xisha Islands).

Diagnosis: Peduncle with well-defined, chitinised constriction. Funnel with 26 chitinised, triangular, pointed radii. Verticil sitting in funnel, consisting of 3 valvular spines with rounded tips, curving inwards, without lateral, internal, external or basal spinules. Spines unequal in size, the dorsal being the largest, all thicker at base and almost triangular in cross-section, apparently meeting at base.

Remarks: *Hydroides trilobula* having an opercular verticil with only three valvular spines differs from all other taxa in the *H. tuberculata*-complex **sensu lato** in the shape of the operculum. For a more general comparison see remarks *H. tuberculata*-complex **sensu lato**.

Distribution: South China Sea.

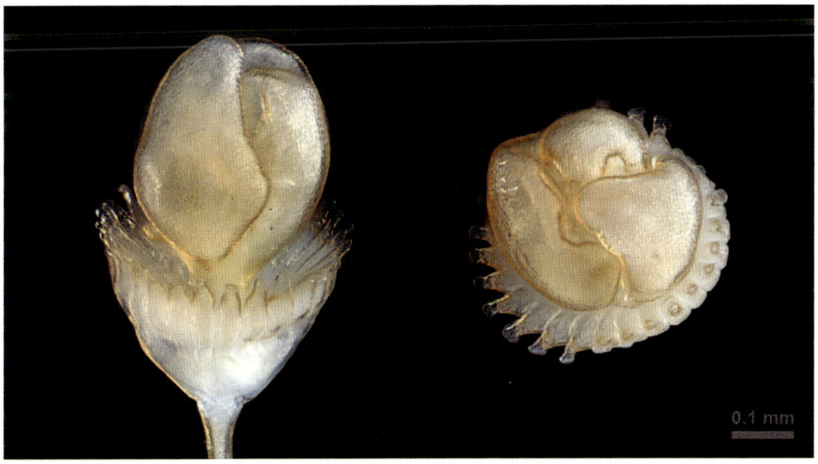

Fig. 112. *Hydroides trilobula*, various views of the same operculum, holotype MBMCAS, Xisha Island, Hainan, China (photo X. Wu).

99. *Hydroides trivesiculosa* Straughan, 1967b

Figure 113

Type information: Heron Island, Qld, Australia, holotype AM W.4111.

Etymology: Not given by Straughan (1967b: 33–34), but evidently refers to the three lobes of the enlarged dorsal verticil spine (Latin *tri* = three (in combinations), *vesicula* = vesicle).

Selected synonymy

Hydroides trivesiculosus Straughan, 1967b: 33–34, fig. 3h–j (diagnosis [original]; Heron Island; material studied]; ten Hove 1990: 119 (Discussion); 1994: 108 (Seychelles); Fiege & Sun 1999: 113–114, table 1 (differential diagnosis); Sun *et al.* 2012b: 540–545 (DNA data, phylogeny; Qld, Australia); Tovar-Hernández *et al.* 2016: table 2, fig. 6 (name, DNA, phylogeny; Qld).

Hydroides trivesiculosa – Kupriyanova *et al.* 2015: 293–294, fig. 10 (diagnosis; Lizard Island, Qld); Sun *et al.* 2015: 85–87, fig. 28, map 25 (diagnosis, synonymy, discussion; WA, NT, Qld, Australia); 2018: 733, 738, 742, figs 1, 3, 5, 6 (DNA, phylogeny; Ashmore, Mermaid and Imperieuse Reefs, WA; Hicks Reef, Lizard Island and Broomfield Reef, Qld, Australia); Sun 2017: 13, 16, 19, 27, 122, 125, 132, 157, 183, 281, 295, figs 28, 2I, 5A, map 25 (diagnosis, DNA; Big Vicki's reef, Lizard Island, Broomfield Reef, Hicks Reef, all Qld; Ashmore Reef, Imperieuse Reef, Mermaid Reef, WA, Australia).

?*Hydroides exaltatus* var. *vesiculosus* Fauvel, 1919 (*nomen inquirendum*, **fide** Read *et al.* 2017.

Diagnosis: Peduncle with lightly chitinised constriction. Funnel with 21 (18–25) truncated to T- or anchor-shaped chitinised radii. Verticil sitting in funnel, bilateral symmetrical, with 6 (4–7 spines), curving outwards from base of large dorsal vesicular spine and ending in rectangular to T-shaped tips; no accessory spinules nor central disk. Very large dorsal spine consisting of a bulbous median part and two latero-dorsal extensions.

Remarks: *Hydroides trivesiculosa* can usually be distinguished from *H. albiceps* by the exceptionally large size of the dorsal bulbous verticil spine, and the lower number of verticil spines, but there are (few) transitional opercular morphologies. A so called *H. 'prisca'*-type operculum, developmental stage, has been recorded (ten Hove & Ben-Eliahu 2005: 18). A dirty, uncleaned operculum of *H. trivesiculosa* might be mistaken for *H. steinitzi*, but this taxon only has a verticil bulb, no verticil spines. Phylogenetically *H. trivesiculosa* was nested within the two main clades of *H. albiceps* (see Sun *et al.* 2018: 738, 742, figs 2, 3, Fig. 13 here), though with low nodal support. For a full discussion and the similarity with forms as *Hydroides multispinosa* var. *ternatensis*, *Hydroides exaltata* var. *vesiculosa* see Remarks under *H. albiceps*. Like *H. albiceps*, the large geographical distribution of nominal *H. trivesiculosa* might suggest that it is a complex of species.

Distribution: Tropical Australia (from Qld to WA), Indonesia, Red Sea, Tanzania, Seychelles.

Fig. 113. *Hydroides trivesiculosa*, various views of opercula, Chevalier Bay, Seychelles (left two) AM W.28450; Heron Island, Queensland, Australia (right two) AM W.4086 (photo E. Wong).

100. *Hydroides trompi* Bastida-Zavala & ten Hove, 2003b

Figure 114

Type information: Miraflores Locks, Western Panama, holotype USNM 58605, paratypes USNM 1013587 and ZMA V.Pol. 3358.

Etymology: named after Mr Jossy S. Tromp, a student supervised by H.A. ten Hove.

Selected synonymy

Hydroides trompi Bastida-Zavala & ten Hove, 2003b: 104–108, figs 22A–Q, 24 (description [original], synonymy, discussion; Baja California, Pacific Mexico); Bastida-Zavala 2008: 30, fig. 6U (name; Oaxaca, Mexico); Bastida-Zavala *et al.* 2016: 427–428, figs 6, 12G (name, synonymy, distribution; Puerto Peñasco, Gulf of California and Guerrero, Mexico); Sun 2017: 153, 183, 288, fig. 1S (diagnosis, DNA, phylogeny; Sinaloa, Oaxaca, Mexico); Sun *et al.* 2018: 738, figs 2, 3, 5, 6 (DNA, phylogeny; Sinaloa and Oaxaca, Mexico).

?*Eupomatus recurvispina* **sensu** Kudenov 1975: 228 (name; Puerto Peñasco, Mexico, Gulf of California) [**non** Rioja, 1941a]. Reidentified by Bastida-Zavala & ten Hove (2003b: 105).

Diagnosis: Peduncle with well-defined constriction. Funnel with 18 (16–22) acutely pointed, rarely narrowly truncated radii, lightly chitinised. Verticil on short basis in funnel, radially symmetrical, with 8 (5–10) spines, incurving, with sharp tips twisted clockwise. All spines similar, simple, smooth, without lateral spinules, but with a knob- to wart-like basal spinule. Small central disk without tooth.

Remarks: The overall appearance of the operculum is the same as that of *Hydroides gairacensis*, however, the funnel radii of this species are T or crescent shaped. The main difference between *H. trompi* and *H. recurvispina* is the presence of a pronounced external knob on the bend of the verticil spines in the latter, absent in the former.

Distribution: Baja California to Panama.

Fig. 114. *Hydroides trompi*, various views of opercula, Miraflores Locks, Panama (left two) ZMA V.Pol.3358; Gulf of California, Mexico (right two) ZMA V.Pol. 3696 (photo E. Wong).

101. *Hydroides tuberculata* Imajima, 1976a-complex sensu stricto

Figure 115

Type information: Urata, Tanega-shima, Japan, holotype NSMT–Pol. H120, 13 paratypes NSMT–Pol. P121.

Etymology: Not given by Imajima (1976a: 132–133), but refers to external swellings (Latin = *tuberculum*) at the shoulder of the verticil spines.

Selected synonymy

Hydroides tuberculata Imajima, 1976a: 132–133, fig. 7a–j (description [original]; south-west Japan); 1987: 16 (distribution in Japan); Imajima & ten Hove 1984: 44–45 (synonymy, extensive discussion; Truk, Ponape Island; Okinawa, experimental fouling; Lizard Island, Qld, experimental fouling, Noosa Heads, Heron Island, Qld, Australia); Bailey-Brock 1985: 210–211, fig. 12a–c (diagnosis; Fiji); Sun *et al.* 2015: 88–91, fig. 29, map 26 (diagnosis, remarks; WA, NT, Qld, Australia); 2016: 42, 46–47, fig. 2, tab. 1 (name, DNA, phylogeny; Lizard Island, Australia); 2018: 738, 739, 741–742, figs 1–3, 5, 6 (DNA, phylogeny; Hibernia and Mermaid Reefs, WA; Broomfield Reef, Dunk Island, Linnet Reef, Lizard Island, Qld, Australia; Gizo, Solomon Islands; Okinawa, Japan).

Hydroides perezi **sensu** Straughan 1967a: 219–220, fig. 6o (diagnosis; Havannah Island, Qld, Australia) [**non** Fauvel, 1918, reidentified by Imajima & ten Hove (1984: 44) and Sun (2017: 126–127)]; **sensu** Bailey-Brock & Magalhães 2013: 4–6, fig. 4 (description, discussion; Guam, Mariana Islands; reidentified herein to *H. tuberculata*-complex [**non** Fauvel, 1918], see remarks under *H. perezi.*

Hydroides brachyacantha **sensu** Dew 1959: 28, fig. 7A–G (diagnosis; Qld, Australia) [**non** Rioja 1941a, reidentified in part, the material from Queensland only, by Imajima & ten Hove (1984: 44, 50)].

Hydroides tuberculatus – ten Hove 1994: 108 (bioperculate chimaera; Seychelles); Ishaq & Mustaquim 1996: 170–172, fig. 6A–H (diagnosis; Pakistan); Fiege & Sun 1999: 121–123, fig. 11A–C (diagnosis, remarks; Hainan, South China Sea); Kupriyanova *et al.* 2008: 428, 430–431, fig. 1B (name; DNA; Lizard Island, Qld, Australia); Bailey-Brock *et al.* 2012: 969, 973–974, fig. 4B, C (description, discussion; Rongelap, Utirik, Marshall Islands); Sun *et al.* 2012a (discussion); Sun & Yang 2014: 257–259, figs 143A–K, 138 (diagnosis in Chinese; Gulei, Dongshan, Fujian; Weizhou Island, Guangxi; Xincun, Longya Bay, Hainan, China); Tovar-Hernández *et al.* 2016: table 2, fig. 6 (name, DNA, phylogeny; Qld, Australia).

Diagnosis: Peduncle with well-defined constriction. Funnel with 21 (18–25) chitinised, sharply tipped radii. Verticil sitting or on short basis in funnel, bilaterally

Fig. 115. *Hydroides tuberculata*-complex, various views of opercula, Top row: Mackay, Queensland, Australia, AM W.3971; bottom row: North Platte Island, Seychelles, AM W.28443 (photo E. Wong).

symmetrical, including dorsal hook with 5(–6) triangular, incurving spines, with a small external knob on the bend. Dorsal hook larger than others, with long, sharp, incurving tip. Spines all thick, almost triangular in cross-section. No lateral, internal, external or basal spinules. Bases of verticil spines almost meeting in centre, only leaving a small central disk, no central tooth. Tube often (but not always) bluish. Bayonet of collar chaetae with 2 teeth and some accessory ones.

Remarks: The above synonymy must be regarded as tentative, and certainly contains material that does not belong to *Hydroides tuberculata* in the strict sense. Many Indo-West Pacific nominal species of the genus *Hydroides* are characterised by a low number of rather valvular verticil spines: *H. adamaformis*, *H. dafnii*, *H. kimberleyensis*, *H. perezi*, *H. rhombobula*, *H. tuberculata*, *H. uniformis* and *H. xishaensis*. The first, *H. adamaformis* differs by its funnel radii with crescent-shaped tips as opposed to pointed or triangular ones in all the other mentioned taxa; genetically too it is widely different (e.g. Sun *et al.* 2018, figs 2, 6, Fig. 13). The remaining series of names may be collectively indicated as *Hydroides tuberculata*-complex **sensu lato**.

Judging by the original description, *Hydroides tuberculata* can be distinguished from *H. kimberleyensis*, *H. rhombobula*, *H. uniformis* and *H. xishaensis* by the presence of an enlarged dorsal verticil spine with external knob, while all verticil spines are equally sized and apparently smooth in the other nominal taxa. However, the two specimens of *H. kimberleyensis* differ between themselves, holotype without dorsal hook, paratype with sharply incurving hook; moreover, a high magnification study of the holotype of *H. dafnii* in Paris showed a sharp incurving tip in the dorsal hook, not mentioned in the original description (Ben-Eliahu, unpublished sketches). All these nominal species differ from *H. perezi* by the cross-section of the verticil spines, subtriangular in the former [as well as in *H. dafnii*], but flat in *H. perezi* (ten Hove 1970; Imajima & ten Hove 1984).

Most members of the *Hydroides tuberculata*-complex **sensu lato** have collar chaetae with two teeth only; however, these chaetae in the nominal species *H. tuberculata* and *H. xishaensis* have accessory teeth as well, and 1 or 2 teeth are figured for *H. kimberleyensis*, possibly with an accessory tooth.

The morphological differences as used here to differentiate between various taxa may not be as clear-cut as presented, ten Hove (1994: 108) mentions a specimen from the Seychelles with two opercula, one of the *tuberculata*-type, the other *uniformis*-type. A bluish colour of the tube has been mentioned for *H. perezi*, *H. rhombobula*, *H. tuberculata* and *H. uniformis*, but not in all tubes. In the past, material even has been mistakenly identified as *H. brachyacantha* (see selected synonymy, above), but can be distinguished from this taxon by its valvular verticil spines without subapical hook (Imajima & ten Hove 1984: 50).

In their phylogenetic analyses, Sun *et al.* 2018 (figs 2, 3. 5–7, Fig. 13 here) recovered the nominal taxon *Hydroides tuberculata* together with *H. kimberleyensis* and *H. uniformis* in a single, well supported clade (e.g. Fig. 2B, lower lineage). This clade can be regarded to form the *H. tuberculata*-complex **sensu stricto**. However, the morphospecies *H. tuberculata* is paraphyletic: depending the genes used in the analyses, this name covers up to seven molecular lineages of specific level, even excluding the nominal taxa *H. kimberleyensis* and *H. uniformis*. The wide distribution of the nominal species, from the Solomon Islands to the Seychelles, is anyhow suspect in view of recent analyses of serpulid taxa with wide distributions (e.g. Sun *et al.* (2016) on *H. brachyacantha*, Sun *et al.* (2017b) on *H. operculata*, Simon *et al.* (2019) and Pazoki *et al.* (2020) on *Spirobranchus kraussii*, Palero *et al.* (2020) on *Spirobranchus tetraceros*) which dissolved in complexes of species when anthropogenic translocations are not involved.

Distribution: Indo-West Pacific (Micronesia, Melanesia, Southern Japan; NT, Qld, WA, Australia); even mentioned from the Seychelles, Indian Ocean.

102. *Hydroides uniformis* Imajima & ten Hove, 1986

Figure 116

Type information: Keaso, Solomon Islands, holotype NSMT–Pol. H195, two paratypes NSMT–Pol, P196.

Etymology: not given by Imajima & ten Hove (1986: 5–8), but evidently refers to the uniform verticil spines.

Selected synonymy

Hydroides uniformis Imajima & ten Hove, 1986: 5–8, fig. 1a–o (description [original], discussion; Solomon Islands); Sun 2017: 68–70, 130, 152, 183–184, 295–296, 319, figs 1D, 2, 3 (name; phylogeny, discussion); Sun *et al.* 2018: 742, figs 2, 3, 5, 6 (DNA, phylogeny; Hideaway Island, Vanuatu; Gizo, Solomon Islands).

Diagnosis: Peduncle with well-defined constriction. Funnel with 19–29 chitinised, slender and sharply tipped radii. Verticil sitting in funnel, with 7–8 uniform spines, straight, equal-sized with rounded tips, each with basal spinule and subtriangular in cross-section. Bases of verticil spines almost meeting in centre, only leaving a small central disk, no central tooth. Interior of tube bluish. Bayonet of collar chaetae with 2 teeth, no accessories.

Remarks: *Hydroides uniformis* is part of the *Hydroides tuberculata*-complex **sensu lato**. In the phylogenetic analyses by Sun *et al.* 2018 (figs 2, 3, 5–7, 13) it was recovered together with *H. tuberculata* and *H. kimberleyensis* in a single, well supported clade (e.g. fig. 2B, lower lineage) which might be called the *Hydroides tuberculata*-complex **sensu stricto**. Apparently the nominal taxon is very similar to *H. kimberleyensis*, in 2001 one of us (HAtH) identified the material of *H. kimberleyensis* described in Sun *et al.* (2015: 44–46) as *H. uniformis*. Material from Indonesia (ZMA V.Pol. 4632, det. ten Hove in 1985) probably is a yet undescribed species (for a full discussion we refer to Sun *et al.* 2015, and to the remarks under *H. tuberculata*, above).

With 7–8 verticil spines, *Hydroides uniformis* can easily be distinguished from *H. xishaensis* and *H. rhombobula*, with 5(4?) spines, moreover more pointed than in *H. uniformis*. ten Hove (1994: 108) mentions a bioperculate chimaera from the Seychelles, one operculum of the *tuberculata*-type, the other *uniformis*-type; however, in view the multiple possible species as evidenced by DNA in material identified previously as *H. tuberculata* from the Indo-West Pacific (see above), it seems unlikely that the distribution of *H. tuberculata*-complex **sensu stricto** would extend to the Indian Ocean.

Distribution: Vanuatu, Solomon Islands.

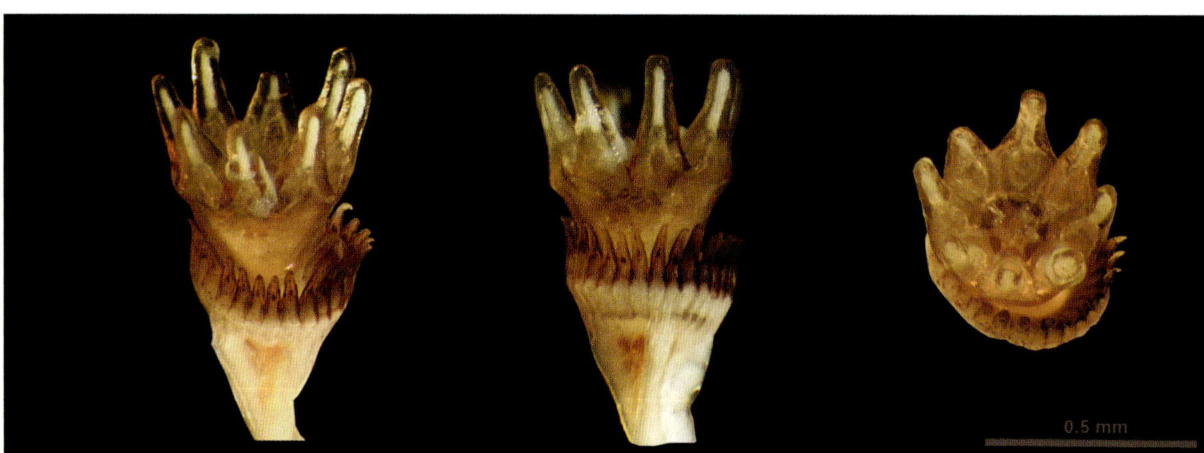

Fig. 116. *Hydroides uniformis*, various views of the same operculum, Vanuatu, AM W.47092 (photo L. Gunton).

103. *Hydroides vizagensis* Lakshmana Rao, 1969

Figure 117C

Type information: Naval Base, Visakhapatnam Harbour, types in Andhra University, India, apparently lost.

Etymology: Not given by Lakshmana Rao, 1969, but evidently named after its type locality.

Selected synonymy

Hydroides vizagensis Lakshmana Rao, 1969: 9, pl. 6a–h (diagnosis [original]; Visakhapatnam Harbour, India, Gulf of Bengal); Sun 2017: 184, 309 (diagnosis, type missing; same).

Diagnosis: Peduncle with ill-defined constriction. Funnel with 25 chitinised, pointed radii. Verticil bilaterally symmetrical, sitting on funnel, with 9–10 slender spines, all pointing in ventral direction and with wings [with broad proximal part] extending for more than half of spine length. Two dorsal spines curving inward with external knee [blunt external spinule] and curving inwards, the others with pointed spinule at 2/3 of their length and pointing ventrally and thus outwards spinules. Central disk not mentioned [nor visible in original drawing], no central tooth.

Remarks: This diagnosis is based on original description and original drawing, which is somewhat in contradiction with original diagnosis. For the time being we have treated this taxon as full species, though we mentioned it as questionable synonym of *Hydroides sanctaecrucis*. The species is only known from the original record in India, its author says: '*H. vizagensis* differs from other members of this group in possessing a single outwardly directed process on the outer surface of all the spines'. Thus, the nominal species *H. vizagensis* is very similar to *H. sanctaecrucis*, and numerical details of funnel radii as well as verticil spines fall within the ranges in *H. sanctaecrucis*. It most probably is a synonym of the latter, but this hypothesis should be corroborated by DNA analysis. Material of *H. sanctaecrucis* (accompanied by DNA sequence data) in the meantime has been reported from the west coast of the subcontinent India (Sun *et al.* 2018, supplementary table S1).

Distribution: East coast of India.

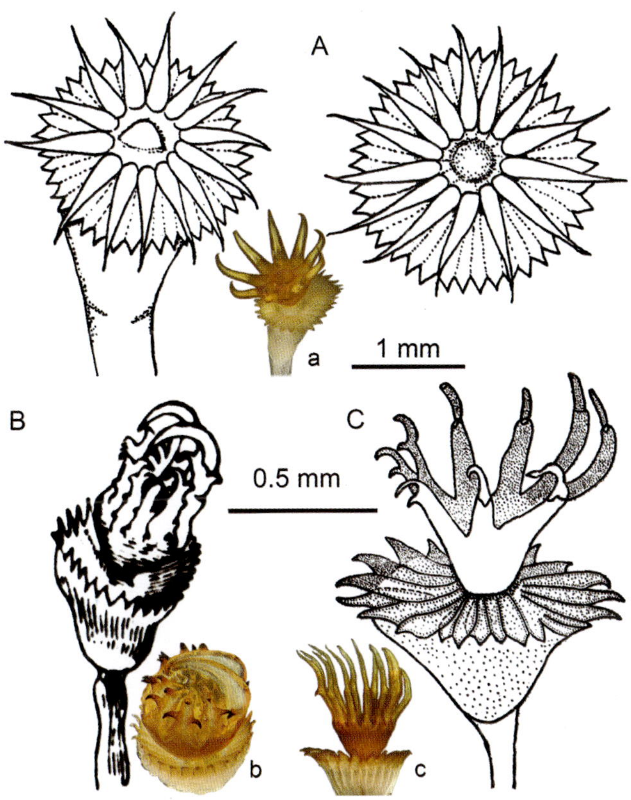

Fig. 117. Original drawings of *Hydroides* species for which material was not available and photos of similar species for comparison. A – *H.huanghaiensis* from Sun & Yang (2000), a – photo of *H. gracilis*; B – *H. ralumiana* from Augener (1927), b – photo of *H. externispina*; C – *Hydroides vizagensis* from Lakshmana Rao (1969), c – photo of *H. sanctaecrucis*.

104. *Hydroides xishaensis* Chen & Wu, 1978

Figure 118

Type information: Xisha Islands, China, holotype MBMCAS.

Etymology: named after its type locality.

Selected synonymy

Hydroides xishaensis Chen & Wu, 1978: 140–141, 145, pl. 1 figs 1–9 (diagnosis, in Chinese [original], shorter in the English abstract; Xisha Islands, China); Sun & Yang 2014: 260–261, fig. 145A–H; copy of Chen & Wu); Sun 2017: 71, 130, 152, 184 fig. 1G (diagnosis; Xisha Islands).

Diagnosis: Peduncle with ill-defined constriction. Funnel with 26–30 rather triangular, bluntly rounded radii. Verticil on short basis in funnel, with 5 smoothly tapering and slightly incurving spines, without lateral, internal, external or basal spinules. Spines equal in size, thick, almost triangular in cross-section, apparently meeting at base. Bayonet collar chaetae with 2 teeth and up to 3 accessories.

Remarks: *Hydroides xishaensis* resembles *H. rhombobula* in the shape of the operculum. However, the latter has rhomboid valves in its verticil, straight or slightly outcurving and bayonet chaetae with 2 teeth only, while *H. xishaensis* has triangular incurving valves, and bayonet chaetae with 2 teeth and up to 3 accessories. For a more general comparison see Remarks for *H. tuberculata*-complex **sensu lato**.

Distribution: Hainan, China.

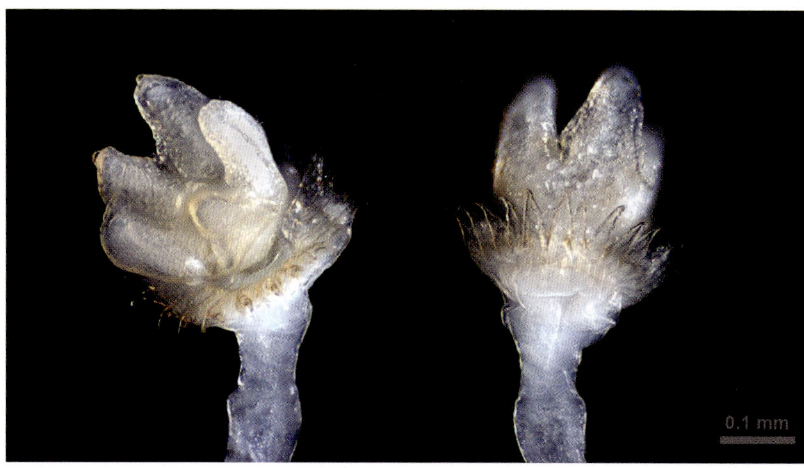

Fig. 118. *Hydroides xishaensis*, various views of the same operculum, Xisha Island, China, holotype MBMCAS (photo X. Wu).

105. *Hydroides zibrowii* nov. spec. ten Hove & Bastida-Zavala

Figure 119

Type information: Santa Marta, near the Institute for Marine and Coastal Research, Colombia (Atlantic side), from experimental (Eternit) fouling panels, legit, don. C.B. García (original identification by CBG cf. *Hydroides gairacensis*, around 1990 reidentified by HAtH *H.* cf. *brachyacanthus*). Holotype ZMA V.Pol. 3742.1, and four paratypes ZMA V.Pol. 3742.2–5.

Etymology: Named after Dr. Helmut Zibrowius, who described several new serpulid species, including 10 (and one new subspecies) in the genus *Hydroides*. He was the first to record this taxon from Brazil, as '*H. brachyacantha*', while mentioning that his specimens were different in some details from Rioja's description.

Synonymy

Hydroides cf. *brachyacanthus* – Bastida-Zavala & ten Hove 2003a: 154–157, figs 29D–K, 31A–D, 33 (extensive description, probably belonging to a species complex; Curaçao, Netherlands Antilles, and Venezuela); questionably fig. 29A–C, from Grenada, see remarks; Bastida-Zavala *et al.* 2017: 24–25, figs 4b, 5 (diagnosis, discussion; Biscayne Bay, Pensacola Bay, Florida, USA).

Hydroides brachyacantha **sensu** Zibrowius 1971b: 6, pl. 1 fig. 4a, b (short description, discussion; São Sebastião, Ubatuba, Brazil; material studied) [**non** Rioja, 1941a]; however, not pl. 1, fig. 3 which is redrawn from Rioja; Rullier & Amoureux 1979: 193–194 (diagnosis; off Itabapoana, 21°38′S, 40°45′W, Brazil); Knight-Jones & Knight-Jones 1991: 581 (name; Guanabara Bay, Brazil); Díaz-Díaz & Liñero-Arana 2001: 11–12, fig. 2g–m (diagnosis; Cariaco Gulf, Venezuela; on PVC dock pilings).

Material additional to that mentioned in Bastida-Zavala & ten Hove (2003a; 154).

Colombia, Santa Marta (1 small specimen, not included in type series since either a juvenile of *priscus* type or its dorsal hook is broken off and missing).

Surinam, shelf, 06°21.8′N, 55°00.3′W, 31 m, 26 April 1969. Agassiz trawl, HNLMS *Luymes*, O.C.P.S. stn I 93A; same, 06° 30.3′N, 54° 59.7′W, 36–37.5 m, 25 April 1969. Soft, sandy mud with inclusions of hard clay, small pebbles and Bryozoa, stn I 120; same, 06°28.3′N, 55°00.2′W, 32.5–33 m, 25 April 1969. Coral reef and rock fragments, stn I 121 (ZMA V.Pol. 5005–5007, 6 specimens and 2 anterior fragments, 1 with 2 opercula).

Brazil, São Paulo, Caraguatatuba Bay, Martim de Sa and Barra do Una, legit J.M. de Matos Nogueira and students (8 operculate samples, ZMA V.Pol. 5184–V.Pol. 5191).

Diagnosis: Peduncle with ill-defined constriction. Funnel radii with 24 (17–30) hardly chitinised pointed radii. Verticil sitting in funnel, bilaterally symmetrical, with (7–)8 incurving spines with pointed tips. Dorsal hook almost smoothly curved, may be accompanied by a pair of smaller hooks, larger than the other spines which may be geniculate and decreasing in size ventrally. All verticil spines with short basal spinule over small central disk. Central tooth absent.

Description

Tube: White, about 0.8 mm wide (up to 1.2, in holotype (h) 1.1 mm, n = 7), with lumen of about 0.6 mm (up to

Fig. 119. *Hydroides zibrowii* nov. spec., various views of operculum, Santa Marta, Colombia (Atlantic Ocean side) ZMA V.Pol. 3742.1 (photo H.A. ten Hove).

Table 4. Main differences between *Hydroides zibrowii* nov. spec., *H. similoides* and *H. deleoni*

Character	*H. zibrowii* nov. spec.	*H. similoides*	*H. deleoni*
Longitudinal keels of tube	0 or 2	(2 or) 3	0 or 2
Number of radii	24 (19–30)	17 (12–22)	36 (30–43)
Tips of radii	pointed	blunt	pointed
Number of verticil spines	7–8	8–13	7–11
Large dorsal verticil spines	1, others ventrally decreasing in size	2–3, others ventrally decreasing in size	1, others almost the same in size
Peduncle length (mm)	1.4–3.2	1.3–3.6	2–4
Distribution	Lesser Antilles to Souhern Brazil	Caribbean to Brazil	Baja California Sur to Ecuador

0.9 mm (h)); some tubes missing, the rest fragmentary and/or covered by epibionts, six anterior ends lack peristomes; six tubes with transversal ridges (Bastida-Zavala & ten Hove 2003a fig. 29A); holotype and one other showing two longitudinal ridges, and three none.

Radiolar crown: With 8 (7–10, h 6) radioles (n = 8). Terminal filament long in four specimens (50%), short in other four (50%).

Peduncle: Insertion left (n = 3) or right (n = 6); smooth, with ill-defined constriction (Bastida-Zavala & ten Hove 2003a fig. 29B, D). Length including operculum 2.2 (1.4–3.2) mm (n = 8). Pseudoperculum present in six specimens and lacking in one.

Operculum: Verticil sitting in funnel, bilaterally symmetrical; seven to eight spines (h 7, n = 17), strongly curving inwards (fig. 29B, D–F), sometimes geniculate; colour yellow to brown. One dorsal verticil spine larger than the others, the rest decreasing gradually in size (Bastida-Zavala & ten Hove 2003a, fig. 29B, D–F) although the pair next to the hook may be larger than the others. Tip of spines pointed. Spines with one basal internal spinule (Bastida-Zavala & ten Hove 2003a, fig. 29D, F); without external and lateral spinules and/or wings. Verticil without central tooth (Bastida-Zavala & ten Hove 2003a, fig. 29F). Funnel with 24 ((17)19–30) radii (h 28, n = 13) with pointed tip (Bastida-Zavala & ten Hove 2003a, fig. 29B–F). Interradial grooves 1/3 of funnel length in three opercula, 2/5 in four and 1/2 in one. Length and width of operculum up to 0.7 mm (n = 11).

Collar and thoracic membranes: Well developed, the latter forming apron ventrally across anterior abdominal chaetigers.

Thorax: With collar chaetiger and 6 uncinigerous chaetigers. Collar chaetae bayonet chaetae with two blunt, short teeth, distal blade smooth (Bastida-Zavala & ten Hove 2003a, fig. 29G); capillary chaetae present (Bastida-Zavala & ten Hove 2003a, fig. 29H). Subsequent chaetigers with limbate chaetae (Bastida-Zavala & ten Hove 2003a, fig. 29I) of two sizes, and saw-shaped uncini (Bastida-Zavala & ten Hove 2003a, fig. 29J, K)

Abdomen: With up to 69 chaetigers (h ~40, 2 or 3 posterior chaetigers missing). Anterior and middle abdominal chaetigers with flat trumpet-shaped chaetae. Posterior chaetigers with 'capillary' chaetae. Anterior uncini saw-shaped, posterior uncini rasp-shaped.

Size: Length up to 13 (h 8) mm, only three complete specimens). Width of thorax 0.6 mm (0.5–0.8, n = 7). Radioles (including operculum) accounting for 1/6 of entire length.

Colour: Body yellow.

Remarks: The description above is based largely on that in Bastida-Zavala & ten Hove (2003a: 154–157), though without the three specimens from Grenada, which showed a considerably larger spine than the others and was regarded to be a possible second taxon by Bastida-Zavala *et al.* (2017: 24–25). Some relevant measurements of the holotype (h) have been added. *Hydroides zibrowii* nov. spec. is yet another species previously identified as *Hydroides brachyacantha* **sensu auctores**. This large species complex is including an unknown number of species that are superficially similar in having a verticil with a large hook covering other verticil spines, see Remarks under *H. brachyacantha*. Rioja (1941a) described his species with one large dorsal verticil spine, almost a hook, and up to 10 additional spines clearly geniculate, with a well-developed knob protruding above subapical incurving tip. *Hydroides zibrowii* nov. spec. does not show such well-developed knobs, though numbers of verticil spines (7–8 *v* 7–11) and radii (19–30 *v* 14–39) overlap widely with *H. brachyacantha* **sensu stricto**. A comparison of

H. zibrowii nov. spec. with two other American members of the *brachyacantha*-complex is given in the Table 4.
Paratype 3, a specimen of only 2.5 mm long, shows two opercula, a primary '*Serpula*'-type, and a small secondary operculum typical for *H. zibrowii* nov. spec. A *Hydroides* '*prisca*'-type operculum, developmental stage, has been recorded for this taxon (ten Hove & Ben-Eliahu 2005: 18).
Distribution: Lesser Antilles, Venezuela and Southern Brazil.

NOMINA INQUIRENDA AND REJECTED NAMES, OCCASIONALLY STILL IN USE

1. *Hydroides affinis* (Marion, 1875a), *nomen inquirendum*

Figure 120A

Type information: no type known, no locality given though Gourret (1902: 689) gives les Goudes (Marseille) as locality for his specimen, which most likely was Marion's original material (Zibrowius 1968: 115).

Etymology: not given by Marion (1875a: 311), but Latin *affinis* = related, in Marion's view it was very similar to –what we nowadays regard to be undeterminable– representatives of the genus *Hydroides*.

Selected literature records:

Eupomatus affinis Marion, 1875a: 310–311 (diagnosis, specimen with two opercula; coralligène).

Hydroides affinis: Gourret 1902: 688, fig. 2a–g (diagnosis based on Marion's specimen; les Goudes, Marseille, France); Zibrowius 1968: 115, pl. 13, fig. 29 (reworked description of Gourret); Zibrowius 1971a: 694, 713–714 (no new data, but a long discussion concluding that it probably is an aberrant specimen of *H. helmata*).

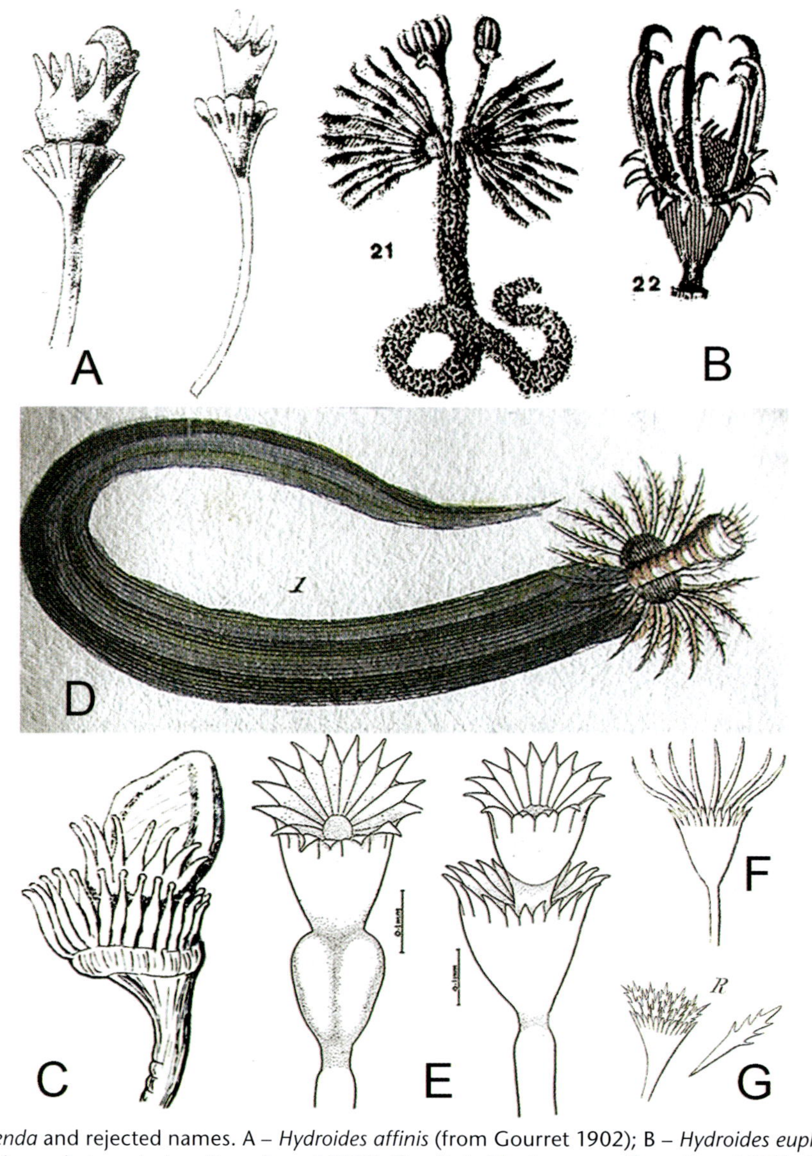

Fig. 120. *Nomina inquirenda* and rejected names. A – *Hydroides affinis* (from Gourret 1902); B – *Hydroides euplaeana* (from Delle Chiaje 1822); C – *Hydroides exaltata vesiculosa* (from Fauvel 1919); D – *Hydroides hexagona* (from Bosc 1802); E – *Hydroides prisca* (from Pillai 1971); F – *Hydroides uncinata* (from Philippi 1844); G – *Hydroides pectinata* (from Philippi 1844).

Nomenclature: *Hydroides affinis* was previously annotated as a *nomen oblitum* in WoRMS (Read & Fauchald 2020), now as *nomen inquirendum*, as discussed in Read *et al.* (2017: 7). We have carefully revised all available literature records, most just names in lists. Marion's remark that the abdominal chaetae are very distinctive by a large hook supporting its pectinate blade ('un fort crochet soutenant la lame pectinée') is, in absence of a figure, disputable. The only figures of the two opercula are those by Gourret (1902 fig. 2a, b), showing a verticil with 6 equal spines, reminiscent of the secondary opercula occurring in several species before the operculum reaches its final shape (for a full discussion of this opercular stage, see ten Hove & Ben-Eliahu 2005 and the rejected name *H. prisca*, below). The second, fully grown, operculum shows in addition 1 larger hook with 2 side spinules, explicitly mentioned by Gourret, the latter absent in *H. helmata*. The seven spines, including large hook, make the operculum indeed reminiscent of that of *H. helmata*. However, in the latter taxon, the verticil spines all are thick, blunt, rounded, not visibly narrowing towards the tips, as opposed to the reasonably thin and pointed verticil spines depicted for *H. affinis*. These differences might be explained by lack of draftsmanship, but seem real enough. Therefore, contrary to Zibrowius (1971a), we cannot synonymise both taxa, and the status of *H. affinis* remains that of a *nomen inquirendum*.

Distribution: Marseille, Mediterranean France.

2. *Hydroides diplochone* (Grube, 1878a), *taxon inquirendum*

Type information: Askold Island, outer Peter the Great Gulf, North Japan Sea, type apparently lost **fide** Zibrowius (1978: 144), but locality inferred from known sampling by Grube (Read *et al.* 2017: 14).

Etymology: Latinised derivation from the Greek διπλοος = double, with two storeys, and the feminine Greek noun χοάνη = funnel, thus double funnel, unfortunately a character applicable to 99% of all *Hydroides* opercula.

Selected literature records:

Serpula (Hydroides) diplochone Grube, 1878a: 105 (very short diagnosis [original]; northern Sea of Japan).

Hydroides diplochone – Zachs 1933: 135 (diagnosis in Russian, restudied by Uschakov [1955] 1965: 400–401 and reidentified by him as *H. ezoensis* Okuda, 1934, a younger and better known name. In this he was followed by Zibrowius (1978: 144), who restudied Zachs' specimen).

Diagnosis (free translation of Grube 1878a: 105): The opercular funnel does not carry a starlet of spines, but a complete second funnel [the verticil in present day use], both have serrated margins, but the radii of the lower funnel (about 46) are three times as long as the 31 of the upper [verticil]. Length of the entire animal about 27 mm.

Remarks:

Hydroides diplochone was originally described but not illustrated by Grube (1878a) and the name was used as valid name in the literature by Zachs (1933), Annenkova (1938) and Uschakov [1955] (1965). Grube's description is short, freely translated in 'Diagnosis', above. Although numbers and sizes of Grube's description agree with what we perceive to be *H. ezoensis* (and not with the sympatric *H. fusicola* Mörch, 1863), the situation that radii of funnel are three times the length of the verticil teeth is absolutely the reverse of what is known of *H. ezoensis* where the teeth of the lower funnel (~40–55) are three times as short as those of the upper verticil (22–34). In absence of the types of Grube (not in Musea Wroclaw or Berlin where most of his material can be found, **fide** Wiktor 1980, Hartwich 1993), we cannot be certain whether the material is the same as *H. ezoensis*, has been described correctly, or if upper and lower verticil were inadvertently swapped by Grube (as surmised by Zibrowius 1978: 144). Zachs's description of *H. diplochone* is equally short, only adding the character collar chaetae, with a single tooth instead of the usual two. Annenkova (1938) is not very useful, but the bayonet collar chaetae with one tooth indeed might refer to *H. ezoensis*: 'Bayonet collar chaetae are distinct in having only one boss, not two. Operculum double. Tube trihedral to hexagonal. This species is absent in our material.' Uschakov [1955] (1965) restudied Zachs's material and concluded it to be *H. ezoensis*. Unfortunately, there is no guarantee that Zachs had the same taxon before him as Grube. It is possible that there are two very similar morphologically species in the Peter the Great Bay, a hypothesis that can be tested in future studies. Here we follow Uschakov [1955] (1965) and Zibrowius (1978), and leave *H. diplochone* as a *taxon inquirendum* rather than synonymise it with *H. ezoensis*.

Distribution: Askold Island, North Japan Sea.

3. *Hydroides euplaeana* (Delle Chiaje, 1822), *nomen inquirendum*

Figure 120B

Type information: Caiola (= Gaiola/Euplaea) Island, Naples, Italy, Tyrrhenian Sea, Mediterranean.

Etymology: Not stated, but *Sabella euplaeana* was evidently named after Euplaea, the old Roman name for Caiola Island (nowadays Gaiola), Naples, where delle Chiaje mentions it was collected.

Selected literature records:

Sabella euplaeana Delle Chiaje, 1822, pl. 48, figs 21, 22; 1828: 219, 226–227 (short diagnoses in Italian (*eupleana*) and Latin (*euplaeana*), based on the figures already published in 1822; Gaiola, Naples, Italy); Hartman 1959: 559 (name in Catalogue only and referred to *Eupomatus euplaeanus*).

Hydroides (*Eupomatus*) *euplaeana* – Mörch 1863: 372–373, 445 (diagnosis taken from the literature, discussion).

Eupomatus euplaeanus – Hartman 1959: 574 (name in Catalogue only, but see Nomenclature, below, as well); Jones 1962: 207 ('description not sufficiently detailed to be determined').

Hydroides euplaeana – Zibrowius 1972d: 117 (name, extensive discussion of the status of the taxon: indeterminable).

Nomenclature: There has been some confusion on the date of publication. Stefano delle Chiaje published his thesis in separate volumes, starting with the plates (1822). Figures are fully named herein, this volume with plates thus should be regarded to be the publication date of the names. Philippi (1844: 195) thought the taxon to be very similar to his *Eupomatus uncinatus* (see below), and questioned the tube built of sand grains as mentioned by delle Chiaje. Mörch (1863: 372) commented (in Danish): Grube is, maybe with reason, of the opinion that this species is the same as *Serpula* (*Eupomatus*) *uncinata*. This opinion was shared by Iroso (1921: 52, as *Sabella 'euphlaema'* [*sic*]) and followed by Zibrowius (1968: 109). Hartman (1959: 559, 574) added to the confusion by differentiating between the original *S. euplaeana* and the questionable attribution of *S. euplaeana* by Wagner in McIntosh (1923: 279) to *Hydroides norvegica*. However, Wagner (1832: 654, pl. X, fig. 9) is a straight copy of delle Chiaje, in the Isis of Oken (1832), a precursor of the Zoological Record. The specific name *euplaeana* now is in disuse and representing a *species inquirenda* that has been compared to *H. pseudouncinata*, but is also superficially similar to other species found in the Mediterranean, such as *H. niger* and *H. dianthus* by Zibrowius (1971a: 693). It is not a candidate *nomen oblitum* (since used as valid by Zibrowius (1968: 109); the latter author ((1972d: 116–117) regarded both *H. euplaeana* and Philippi's *H. uncinata* to be 'indeterminable' (see separate paragraph, below). If suppression would be desirable, prevailing usage of *H. pseudouncinata* would be maintained (ICZN 1999, Recommendation 23A; see Read *et al.* 2017).

Distribution: Naples, Italy.

4. *Hydroides hexagona* (Bosc, 1802), *nomen inquirendum*

Figure 120D

Type information: No type known, type locality Charleston Harbour, Charleston, South Carolina, USA.

Etymology: Bosc described the tube of *Serpula hexagona* as 'montrant la moitié d'un prisme hexagone …' ('showing half of hexagonal prism'), and the name is a New Latin adjectival form for six-sided, modified from Greek. Bosc's figure shows two ridges so the tube cross-section would be trapezoidal, not literally hexagonal as named, but half (la moitié) of that.

Selected literature records:

Serpula hexagona Bosc, 1802: 176, pl. VII, fig. 1 (diagnosis; Charleston Harbour, South Carolina, USA); regarded to be indeterminable by Hartman (1959: 595), and to be a *species inquirenda* by Read *et al.* (2017: 22–23)

Not *Hydroides hexagona/us* **sensu** Pratt 1916: 302 (colour note; Florida to Cape Cod, Massachusetts, USA, very common) [**non** (Bosc, 1802)]; Grave & Oliphant 1930: 234–236 (as *H. hexagonis* [*sic*], longevity of eggs and sperm; Woods Hole, Massachusets, USA); Costello *et al.* 1957: 77–80 (name, larval culture methods; Woods Hole, Massachusetts, USA); see below.

Nomenclature: The original description and figure are rudimentary, clearly refer to the genus *Hydroides*, but the species Bosc saw will remain indeterminable unless original specimens are found (unlikely). The name has been disused by taxonomists, in their opinion representing a *species inquirenda*. Nevertheless, there are multiple modern citations in manuals of invertebrates (e.g. Pratt 1916: 302) and of research on *Hydroides* sperm (e.g. Grave & Oliphant 1930) in which the name appeared; it thus cannot be a *nomen oblitum*, nor can the original name simply be synonymised with the junior name *H. dianthus* (as suggested by Zibrowius 1971a: 697; Bastida-Zavala & ten Hove 2003a: 108). Later uses of the name *H. hexagona/us* should be regarded as misidentifications of *H. dianthus* (**fide** Read *et al.* 2017: 22–23).

Distribution: South Carolina, USA.

5. *Hydroides pectinata* (Philippi, 1844), *nomen inquirendum*

Figure 120G

Type information: no type known. Type locality unspecified Mediterranean, but can be narrowed to the Tyrrhenian Sea coast of Italy as Philippi's activities were in western Italy, and plausibly to Naples, where he was based there prior to 1844.

Etymology: Philippi's brief Latin description of *E. pectinatus* describes the opercular spines as '*utrinque pectinatis*' (pectinate both sides) with three sharp teeth. The Latin adjective *pectinatus* indicates a comb-like structure.

Selected literature records:

See discussion, below.

The name *Eupomatus pectinatus* Philippi, 1844 (p. 195, pl. 6, fig. Q) has rarely been used in the literature (e.g. Marion 1875a, b; Grube 1862; Saint-Joseph 1895; Stossich 1882). As *Hydroides pectinata* or *-us* it was mentioned in about 20 papers between 1863 and 1965, all from the Mediterranean coasts of France and Italy. Incidentally it was attributed to the genus *Serpula* (by e.g. Grube 1872) or *Vermilia* (by Quatrefages 1866). It was considered to be a synonym of *Hydroides norvegica* e.g. Fauvel (1927: 356) and fell into disuse. Zibrowius (1971a: 717–725) separated *Hydroides elegans* from *H. norvegica* **sensu auctores**, and was able to attribute many of the 25 records under the specific name *pectinata*, *-us* (above) to either of these two taxa. However, for the original record he explicitly (Zibrowius 1971a: 720) stated, freely translated from the French: 'the attribution of *Eupomatus pectinatus* Philippi, 1844 either to *H. norvegica* or to *H. elegans* remains uncertain; bayonet chaetae of *E. pectinatus* are unknown and its provenance is imperfectly known'. We agree with the conclusion by Read *et al.* (2017: 31–32): name now disused and representing a *species inquirenda*. It is not eligible as a candidate *nomen oblitum* (used in taxonomy by e.g. Iroso 1921: 49, Naples), but is indeterminable unless original specimens are found.

Distribution: Mediterranean.

6. *Hydroides prisca* Pillai, 1971, rejected name

Figure 120E

Type information: Holotype BMNH 1968.143, Paratype BMNH 1968.144, Panadura River, Sri Lanka.

Etymology: Not given by Pillai, but the Latin adjective *priscus, -a, -um* is amongst other meanings: primitive, from old times, obviously refers to the undifferentiated opercula.

Selected literature records:

See discussion, below.

The nominal species *Hydroides prisca, -us* has been unmasked as being composed of juvenile specimens with a developing operculum, not to be assigned to just a single known species of the genus *Hydroides* – and thus as a not valid, a rejected species name – by ten Hove & Ben-Eliahu (2005). Nevertheless, it has been mentioned as valid name in a few later checklists and similar literature overviews (e.g. Liu 2008, Yan *et al.* 2014, Sun & Yang 2014; Sun *et al.* 2018). Ten Hove & Ben-Eliahu (2005) introduced the concept 'bioperculate chimaera' for what they recognised to be a developmental stage in the opercular ontogeny of some *Hydroides* species, showing two different *Hydroides* type opercula: one radiolar lobe showing a small, undifferentiated operculum with two superimposed tiers of simple spines, a *H. 'priscus'*-type; the other radiolar lobe showing an operculum characteristic for *Hydroides* cf. *albiceps*, or *H.* cf. *brachyacantha*, *H. dirampha*, *H. homoceros*, *H. minax* and *H. trivesiculosa*. All these nominal taxa occur in Sri Lanka, but none in Panadura River together with the two type specimens of *H. prisca*, so it is impossible to attribute the record to a specific taxon. An additional and presumably more advanced bioperculate chimaera was mentioned by ten Hove (1994: 108) from the Seychelles, one operculum characteristic for *H.* cf. *albiceps*, the replacement operculum at the other side characteristic for *H. ancorispina/malleolaspina*. Such rather simple, undifferentiated opercula have not only been identified as *H. prisca*, but also as *H. novaepommeraniae* (**sensu** Dew 1959 and Straughan 1967a, both reidentified as *H. recta* by Imajima & ten Hove 1984: 47) and **sensu** Ben-Eliahu (1991: 527), the latter proved to have a second operculum typical for *H. homoceros* (see ten Hove & Ben-Eliahu 2005: 8). The co-occurrence of '*Hydroides prisca*' with *H. ezoensis* and *H. fusicola* as mentioned by Sun & Yang (2000, and again in 2014), both characterised by opercula with simple, unadorned spines, points to a misidentification of juvenile specimens belonging to the latter taxa (ten Hove & Ben-Eliahu 2005: 9–10). In a similar way the rather small (not yet fully developed?) specimens from the Mediterranean (fouling navy ships in Toulon) and Sri Lanka (original description, specimens 5.5 mm long with five pairs of radioles only), identified as *Hydroides grubei*, were regarded to be undeterminable juveniles (Ben-Eliahu & ten Hove 1992: 42, ten Hove & Ben-Eliahu 2005: 8). See discussion below, on the *nomen inquirendum Hydroides uncinata* as well.

7. *Hydroides uncinata* (Philippi, 1844), *nomen inquirendum*

Figure 120F

Type information: No type known. Type locality unspecified Mediterranean, but can be narrowed to the Tyrrhenian Sea coast of Italy as Philippi's activities were in western Italy, and plausibly to Naples, where he was based prior to 1844.

Etymology: Not stated, but Philippi described the verticil spines as with '*cornubus octo, apice incurvo uncinatis.* (Fig. Q)' (eight horns, hooked incurving tip), thus the name refers to the hooked spines, from the Latin adjective *uncinatus* 'hooked'.

Selected literature records:

See discussion, below.

Hydroides uncinata is one of the most confused, and still confusing, names in serpulid taxonomy. The original diagnosis by Philippi (1844: 195, pl. 6, fig. Q, as *Eupomatus uncinatus*) is too short to identify his material with certainty, but in view of what we presently know of Mediterranean serpulids most probably was based on either *H. pseudouncinata* or *H. dianthus*. The taxon, in absence of type material, thus formally has been classified as *nomen inquirendum*.

Zibrowius (1971a) diagnosed 12 species of *Hydroides*, mostly valid, at one time or another included in '*uncinata*': *H. affinis* [as synonym of *H. helmata*], *H. alatalateralis*, *H. arnoldi*, *H. dianthus*, *H. dipoma*, *H. fusicola*, *H. gairacensis*, *H. gracilis*, *H. heterocera*, *H. operculata*, *H. pseudouncinata*, *H. sanctaecrucis*, and *H. protulicola*. Following the use of the name by for instance Fauvel (1927), Monro (1933) and Rioja (1959), the next listed species could also be classified as '*H. uncinata*': *H. inornata*, *H. nigra*, *H. spongicola*, *H. elegantula*, *H. decora* and *H. plateni*. In total, Zibrowius (1971a) thus gave 19 names that had been, or could have been, included in the nominal taxon *H. uncinata*. Later, the list of *Hydroides* species with long, slender and rather undifferentiated verticil spines confused with *H. uncinata* was expanded with taxa as *H. augeneri*, *H. capensis*, *H. elegans* (N.B.), *H. floridana*, *H. monroi*, *H. novaepommeraniae*, '*priscus*', *H.* cf. *recta*, bringing the total of names at a conservative estimate of 27, almost a one fifth of all *Hydroides* names. Although Zibrowius (1968) with the introduction of *H. pseudouncinata* tried to give a well-defined alternative for the questionable name *H. uncinata*, the use of this ill-defined specific name persisted. Some of these (in the light of our present knowledge mis-) identifications have been mentioned in the selected synonymies given for various taxa, but our listings certainly are not exhaustive. Most recently the name was used in various checklists and discussions such as by Pillai (1972), Uchida (1978), Liu (2008), Dean (2012), Joydas *et al.* (2012), Amaral *et al.* (2013), Schwan *et al.* (2016), and Rodrigues *et al.* (2020). For a nomenclatural discussion see Read *et al.* (2017: 40).

Abbreviations of institutions where type material is deposited

AHF – Allan Hancock Foundation (AHF), material now curated by the Natural History Museum of Los Angeles County (LACM)
AM – Australian Museum, Sydney, Australia
AMNH – American Museum of Natural History, New York, USA
ANC – Acuario Nacional de Cuba
BMNH – British Museum (Natural History), now Natural History Museum, London, UK (NHML)
BPBM – Bernice Pauahi Bishop Museum, Hawaii, USA
ECOSUR – El Colegio de la Frontera Sur, Chetumal, Mexico (reference collection)
FSBC I – Florida Marine Research Institute, Invertebrate collection, St Petersburg, Florida, USA
HUJ – Hebrew University of Jerusalem, Section of Invertebrates of the National Natural History Collections
IOCAS – Institute of Oceanology of the Chinese Academy of Sciences, Qingdao, China (same as MBMBAC)
MAGNT – Museum and Art Gallery of Northern Territory, Darwin, Australia
MBMCAS – Marine Biological Museum of the Chinese Academy of Sciences, Qingdao, China
MCZ – Museum of Comparative Zoology, Cambridge, Massachusetts, USA
MESC – Marine Environmental Science Consortium, Dauphin Island, Alabama, USA
MNHN – Muséum National d'Histoire Naturelle, Paris, France
MNHW – Museum of Natural History, Wroclaw University, Poland
MRAC – Musée Royal de l'Afrique Centrale, Tervuren, Belgium
Naturalis – short for Naturalis Biodiversity Center, formerly known as RijksMuseum van Natuurlijke Historie, Leiden
NHML – Natural History Museum, London, UK
NHMW – Natural History Museum Wien, Austria
NSMT – National Science Museum, Tokyo, Japan
QM – Queensland Museum, South Brisbane, Queensland, Australia
RMNH – collection number of the RijksMuseum van Natuurlijke Historie, Leiden, now Naturalis Biodiversity Center
SAM – South Australian Museum, Adelaide, Australia
SMF – Senckenberg Research Institute and Natural History Museum Frankfurt/M, Germany
USNM – collection number of the Smithsonian National Museum of Natural History (NMNH), Washington, D.C., USA, formerly United States National Museum
YPM – Yale Peabody Museum, New Haven, Connecticut, USA
ZINRAS – Zoological Institute and Museum of the Russian Academy of Sciences, St Petersburg, Russia
ZMA – Zoological Museum of Amsterdam, now incorporated in the Naturalis Biodiversity Center, Leiden, the Netherlands
ZMB – Zoologisches Museum der Humboldt-Universität, Berlin, Germany
ZMC – Zoological Museum, Copenhagen, Denmark
ZMH – Zoologisches Institut und Zoologisches Museum, Hamburg, Germany

Glossary

abdomen (Sabellidae and Serpulidae): body region posterior to the thorax; recognised by notopodial (dorsal) uncini and neuropodial (ventral) chaetae (Fig. 1)

aff.: *affinis* (Latin) – related to, in taxonomy usually indicates having affinity with but not identity with

apron: membranous flap formed by thoracic membranes joined ventrally past the last thoracic chaetigers (Fig. 1)

auriculate: having one or more structures shaped like an ear or earlobe

basal spinule: spinule located at the base of verticil **spine** (Fig. 4, bs)

bayonet chaetae: collar chaetae with one or two (sometimes more) large proximal bosses (or 'teeth') at the base of a distal limbate zone (Fig. 5A–C)

bifurcate: divided into two parts or branches

bilobed: having or consisting of two lobes

binomen: a scientific name at the rank of species, with two terms: a generic name and a specific name

bioperculate chimaera: the ontogenetic stage in the genus *Hydroides* having two different types of two-tier opercula

blade: distal, seemingly flat portion of a chaeta (see, however, **capillary** or **limbate** chaeta)

branchiae: gills or similar organs of respiration

boss: small projection or knob-like process in collar chaetae of *Hydroides* (Fig. 5A–C)

capillary chaeta: slender, often long, chaeta tapering to a fine point; a collective term for elongate, needle-like chaetae of otherwise variable shape and ontogeny (Fig. 6A, B).

cf.: *confer* (Latin) – to compare, or to be compared with

chaeta (*pl.* chaetae): chitinous bristle protruding from an epidermal pocket in the body wall.

chaetal inversion: in serpulids the thorax bears chaetae dorsally and uncini ventrally; in the abdomen the position of chaetae and uncini is reversed

chaetiger: the segment bearing the chaetae

collar: an encircling membranous fold or flap covering the base of the radiolar crown (Fig. 1)

collar chaetae: the most anterior (paired) thoracic bundle of chaetae; see **collar segment**

collar segment: with an anterior encircling fold or flap covering the base of the radiolar crown; the first chaetiger bears an anterior collar and notochaetae (**collar chaetae**), but lacks uncini

compensatory regeneration: the development of functional opercula from regeneration of the pseudoperculum after the functional operculum has been amputated or shed

compensatory regulation: regulation of **compensatory regeneration**

constriction: narrowing of the opercular peduncle or a transverse groove, at the base of the funnel

det.: *determinavit* (Latin) – identified

dorsal: pertaining to, or situated at the back, or dorsum

duplicity: occurrence of bioperculate specimens in species of *Crucigera*, *Hydroides* and *Serpula* (Fig. 11), very rarely in other genera

eucosmopolitan: relating to a species with natural, truly cosmopolitan natural distribution

fang: sharply pointed anterior tooth of uncinus (e.g. Fig. 6C, D, H)

fide: trust (Latin)

flat trumpet-shaped chaeta: a chaeta with a thin, flat distal part with small teeth on its edge (Fig. 6E, F)

funnel: a descriptive term used to indicate the inverted cone-like proximal part of the operculum in *Hydroides* (Fig. 1)

geniculate: bent abruptly at an angle like a bent knee

holotype: the single specimen designated as the type of a species by the original author at the time the species' name and description were published

in litteris: in the document (Latin)

infraspecific: the scientific name for any taxon below the rank of species (i.e. infraspecies)

inter-radiolar membrane: thin membrane connecting basal parts of radioles in some Serpulidae

IOM: insoluble (in EDTA) organic matter in serpulid tube

keel: longitudinal ridge running along the tube length (Fig. 7, k)

lapsus calami: Latin for 'lapse, slip, error', an involuntary mistake made while writing or speaking

lateral: located on the side

lateral spinule: spinule on the side on verticil spine (Fig. 4, ls)

legit: collected (Latin).

limbate chaetae: stiff, elongate chaetae made of densely packed fibrils; under a compound microscope they seem to have a distal limbus or flat blade, which on inspection with a SEM is an outer layer where the fibrils are packed less tightly than in the central shaft (Fig. 6A, B).

limbus: the seemingly flattened distal border of chaetae, longitudinal flange; however, see **limbate chaetae**

mouth: the anterior opening of the alimentary canal; in serpulids it is also used as an anterior opening of the tube

neotype: the specimen that is chosen to act as the type material in cases when the original type was not designated or has been subsequently lost

neocosmopolitan species: species that achieve extensive geographical ranges only through anthropogenic dispersal

neuropodial: relating to (chaetae of the) ventral branch of a parapodium

nomen dubium: Latin for 'doubtful name', plural *nomina dubia*; a taxonomic name that cannot be assigned with certainty to any taxonomic group because the description is insufficient for identification and the original specimens no longer exist

nomen inquirendum: a species of doubtful identity requiring further investigation

nomen novum: new name

***nomen oblitum* (*pl. nomina oblita*)**: Latin for 'forgotten name'

nomen protectum: zoological name that is protected against the reuse of a *nomen oblitum* (Latin)

nominotypical: International Code of Zoological Nomenclature (https://code.iczn.org): 'Article 37. Nominotypical taxa. 37.1. Definition. When a family-group taxon is subdivided, the subordinate taxon that contains the type genus of the superior taxon is denoted by the same name (except for suffix) with the same author and date [Art. 36.1]; this subordinate taxon is termed the "nominotypical taxon".' See further Articles 44 and 47.

notochaetae: chaetae of the dorsal branch of a parapodium (the notopodium)

operculum (*pl.* opercula): the transformed tip of modified radiole used to plug the tube when the worm is retracted (e.g. Fig. 1)

paraphyletic: relating to a taxonomic group that does not include all descendants of a common ancestor

parapodium (*pl.* parapodia): fleshy lateral projection from a body segment that usually bears chaetae

paratype: a specimen collected at the same time as the holotype and designated or indicated as such by the original author

pectinate: comb-like; with series of projections like the teeth of a comb; an arrangement of radioles in serpulids

peduncle: modified radiole lacking pinnules and bearing the operculum (Fig. 1)

peripheral flange: a flattened tube rim completely attached to substrate for increasing area of tube attachment (e.g. Fig. 7, pf)

peristome: collar-like widening of tube (former tube-mouth)

peristomium: anterior body region surrounding the mouth and located posterior to and/or below the prostomium

pinnules: small paired side branches of the radioles, giving each radiole a feathery appearance (Fig. 1)

polyphyly: a group of organisms whose last common ancestor is not a member of the group

pro parte: a Latin phrase meaning in part, usually referring to a situation in which only part of a taxon possesses a given feature; often abbreviated as p.p.

prostomium: anteriormost presegmental region of body, usually bearing radioles and sensory organs such as palps, antennae, nuchal organs and eyes

proximal rasp: in bayonet chaetae, a denticulate zone proximal from the two or three large teeth of the bayonet or instead of the bayonet (Fig. 5 D, E, pr)

pseudocosmopolitan: describes species with reported broad distribution that is a reflection of its uncertain taxonomic status

pseudoperculum (*pl.* pseudopercula): an underdeveloped operculum on a modified radiole (usually the second dorsal one), generally shortened and without pinnules in *Hydroides* (Fig. 1) and closely related *Serpula*; can develop into a new functional operculum when the latter is lost

pygidium: the post-segmental terminal body-part surrounding the anus

radiolar crown: a bilaterally symmetrical branched structure consisting of two separate lobes, each with a number of radioles bearing pinnules; often including an operculum with peduncle on one lobe and sometimes also a pseudoperculum on the opposite lobe (Fig. 1)

radiolar eyes: ocelli found in the radiolar crown that can vary in number, arrangement and structure

radiolar lobes: the proximal part of the radiolar crown attached to the anterior end of the body; generally, as two semicircles located on both sides of the mouth, but forming spirals in some species (Fig. 1, rl)

radioles: pinnulate filaments of radiolar crown, used for respiration and feeding (Fig. 1)

radius (*pl.* radii): the radial projection of the funnel in *Hydroides* and *Serpula* (Figs 1 and 4)

rasp-shaped uncinus: with two or more rows of teeth (Fig. 6H)

reversible asymmetry: alternating development of functional opercula and regeneration of pseudopercula, leading to an alternation of position (left or right) between them

saw-shaped uncinus: with only one row of teeth (Fig. 6C, D)

sedentary: attached to a surface and not moving freely

segment: one of the serially repeated units comprising the annelid body; often separated internally by septa or dissepiments

sensu auctores: in the sense as given in the literature, not necessarily as defined by the original author (Latin)

sensu lato: in the broad sense (Latin)

sensu stricto: in the strict/stricter/strictest sense (Latin)

semaphoront: an organism as seen in a specific time, as the object of identification or basis for systematics

shaft: the proximal smooth capillary part of a chaeta, partly embedded in the tissue

species inquirenda: names which are otherwise valid but require taxonomic clarification due to inadequate original descriptions

spine: see **verticil spine**

spinule: each of the tubercular or tooth-like projections of a spine in the verticil of the genus *Hydroides*.

SOM: soluble (in EDTA) organic matter in serpulid tube

syntype: any of two or more specimens listed in the original description of a taxon when a holotype was not designated

thoracic membranes: thin folds on both sides of thorax, extending from dorsal part of collar to lateral and/or ventral side of posterior thorax (Fig. 1)

thorax: anterior region of the body behind the head (Fig. 1)

topotypical material: a specimen/specimens found in the type locality of a taxon to which it is thought to belong, but that is not necessarily of that type series

torus (*pl.* tori): transverse elevation of parapodium surrounding the uncini (Fig. 6H)

transversal ridge: annular elevation of tube, less pronounced than peristome

triangular depression: the area between thoracic uncinigerous tori when gradually approaching and almost touching one another posteriorly and ventrally

trumpet chaetae: abdominal chaetae in *Hydroides*, formerly thought to be hollow; however, more recently proven to be flat, not hollow. See the preferred term **flat trumpet-shaped chaeta** (Fig. 6E, F)

tube: protective structure made of calcium carbonate and completely enclosing the body (Fig. 7)

tweezing: a technique used to open verticil by pressing the basis of the verticil with fine tweezers

type: a zoological object that serves as the base for the name of a taxon

uncinigerous: bearing uncini

uncinus (*pl.* uncini): a deeply embedded comb-shaped chaeta with only its dentate edge protruding from the body wall; uncini are usually arranged in tori, elevated rows transverse to the axis of the animal (Fig. 6C, D)

unsculptured tube: usually circular in cross-section and not bearing any sculptural elements such as keels, ridges, alveoli, or spines (Fig. 7A)

ventral: lower or underside of the body; side of the polychaete body bearing the mouth

verticil spine: any of the radial elements, generally around a central disk, together forming the **verticil** in *Hydroides* (Fig. 1)

verticil: the distal part of operculum in *Hydroides*, usually made of chitinous spines

wing: lateral flat expansion of verticil spine of the operculum (Fig. 4A, w)

WoRMS: World Record of Marine Species (https://www.marinespecies.org); a taxonomic database that aims to provide an authoritative and comprehensive list of names of marine organisms

zygomorphy: having bilateral symmetry

References

Abeloos M (1949) Sur la régénération compensatrice de l'opercule du serpulide *Hydroides norvegica* (Gunn.). *Comptes Rendus Hebdomadaires des Seances de l Academie des Sciences* **228**, 1157–1159.

Abeloos M (1952a) Sur le développement de l'opercule du serpulide *Hydroides norvegica*. *Comptes Rendus de l'Académie des Sciences* **146**, 459–461.

Abeloos M (1952b) Régénération simultanée des opercules du serpulide *Hydroides norvegica* (Gunn.). *Comptes Rendus des Seances de la Societe de Biologie et des ses Filiales (Paris)* **146**, 734–736.

Abeloos M (1952c) Nature des corrélations entre les opercules du serpulide *Hydroides norvegica* (Gunn.). *Comptes Rendus de l'Académie des Sciences (Paris)* **234**, 2489–2491.

Abeloos M (1952d) La 'régénération compensatrice' chez *Hydroides norvegica* (Gunn.) et les mécanismes de la morphogénèse. *Bulletin de la Société Zoologique de France* **77**, 462–470.

Abeloos N (1954a) Morphogénèse de l'opercule du serpulide *Hydroides norvegica* (Gunn.). *Comptes Rendus des Seances de la Societe de Biologie et des ses Filiales (Paris)* **148**, 1440–1441.

Abeloos M (1954b) Sur la régénération des Sabelles. *Bulletin du Laboratoire Maritime de Dinard* **40**, 1–3.

Albano MJ, Obenat SM (2019) Fouling assemblages of native, non-indigenous and cryptogenic species on artificial structures, depths and temporal variation. *Journal of Sea Research* **144**, 1–15.

Alcázar J, San Martín G (2016) Serpúlidos (Annelida, Serpulidae) colectados en la Campaña Oceanográfica 'Fauna II' y catálogo actualizado de especies íbero-baleares de la familia Serpulidae. *Graellsia* **72** (2), e053.

Al-Kandari M, Sattari Z, Hussain S, Radashevsky VI, Zhadan A (2019) Checklist of intertidal polychaetes (Annelida) of Kuwait, Northern part of the Arabian Gulf. *Regional Studies in Marine Science* **32**, 100872.

Allen FE (1953) Distribution of marine invertebrates by ships. *Australian Journal of Marine and Freshwater Research* **4**, 307–316.

Allen FE, Wood EJF (1950) Investigations on underwater fouling. II. The biology of fouling in Australia: results of a year's research. *Australian Journal of Marine and Freshwater Research* **1**, 92–105.

Allen RM, Marshall D (2014) Egg size effects across multiple life-history stages in the marine annelid *Hydroides diramphus*. *PloS One* **9**, e102253.

Amaral ACZ, Nallin SAH, Steiner TM, Forroni TO, Gomes-Filho D (2013) *Catálogo das espécies de Annelida Polychaeta do Brasil*. Unicamp, Campinas.

Amoureux L (1974) Annélides polychètes de Madagascar receuillies par G. Cherbonnier en 1960. *Bulletin du Muséum National d'Histoire Naturelle, Paris, 3e série, Zoologie* **145**, 425–462.

Amoureux L (1978) Annélides polychètes recoltes par J. Stirn en 1969, sur les côtes marocaines du détroit de Gibraltar. *Cuadernos de Ciencias Biol*ógicas **5**, 5–33.

Amoureux L, Rullier F, Fishelson L (1978) Systématique et ecologie d'Annélides Polychètes de la presqu'îl du Sinai. *Israel Journal of Zoology* **27**, 57–163.

Annenkova NP (1938) Polychaeta of the North Japan Sea and their horizontal and vertical distribution. *Reports of the Japan Sea Hydrobiological Expedition of the Zoological Institute of Academy of Sciences of the USSR in 1934* **1938**, 81–236.

Arakawa KY (1971) Notes on a serious damage to cultured oyster crops in Hiroshima caused by unique and unprecedented outbreak of a serpulid worm, *Hydroides norvegica* (Gunnerus) in 1969. *Venus* **30**, 75–82.

Arenas-Mena C (2007) Sinistral equal-size spiral cleavage of the indirectly developing polychaete *Hydroides elegans*. *Developmental Dynamics* **236**, 1611–1622.

Arenas-Mena C, Li A (2014) Development of a feeding trochophore in the polychaete *Hydroides elegans*. *International Journal of Developmental Biology* **58**, 575–583.

Arndt W (1927) Kalk- und Kieselschwämme von Curaçao. *Bijdragen tot de Dierkunde* **25**, 133–158.

Arumugam S (2012) Toxic effects of flucloxacillin on the early development of the polychaete *Hydroides elegans*. *Pune India* **2**, 135–142.

Augener H (1914) Polychaeta, II., Sedentaria. In *Die Fauna Südwest-Australiens* (Eds W Michaelsen, R Hartmeyer) pp. 1–72. Ergebnisse der Hamburger Südwest-Australischen Forschungsreise 1905. Volume 5. Gustav Fischer, Jena.

Augener H (1918) Polychaeta. *Beitrage zur Kenntnis der Meeresfauna Westafrikas* **2**, 67–625.

Augener H (1922) Über litorale Polychaeten von Westindien. *Sitzungsberichte Gesellschaft Naturforschender Freunde zu Berlin* **1922**, 38–53.

Augener H (1925a) Über westindische und einige andere Polychaeten-Typen von Grube (Oersted), Krøyer, Mörch und Schmarda. *Publikationer fra Universitetets Zoologiske Museum, København, København* **39**, 1–47.

Augener H (1925b) Die Polychaeten der Südsee-Expedition der Hamburgischen wissenschaftliche Stiftung 1908–1909. *Mitteilungen der Zooologisches Staatinstitut und Zoologisches Museum, Hamburg* **41**, 53–70.

Augener H (1927) Polychaeten von Neu-Pommern. *Sitzungsberichte der Gesellschaft der naturforschende Freunde zur Berlin* **1926**, 119–152.

Augener H (1934) Polychaeten aus den Zoologischen Museen von Leiden und Amsterdam. IV Schluss. *Zoologische Mededeelingen s'Rijks Museum van Natuurlijke Historie Leiden* **17**, 67–160.

Bailey-Brock JH (1985) Polychaetes from Fijian coral reefs. *Pacific Science* **39**, 195–220.

Bailey-Brock JH (1991) Tubeworms (Serpulidae, Polychaeta) collected from sewage outfalls, coral reefs and deep waters off the Hawaiian Islands, including a new *Hydroides* species. *Bulletin of Marine Science* **48**, 198–207.

Bailey-Brock JH, Magalhães WF (2012) A new species and record of Serpulidae (Annelida: Polychaeta) from Cross Seamount in the Hawaiian Chain. *Zootaxa* **3192**, 49–58.

Bailey-Brock JH, Magalhães W (2013) New records of *Hydroides* (Polychaeta: Serpulidae) from Guam, Mariana Islands. *Marine Biodiversity Records* **6**, 1–7.

Bailey-Brock JH, Magalhães WF, Brock RE (2012) Coral reef inhabiting tubeworms (Polychaeta: Serpulidae) from Enewetak, Kwajalein, Rongelap and Utirik Atolls, Marshall Islands. *Journal of the Marine Biological Association of the United Kingdom* **92**, 967–988.

Bakken T, Fremstad E, Aagaard K (2011) Bishop Gunnerus as naturalist: his impact on biodiversity research today. *Det Kongelige Norske Videnskabers Selskabs Skrifter* **2011**, 109–128.

Bao W-Y, Lee O-O, Chung H-C, Li M, Qian P-Y (2010) Copper affects biofilm inductiveness to larval settlement of the serpulid polychaete *Hydroides elegans* (Haswell). *Biofouling, The Journal of Bioadhesion and Biofilm Research* **26**, 119–128.

Barnes DH (1982) The colonization of an artificial reef in the north central Gulf of Mexico by polychaetes. MSc thesis. University of Southern Mississippi, USA.

Barroso R, Klautau M, Solé-Cava AM, Paiva PC (2010) *Eurythoe complanata* (Polychaeta: Amphinomidae), the 'cosmopolitan' fireworm, consists of at least three cryptic species. *Marine Biology* **157**, 69–80.

Bastida-Zavala JR (1993) Taxonomía y composición biogeográfica de los poliquetos (Annelida: Polychaeta) de la Bahía de La Paz, B.C.S., Mexico. *Revista de Investigaciones Científicas* **4**, 11–39.

Bastida-Zavala JR (2008) Serpulids (Annelida: Polychaeta) from the Eastern Pacific, including a brief mention of Hawaiian serpulids. *Zootaxa* **1722**, 1–61.

Bastida-Zavala JR, de León-González, JA (2002) A new species of *Hydroides* (Polychaeta: Serpulidae) from western Mexico. *Journal of the Marine Biological Association of the United Kingdom* **82**, 389–393.

Bastida-Zavala JR, Salazar-Vallejo SI (2000) Serpúlidos (Polychaeta: Serpulidae) del Caribe noroccidental: *Hydroides* y *Serpula*. *Revista de Biología Tropical* **48**, 841–858.

Bastida-Zavala JR, ten Hove HA (2003a) Revision of *Hydroides* Gunnerus, 1768 (Polychaeta: Serpulidae) from the Western Atlantic Region. *Beaufortia* **52**, 103–178.

Bastida-Zavala JR, ten Hove HA (2003b) Revision of *Hydroides* Gunnerus, 1768 (Polychaeta: Serpulidae) from the Eastern Pacific region and Hawaii. *Beaufortia* **53**, 67–110.

Bastida-Zavala JR, Rodríguez Buelna AS, de León-González JA, Camacho-Cruz KA, Carmona I (2016) New records of sabellids and serpulids (Polychaeta: Sabellidae, Serpulidae) from the Tropical Eastern Pacific. *Zootaxa* **4184**, 401–457.

Bastida-Zavala JR, McCann LD, Keppel E, Ruiz GM (2017) The fouling serpulids (Polychaeta: Serpulidae) from United States coastal waters: an overview. *European Journal of Taxonomy* **344**, 1–76.

Batzel G, Nedved BT, Hadfield MG (2016) Expression and localization of carbonic Anhydrase genes in the serpulid polychaete *Hydroides elegans*. *The Biological Bulletin* **231**, 175–184.

Beckmann M, Harder T, Qian PY (1999) Induction of larval attachment and metamorphosis in the serpulid polychaetes *Hydroides elegans* by dissolved free amino acids: mode of action in laboratory bioassays. *Marine Ecology - Progress Series* **190**, 181–190.

Behrens EW (1968) Cyclic and current structures in a serpulid reef. *Contributions to Marine Science* **13**, 21–27.

Benedict JE (1887) Descriptions of ten species and one new genus of annelids from the dredging of the U.S. Fish Commission Steamer Albatross. *Proceedings of the United States National Museum* **9**, 547–553.

Ben-Eliahu MN (1972) A description of *Hydroides steinitzi* n. sp. (Polychaeta: Serpulidae) from the Suez Canal with remarks on the serpulid fauna of the canal. *Israel Journal of Zoology* **21**, 77–81.

Ben-Eliahu MN (1976) Polychaete cryptofauna from rims of similar intertidal vermetid reefs on the Mediterranean coast of Israel and in the Gulf of Elat: Serpulidae (Polychaeta Sedentaria). *Israel Journal of Zoology* **25**, 103–119.

Ben-Eliahu MN (1991) Red Sea serpulids (Polychaeta) in the eastern Mediterranean. In *Proceedings of the 2nd International Polychaete Conference*. 18–23 August 1986, Copenhagen (Eds ME Petersen, JB Kirkegaard). *Ophelia Supplement* **5**, 515–528.

Ben-Eliahu MN, Fiege D (1996) Serpulid tube-worms (Annelida: Polychaeta) of the Central and Eastern Mediterranean with particular attention to the Levant Basin. *Senckenbergiana Maritima* **28**, 1–51.

Ben-Eliahu MN, ten Hove HA (1989) Redescription of *Rhodopsis pusilla* Bush, a little known but widely distributed species of Serpulidae (Polychaeta). *Zooligica Scripta* **18**, 381–395.

Ben-Eliahu MN, ten Hove HA (1992) Serpulids (Annelida: Polychaeta) along the Mediterranean coast of Israel – New population buildups of Lessepsian migrants. *Israel Journal of Zoology* **38**, 35–53.

Ben-Eliahu MN, ten Hove HA (2011) Serpulidae (Annelida: Polychaeta) from the Suez Canal – from a Lessepsian migration perspective (a monograph). *Zootaxa* **2848**, 1–147.

Bensari B, Bahbah L, Lounaouci A, Eddinah Fahci S, Bouda A, *et al.* (2020) First records of non-indigenous species in port of Arzew (Algeria: southwestern Mediterranean). *Mediterranean Marine Science* **21**(2), 393–399.

Berkeley E, Berkeley C (1941) On a collection of Polychaeta from Southern California. *Bulletin of the Southern California Academy of Sciences* **40**, 16–60.

Berkeley MJ (1834) Observations on some British Serpulae. *Magazine of Natural History, London* **7**, 420–421.

Bernhardt AM, Manyak DM, Wilbur KM (1985) In vitro recalcification of organic matrix of scallop shell and serpulid tubes. *Journal of Molluscan Studies* **51** (3), 284–289.

Bianchi CN (1981) *Policheti Serpuloidei. Guide per il Riconoscimento delle Specie Animali delle Acque Lagunari e Costiere Italiane. AQ/1/96, 5.* Consiglio Nazionale delle Ricercher, Genova.

Bick A (2006) Polychaete communities associated with gastropod shells inhabited by the hermit crabs *Clibanarius erythropus* and *Calcinus tubularis* from Ibiza, Mediterranean Sea. *Journal of the Marine Biological Association of the United Kingdom* **86**, 83–92.

Bleeker J, van der Spoel S (1992) Catalogue of the polychaeta collected by the Siboga Expedition and type specimens of polychaeta in the Zoological Museum of Amsterdam. *Bulletin Zoologisch Museum* **13**, 121–166.

Boaventura D, Moura A, Leitão F, Carvalho S, Cúrdia J, *et al.* (2006) Macrobenthic colonisation of artificial reefs on the southern coast of Portugal (Ancão, Algarve). *Hydrobiologia* **555**, 335–343.

Boeck A (1865) Oversigt over de ved Norges Kyster jagttagne Copepoder henhorende til Calanidernes, Cyclopidernes og Harpactidernes Familier. *Forhandlinger i Videnskabs-Selskabet i Christiania* **1864**, 226282.

Boltachova NA, Lisitskaya EV, Lebedovskaya MV (2011) New species *Hydroides dianthus* (Verrill, 1873) (Polychaeta: Serpulidae) from the coastal waters of Crimea (the Black Sea). *Marine Ecological Journal, Sevastopol* **10**, 34–38.

Bornhold BD, Milliman JD (1973) Generic and environmental control of carbonate mineralogy in serpulid (polychaete) tubes. *Journal of Geology* **81**, 363–373.

Bosc LAG (1802) *Histoire naturelle des vers contenant leur description et leurs moeurs, avec figures dessinées d'après nature. Tome Premiere.* De l'imprimerie de Guilleminet, Paris.

Breton G, Vincent T (1999) Invasion du port du Havre (France, Manche) par *Hydroides ezoensis* (Polychaeta, Serpulidae), espèce d'origine japonaise. *Bulletin trimestriel de la Société géologique de Normandie et Amis Muséum du Havre* **86**, 33–43.

Bryan PJ, Qian PY, Kreider JL, Chia FS (1997) Induction of larval settlement and metamorphosis by pharmacological and conspecific associated compounds in the serpulid polychaete 1834 *Hydroides elegans. Marine Ecology – Progress Series* **146**, 81–90.

Bryan PJ, Kreider JL, Qian PY (1998) Settlement of the serpulid polychaetes *Hydroides elegans* (Haswell) on the arborescent bryozoans *Bugula neritina* (L.): evidence of a chemically mediated relationship. *Journal of Experimental Marine Biology and Ecology* **220**, 171–190.

Budaeva NE, Jirkov IA, Savilova TA, Paterson GLJ (2014) Deep-sea fauna of European seas: an annotated species check-list of benthic invertebrates living deeper than 2000 m in the seas bordering Europe. Polychaeta. *Invertebrate Zoology* **11**, 217–230.

Bush KJ (1905) Tubicolous annelids of the tribes Sabellides and Serpulides from the Pacific Ocean. *Harriman Alaska Expedition* **12**, 169–346.

Bush KJ (1910) Description of new serpulids from Bermuda with notes on known forms from adjacent regions. *Proceedings of the Academy of Natural Sciences, Philadelphia* **62**, 490–501.

Cao S-M, Xu H, Liu K (2008) Acute toxicity of four heavy metal ions to trochophores of *Hydroides ezoensis. Journal of Dalian Fisheries University* **23**, 73–76.

Cao S-M, Xu H, Liu P-L, Li J, Sun W-S (2009) Effects of temperatures on growth, development and metamorphosis of *Hydroides ezoensis* larvae. *Journal of Dalian Fisheries University* **24**, 531–535.

Cao S-M, Liu R-C, Li X-J, Hao S, Xu H (2013) Effects of salinity on fertility, hatching rate, survival and metamorphosis in *Hydroides ezoensis* larvae. *Journal of Dalian Ocean University* **28**, 133–137.

Carpizo-Ituarte E, Hadfield MG (1998) Stimulation of metamorphosis in the polychaete *Hydroides elegans* Haswell (Serpulidae). *The Biological Bulletin* **194**, 14–24.

Carpizo-Ituarte EJ, Hadfield MG (2003) Transcription and translation inhibitors permit metamorphosis up to radiole formation in the serpulid polychaete *Hydroides elegans* Haswell. *Biological Bulletin* **204**, 114–125.

Carr CM, Hardy SM, Brown TM, Macdonald TA, Hebert PHN (2011) A tri-oceanic perspective: DNA barcoding reveals geographic structure and cryptic diversity in Canadian polychaetes. *PloS ONE* **6**, e22232.

Chamberlin RV (1919) New polychaetous annelids from Laguna Beach, California. *Journal of Entomology and Zoology of Pomona College* **11**, 1–23.

Chan VBS, Li C, Lane AC, Wang Y, Lu X, *et al.* (2012) CO_2-driven ocean acidification alters and weakens integrity of the calcareous tubes produced by the serpulid tubeworm, *Hydroides elegans.* PloS One **7**, e42718.

Chan VBS, Thiyagarajan V, Lu XW, Zhang T, Shih K (2013) Temperature dependent effects of elevated CO_2 on shell composition and mechanical properties of *Hydroides elegans*: insights from a multiple stressor experiment. *PLoS ONE* **8** (11), e78945.

Chan VBS, Toyofuku T, Wetzel G, Saraf L, Thiyagarajan V, *et al.* (2015a) Direct deposition of crystalline aragonite in the controlled biomineralization of the calcareous tubeworm. *Frontiers in Marine Science* **97**, 1–10.

Chan VB, Vinn O, Li LX, Kudryavtsev AB, Schopf JW, *et al.* (2015b) Evidence of compositional and ultrastructural shifts during the development of calcareous tubes in the biofouling tubeworm, *Hydroides elegans. Journal of Structural Biology* **189**, 230–237.

Chapman JW, Carlton JT (1991) A test of criteria for introduced species: the global invasion by the isopod *Synidotea laevidorsalis* (Miers, 1881. *Journal of Crustacean Biology* **11**, 386–400.

Chen M, Wu BL (1978) Two new species of the genus *Hydroides* (Polychaeta, Serpulidae) from the Xisha Islands, Guandong Province, China. *Studia Marina Sinica* **12**, 141–145.

Chen M, Wu BL (1980) Two new species of the genus *Hydroides* (Polychaeta, Serpulidae). *Oceanologia et Limnologia Sinica* **11**, 247–250.

Chung HC, Lee OO, Huang YL, Mok SY, Kolter R, *et al.* (2010) Bacterial community succession and chemical profiles of subtidal biofilms in relation to larval settlement of the polychaete *Hydroides elegans*. *The ISME Journal* **4**, 817–28.

Çinar M (2006) Serpulid species (Polychaeta: Serpulidae) from the Levantine coast of Turkey (eastern Mediterranean), with special emphasis on alien species. *Aquatic Invasions* **4**, 223–240.

Claparède E (1870) Les Annélides Chétopodes du Golfe de Naples. Supplément. *Mémoires de la Société de Physique et d'Histoire Naturelle de Genève* **20**, 365–542.

Colwin AL, Colwin LH (1961a) Fine structure of the spermatozoon of *Hydroides hexagonus* (Annelida), with special reference to the acrosomal region. *Journal of Biophysical and Biochemical Cytology* **10**, 211–230.

Colwin AL, Colwin LH (1961b) Changes in spermatozoon during fertilization in *Hydroides hexagonus* (Annelida). 2. Incorporation with the egg. *Journal of Biophysical and Biochemical Cytology* **10**, 255–274.

Colwin LH, Colwin AL (1961c) Changes in spermatozoon during fertilization in *Hydroides hexagonus* (Annelida). 1. Passage of the acrosomal region through the vitelline membrane. *Journal of Biophysical and Biochemical Cytology* **10**, 231–254.

Costello DP, Davidson ME, Eggers A, Fox MH, Henley C (1957) *Methods for Obtaining and Handling Marine Eggs and Embryos*. Marine Biological Laboratory, Woods Hole.

Cresp J (1964) Études expérimentales et histologiques sur la régénération et le bourgeonnement chez les serpulides *Hydroïdes norvegica*' (Gunn.) et *Salmacina incrustans* (Clap.). *Bulletin Biologique de la France et de la Belgique* **98**, 3–152.

Crisp DJ (1974) Factors influencing the settlement of marine invertebrate larvae. In *Chemoreception in marine organisms*. (Eds PT Grant, AM Mackie) pp. 177–265. Academic Press, London.

Dahms H-U, Qian P-Y (2005) Exposure of biofilms to meiofaunal copepods affects the larval settlement of *Hydroides elegans* (Polychaeta). *Marine Ecology - Progress Series* **297**, 203–214.

Dahms H-U, Harder T, Qian P-Y (2004) Effect of meiofauna on macrofauna recruitment: settlement inhibition of the polychaete *Hydroides elegans* by the harpacticoid copepod *Tisbe japonica*. *Journal of Experimental Marine Biology and Ecology* **311**, 147–61.

Daniel A (1958) Settlement of marine foulers and borers in the Madras Harbour in relation to velocity of water currents. *Journal of Scientific and Industrial Research* **17C**, 18–20.

Darling JA, Carlton JT (2018) A framework for understanding marine cosmopolitanism in the Anthropocene. *Frontiers in Marine Science* **5**, 293.

Day JH (1951) The Polychaet fauna of South Africa. Pt. 1. The intertidal and estuarine Polychaeta of Natal and Mosambique. *Annals of the Natal Museum* **12**, 1–67.

Day JH (1957) The Polychaet fauna of South Africa. Part 4. New species and records from Natal and Moçambique. *Annals of the Natal Museum* **14**, 59–129.

Day JH (1961) The Polychaet fauna of South Africa. 6. Sedentary species dredged off Cape coasts with a few new records from the shore. *Journal of the Linnean Society of London, Zoology* **44**, 463–560.

Day JH (1967) *A Monograph on the Polychaeta of Southern Africa. Part 2. Sedentaria*. British Museum (Natural History), London.

Day JH (1973) 'New Polychaeta from Beaufort, with a key to all species recorded from North Carolina'. Technical Report NMFS Circ-375. US Department of Commerce, National Oceanic and Atmospheric Administration, Seattle.

Day JH, Hutchings PA (1979) An annotated check-list of Australian and New Zealand Polychaeta, Archiannelida and Myzostomida. *Records of the Australian Museum* **32**, 80–161.

Dean HK (2012) A literature review of the Polychaeta of the Caribbean Sea. *Zootaxa* **3596**, 1–86.

De Assis JE, de Araújo Costa D, Christoffersen ML (2009) First record of *Hydroides similoides* (Serpulidae: Polychaeta) from Brazil. *Marine Biodiversity Records* **2**, e27.

Delle Chiaje S (1822) *Memorie sulla storia e notomia degli animali senza vertebre del regno di Napoli*.

Delle Chiaje S (1828) *Memorie Sulla Storia e Notomia Degli Animali Senza Vertebre del Regno di Napoli*. Volume 3.

Devakumar C, Gopalakrishnan H, Chinnasamy A, Subrmanian B, Durai P (2017) Toxicity of silver nanoparticles on fertilization success and early development of the marine polychaete *Hydroides elegans* (Haswell, 1883). *Journal of Basic and Applied Zoology* **78**. doi:10.1186/s41936-017-0001-7

Dew B (1958) Variations in the secondary operculum of the Australian representative of the polychaete worm *Hydroides norvegica* Gunnerus, and notes on a polychaete worm recovered from the aquarium at Taronga zoological park. *Proceedings of Royal Zoological Society of New South Wales* **1956–57**, 52–57.

Dew B (1959) Serpulidae (Polychaeta) from Australia. *Records of the Australian Museum* **25**, 19–56.

Dew B, Wood EJ (1954) Observations on periodicity in marine invertebrates. *Australian Journal of Marine and Freshwater Research* **6**, 469–478.

Díaz-Castañeda V, Almeda-Jauregui C (1999) Early benthic organism colonization on a Caribbean coral reef (Barbados, West Indies): a plate experimental approach. *Marine Ecology* **20**, 197–220.

Díaz-Castañeda V, Cox TE, Gazeau F, Fitzer S, Delille J, *et al.* (2019) Ocean acidification affects calcareous tube growth in adults and reared offspring of serpulid polychaetes. *Journal of Experimental Biology* **222**, jeb196543.

Díaz–Díaz O, Liñero Arana I (2001) Poliquetos asociados a substratos arteficiales sumergidos en la costa nororiental de Venezuela. II. Serpulidae y Spirorbidae. *Boletín del Instituto Oceanográfico de Venezuela, Universidad de Oriente* **40**, 9–20.

Dill LM, Fraser AHG (1997) The worm re-turns: hiding behavior of a tube-dwelling marine polychaete, *Serpula vermicularis*. *Behavioral Ecology* **8** (2), 186–193.

Dobretsov S, Dahms HU, Qian PY (2004) Antilarval and antimicrobial activity of waterborne metabolites of the sponge *Callyspongia pulvinata*: evidence of allelopathy. *Marine Ecology - Progress Series* **271**, 133–146.

Dobretsov S, Dahms H-U, Tsoi MY, Qian P-Y (2005) Chemical control of epibiosis by Hong Kong sponges: the effect of sponge extracts on micro- and macrofouling communities. *Marine Ecology - Progress Series* **297**, 119–129.

Dong Z, Sun T, Wang L (2018) The biogenic reefs formed by the alien polychaete *Hydroides dianthus* (Serpulidae, Annelida) favor the polyp stage of *Aurelia coerulea* (Cnidaria, Scyphozoa) in a coastal artificial lake. *Marine Pollution Bulletin* **129**, 86–91.

Duchassaing de Fontbressin P, Michelotti G (1864) Spongiaires de la mer Caraibe. *Natuurkundige Verhandelingen van de Hollandsche Maatschappij der Wetenschappen te Haarlem* **21**, 1–124.

Dueñas PR (1981) Inventario preliminar de los poliquetos (Annelida) de aguas someras de la Bahía de Cartagena y areas adyacentes. *Boletín del Museo del Mar, Universidad Jorge Tadeo Lozano, Bogotá* **10**, 82–138.

Dürr S, Watson DI (2010) Biofouling and antifouling in aquaculture. In *Biofouling* (Eds S Dürr, JC Thomason) pp. 267–287. Wiley-Blackwell, Oxford.

Ehlers E (1887) Report on the Annelids. Reports on the results of dredging, under the direction of L.F. Pourtalès, during the years 1868–1870, and of Alexander Agassiz, in the Gulf of Mexico, and in the Caribbean Sea, in the U.S. Coast Survey Steamer 'Blake'. *Memoirs of the Museum of Comparative Zoology at Harvard College* **15**, 1–335.

Ehlers E (1913) Die Polychaeten-Sammlungen der deutschen Südpolar- Expedition 1901–1903. *Deutsche Südpolar-Expedition* **13**, 397–598.

Ekman S (1953) *Zoogeography of the Sea*. Sidgwick & Jackson Limited, London.

Eschscholtz F (1825) Bericht über die zoologische Ausbeute während der Reise von Kronstadt bis St. Peter und Paul. *Isis* **1825**, 733–747.

Faasse M, ten Hove H, Paalvast P (2020) The Pacific serpulid tube worm *Hydroides ezoensis* Okuda, 1934 reaches the ports of Rotterdam and Vlissingen (Flushing), the Netherlands (North Sea). *Cahiers de Biologie Marine* **61**, 415–421.

Faimali M, Garaventa F, Geraci S (2002) Laboratory larval growth and settlement of the tube worm *Hydroides elegans* (Polychaeta: Serpulidae) for toxicological bioassays. *Periodicum Biologorum* **104**, 217–223.

Faulwetter S, Simboura N, Katsiaras N, Chatzigeorgiou G, Arvanitidis C (2017) Polychaetes of Greece: an updated and annotated checklist. *Biodiversity Data Journal* **5**, e20997.

Fauvel P (1909) Deuxième note préliminaire sur les Polychètes provenant des campagnes de l'Hirondelle et de la Princesse-Alice, ou déposées dans le Musée Océanographique de Monaco. *Bulletin de l'Institut Océanographique de Monaco* **142**, 1–76.

Fauvel P (1918) Annélides Polychètes des côtes d'Arabie récoltées par M. Ch. Pérez. *Bulletin du Muséum d'Histoire Naturelle, Paris* **24**, 329–344.

Fauvel P (1919) Annélides Polychètes des îles Gambier et Touamotou. *Bulletin du Muséum d'Histoire Naturelle, Paris* **25**, 336–343.

Fauvel P (1927) Polychètes sédentaires. Addenda aux errantes, Archiannélides, Myzostomaires. *Faune de France* **16**, 1–494.

Fauvel P (1930) Annélides polychètes de Nouvelle-Calédonie receuillies par Mme. A. Pruvot-Fol en 1928. *Archives de Zoologie Expérimentale et Générale* **69**, 501–562.

Fauvel P (1936) Annélides polychètes du Japon. *Memoirs of the College of Science, Kyoto Imperial University, Series B* **12**, 41–92.

Fauvel P (1947) Annélides polychètes de Nouvelle-Calédonie et des îles Gambier. *Faune Empire Français* **8**, 1–107.

Ferrario J, Minchin D (2017) Spread of the non-indigenous serpulid *Hydroides sanctaecrucis* Krøyer in Mörch, 1863 in the Pacific Ocean: a new record from Taiwan. *BioInvasions Records* **6**, 33–38.

Ferrero L, Obenat S, Zárate MA (2005) Mid-Holocene serpulid build-ups in an estuarine environment (Buenos Aires Province, Argentina). *Palaeogeography, Palaeoclimatology, Palaeoecology* **222**, 259–271.

Fiege D, Sun R (1999) Polychaeta from Hainan Island, South China Sea. Part I: Serpulidae (Annelida, Polychaeta, Serpulidae). *Senckenbergiana Biologica* **79**, 109–141.

Fischli H (1903) Polychäten von Ternate. *Abhandlungen Herausgegeben von der Senckenbergischen Naturforschenden Gesellschaft* **25**, 90–136.

Fitridge I, Dempster T, Guenther J, De Nys R (2012) The impact and control of biofouling in marine aquaculture: a review. *Biofouling* **28**, 649–669.

Folmer O, Black M, Hoeh W, Lutz R, Vrijenhoek R (1994) DNA primers for amplification of mitochondrial cytochrome c oxidase subunit I from diverse metazoan invertebrates. *Molecular Marine Biology and Biotechnology* **3**, 294–299.

Franzén A (1956) On spermatogenesis, morphology of spermatozoan, and biology of fertilization among invertebrates. *Zoologiska bidrag från Uppsala* **31**, 355–482.

Gaikwad UD, Ranade MR (1979) Studies on the marine fauna of Ratnagiri (West coast, India). II – Polychaeta (Annelida). *Bulletin of Fisheries Faculty of Konkan Agricultural University of India* **1**, 45–51.

Gibbs PE (1971) The polychaete fauna of the Solomon Islands. *Bulletin of the British Museum (Natural History) Zoology* **21**, 101–211.

Glasby TM, Connell SD, Holloway MG, Hewitt CL (2007) Non-indigenous biota on artificial structures: could habitat creation facilitate biological invasions? *Marine Biology* **151**, 887–895.

Gopalakrishnan S, Thilagam H, Raja PV (2007) Toxicity of heavy metals on embryogenesis and larvae of the marine sedentary polychaete *Hydroides elegans*. *Archives of Environmental Contamination and Toxicology* **52**, 171–178.

Goren M (1979) Succession of benthic community on artificial substratum at Elat (Red Sea). *Journal of Experimental Marine Biology and Ecology* **38**, 19–40.

Gourret P (1902) Sur quelques annélides sédentaires (gen. *Hydroides, Pomatoceros* et *Hermella*) du golfe de Marseille.

Comptes rendu de l'Association Française pour l'Avancement des Science **30**, 685–691.

Grave BH (1933) Rate of growth, age of sexual maturity, and duration of life of certain sessile organisms, at Woods Hole, Massachusetts. *The Biological Bulletin* **65**, 375–386.

Grave BH, Oliphant JF (1930) The longevity of unfertilized gametes. *Biological Bulletin* **59**, 233–239.

Gravier C (1906) Sur les annélides polychètes de la Mer Rouge (Serpulides). *Bulletin du Muséum d'Histoire Naturelle* **12**, 110–115.

Gravier C (1908 [1906]) Contribution à l'étude des annélides polychètes de la Mer Rouge. Suite. *Nouvelles Archives du Muséum* **4**, 67–168.

Green J (1972) Opercular regulation in *Hydroides dianthus* (Verrill, 1873). *Dissertation Abstracts International* **32B** (9), 5033.

Grosse M, Pérez R, Juan-Amengual M, Pons J, Capa M (2021) The elephant in the room-first record of invasive gregarious species of serpulids (calcareous tube annelids) in Majorca (Western Mediterranean). *Scientia Marina* **85**, 15–28.

Grube AE (1850) Die Familien der Anneliden. *Archiv für Naturgeschichte Berlin* **16**, 249–364.

Grube AE (1862) Mittheilungen über die Serpulen, mit besonderer Berücksichtigung ihrer Deckel. Jahresb. *JahresBericht und Abhandlungen der Schlesischen Gesellschaft in Breslau* **39**, 53–69.

Grube AE (1868) Beschreibungen einiger von Georg Ritter von Frauenfeld gesammelter Anneliden und Gephyreen des rothen Meeres. *Verhandlungen der kaiserlich-königlichen Zoologisch-botanischen Gesellschaft in Wien* **18**, 629–650.

Grube AE (1870) [1869]. Beschreibungen neuer oder weniger bekannter von Hrn. Ehrenberg gesammelter Anneliden des rothen Meeres. *Monatsberichte der Königlich Preussischen Akademie der Wissenschaften zu Berlin aus dem Jahre* **1869**, 484–521.

Grube AE (1872) Zur kritischen Übersicht der bisher beschriebenen Terebellen und über *Terebellides anguicomus* und einige Serpulaceen. *Jahres-Bericht Der Schlesischen Gesellschaft Fur VaterlandischeCultur* **49**, 48–53.

Grube AE (1878a) Einige neue anneliden aus Japan. *Jahres-Bericht der Schlesischen Gesellschaft für Vaterländische Cultur* **55**, 104–106.

Grube AE (1878b) Annulata Semperiana. Beiträge zur Kenntniss der Annelidenfauna der Philippinen. *Memoires de L'Academie Imperiale des Sciences de St. Petersbourg Ser. 7* **25**, 1–300.

Gruet Y, Hèral M, Robert JM (1976) Premières observations sur l'introduction de la faune associée au naissain d'huîtres japonaises *Crassostrea gigas* (Thunberg), importé sur la côte atlantique française. *Cahiers de Biologie Marine* **17**, 173–184.

Gunnerus JE (1768) Om Nogle Norske Coraller. *Det Kongelige Norske Videnskabers selskabs skrifter* **4**, 38–73.

Hadfield MG (1998) The D P Wilson lecture: Research on settlement and metamorphosis of marine invertebrate larvae: past, present and future. *Biofouling* **12**, 9–29.

Hadfield MG (2011) Biofilms and marine invertebrate larvae: what bacteria produce that larvae use to choose settlement sites. *Annual Review of Marine Science* **3**, 453–470.

Hadfield MG, Paul VJ (2001) Natural chemical cues for settlement and metamorphosis of marine invertebrate larvae. In *Marine Chemical Ecology*. (Eds McClintock JB, Baker BJ) pp. 431–461. CRC Press, Boca Raton.

Hadfield MG, Unabia CC, Smith CM, Michael TM (1994) Settlement preferences of the ubiquitous fouler *Hydroides elegans*. In *Recent Developments in Biofouling Control* (Eds MF Thompson, R Nagabhushanam, R Sarojini, and M Fingerman) pp. 65–74. Oxford & IBH Publishing Company, New Delhi.

Hadfield MG, Nedved BT, Wilbur S, Koehl MAR (2014) Biofilm cue for larval settlement in *Hydroides elegans* (Polychaeta): is contact necessary? *Marine Biology* **161**, 2577–2587.

Harder T, Qian P-Y (1999) Induction of larval attachment and metamorphosis in the serpulid polychaete *Hydroides elegans* by dissolved free amino acids: isolation and identification. *Marine Ecology – Progress Series* **179**, 259–271.

Harder T, Lau SC, Dahms HU, Qian PY (2002) Isolation of bacterial metabolites as natural inducers for larval settlement in the marine polychaete *Hydroides elegans* (Haswell). *Journal of Chemical Ecology* **28**, 2029–2043.

Hartman O (1945) The marine annelids of North Carolina. *Duke University Marine Station Bulletin* **2**, 1–54.

Hartman O (1952) Fouling serpulid worms, new to the Gulf of Mexico. *Texas Journal of Science* **4** (1), 63–64.

Hartman O (1954) Marine annelids from the northern Marshall Islands. *Professional Paper U.S. Geological Survey* **260Q**, 615–644.

Hartman O (1959) Catalogue of the polychaetous annelids of the world. Part II. *Occasional Papers of Allan Hancock Foundation* **23**, 354–628.

Hartmann-Schröder G (1962) Die Polychaeten des Eulitorals. In *Zur Kenntnis des Eulitorals der chilenischen Pazifikküste und der argentinischen Küste Südpatagoniens unter besonderer Berücksichtigung der Polychaeten und Ostracoden*. (Eds G Hartmann-Schröder, G Hartmann). *Mitteilungen aus dem Hamburgischen –Zoologischen Museum und Institut* **60**, *Supplement*, 57–270.

Hartwich G (1993) The types of Polychaeta in the Zoological Museum, Berlin. *Mitteilungen aus dem Zoologischen Museum in Berlin* **69**, 73–154.

Hassan M, (1998) Modification of carbonate substrata by bioerosion and bioaccretion on coral reefs of the Red Sea. PhD thesis. Kiel University, Germany.

Haswell WA (1883) On some new Australian tubicolous annelids. *Proceedings of the Linnean Society of New South Wales* **7**, 633–638.

Hayes KR, Sliwa C (2003) Identifying potential marine pests – a deductive approach applied to Australia. *Marine Pollution Bulletin* **46**, 91–98

Hayes KR, Sliwa C, Migus S, McEnnulty F, Dunstan P (2005) 'National priority pests: Part II. Ranking of Australian marine pests'. Department of Environment and Heritage,

Australian Government Department of Environment and Heritage, Canberra.

Hebert PD, Ratnasingham S, De Waard JR (2003) Barcoding animal life: cytochrome c oxidase subunit 1 divergences among closely related species. *Proceedings of the Royal Society, Biological Sciences* **270**, S96–S99.

Hedley RH (1956a) Studies of serpulid tube formation: I. The secretion of the calcareous and organic components of the tube by *Pomatoceros triqueter*. *Quarterly Journal of Microscopical Science* **97**, 411–419.

Hedley RH (1956b) Studies of serpulid tube formation: II. The calcium-secreting glands in the peristomium of *Spirorbis*, *Hydroides*, and *Serpula*. *Quarterly Journal of Microscopical Science* **97**, 421–427.

Hedley RH (1958) Tube formation by *Pomatoceros triqueter* (Polychaeta). *Journal of Marine Biological Association of the United Kingdom* **37**, 315–322.

Hill MB (1967) The life cycles and salinity tolerance of the serpulid *Mercierella enigmatica* (Fauvel) and *Hydroides uncinata* (Philippi) at Lagos, Nigeria. *Journal of Animal Ecology* **36**, 303–332.

Ho JS (1982) Copepoda associated with echinoderms of Japan. *Annual Report of the Sado Marine Biological Station, Niigata University* **12**, 33–61.

Holm ER, Nedved BT, Carpizo-Ituarte E, Hadfield MG (1998) Metamorphic signal transduction in *Hydroides elegans* (Polychaeta, Serpulidae) is not mediated by a G protein. *Biological Bulletin* **195**, 21–29.

Hou H (2013) Marine biofouling organisms respond to multiple stressors in a changing climate. PhD thesis. University of Hong Kong, Pokfulam.

Huang S, Hadfield MG (2003) Composition and density of bacterial biofilms determine larval settlement of the polychaete *Hydroides elegans*. *Marine Ecology – Progress Series* **260**, 161–172.

Huang Y-L, Dobretsov S, Xiong H, Qian P-Y (2007a) Effect of biofilm formation by *Pseudoalteromonas spongiae* on induction of larval settlement of the polychaete *Hydroides elegans*. *Applied and Environmental Microbiology* **73**, 6284–6288.

Huang Y-L, Dobretsov S, Ki J-S, Yang L-H, Qian P-Y (2007b) Presence of acyl-homoserine lactone in subtidal biofilm and the implication in larval behavioral response in the polychaete *Hydroides elegans*. *Microbial Ecology* **54**, 384–392.

Huang ZG, Mak PMS (1982) Studies on biofouling in Tolo Harbour. In *The Marine Flora and Fauna of Hong Kong and Southern China*. (Eds BS Morton, CK Tseng) pp. 767–787. *Proceedings of the First International Marine Biology Workshop 1980*. University Press, Hong Kong.

Huggett MJ, Nedved BT,Hadfield MG (2009) Effects of initial surface wettability on biofilm formation and subsequent settlement of *Hydroides elegans*. *Biofouling* **25**, 387–399.

Humann P, DeLoach N, Wilk L (2013) *Reef Creature Identification: Florida, Caribbean, Bahamas*. 3rd edn. New World Publications, Jacksonville.

Hung OS, Thiyagarajan V, Rudolf S, Wu S, Qian PY (2005) Effect of ultraviolet radiation on biofilms and subsequent larval settlement of *Hydroides elegans*. *Marne Ecology – Progress Series* **304**, 155–166.

Hung OS, Lee OO, Thiyagarajan V, He H-P, Xu Y, *et al.* (2009) Characterization of cues from natural multi-species biofilms that induce larval attachment of the polychaete *Hydroides elegans*. *Aquatic Biology* **4**, 253–262.

Hutchings PA (2008) Worms. In *The Great Barrier Reef: Biology, Environment and Management*. (Eds PA Hutchings, M Kingsford, and O Hoegh-Guldberg) pp. 246–261. CSIRO Publishing, Melbourne.

Hutchings P, Kupriyanova E (2018) Cosmopolitan polychaetes – fact or fiction? Personal and historical perspectives. *Invertebrate Systematics* **32**(1), 1–9.

Ichikawa A, Takagaki N (1942) The reversible asymmetry in the opercula of *Hydroides ezoensis*. I. Observations on the intact opercula. *Journal of Faculty of Science Hokkaido Imperial University. Zoology* **8**(6), 1–8.

ICZN; International Commission on Zoological Nomenclature; ITZN (1999) *International Code of Zoological Nomenclature*. 4th edn. The International Trust for Zoological Nomenclature, London.

Ilan M, Ben-Eliahu MN, Galil BS (1994) Three deep water sponges from the Eastern Mediterranean and their associated fauna. *Ophelia* **39**(1), 45–54.

Imajima M (1976a) Serpulid polychaetes from Tanega-shima, Southwest Japan. *Memoirs of the National Science Museum Tokyo* **9**, 123–143.

Imajima M (1976b) Serpulinae (Annelida, Polychaeta) from Japan. I. The genus *Hydroides*. *Bulletin of the National Science Museum* **2**, 229–248.

Imajima M (1978) Serpulidae (Annelida, Polychaeta) collected around Nii-jima and O-shima, Izu Islands. *Memoirs of the National Science Museum, Tokyo* **11**, 49–72.

Imajima M (1982) Serpulinae (polychaetous annelids) from the Palau and Yap Islands, Micronesia. *Proceedings of the Japanese Society of Systematic Zoology* **23**, 37–55.

Imajima M (1987) *Catalogue of Polychaetous Annelids (4). Information Retrieval on Species of the Family Serpulidae*. National Science Museum, Tokyo.

Imajima M, Hartman O (1964) The polychaetous Annelids of Japan. Part 2. *Allan Hancock Foundation Occasional Papers* **26**, 239–452.

Imajima M, ten Hove HA (1984) Serpulinae (Annelida, Polychaeta) from the Truk Island, Ponape and Majuro Atoll, with some other new Indo-Pacific records. *Proceedings of the Japanese Society of Systematic Zoology* **27**, 35–66.

Imajima M, ten Hove HA (1986) Serpulinae (Annelida, Polychaeta) from Nauru, the Gilbert Islands (Kiribati) and the Solomon Islands. *Proceedings of the Japanese Society of Systematic Zoology* **32**, 1–16.

Imajima M, ten Hove HA (1989) Two new species of serpulids (Annelida, Polychaeta) from Sesoko Island, Okinawa. *Bulletin of the National Science Museum, Tokyo, Series A* **15**, 11–17.

Ippolitov AP, Desai BG (2015) Ultrastructures of fossil serpulid tubes from the Vinjhan Formation (Miocene): first evidence of preserved lamello-fibrillar structure. In *The Fifteenth Indian Colloquium on Micropalaeontology and Stratigraphy*, 18–20 December, Aurangabad, p. 35.

Ippolitov AP, Vinn O, Kupriyanova EK, Jäger M (2014) Written in stone: history of serpulid polychaetes through time. *Memoirs of Museum Victoria* **71**, 123–160.

Ippolitov AP, ten Hove HA, Kosorukov V, Sun Y, Kupriyanova EK (2016) Ultrastructural and mineralogical diversity of *Hydroides* (Serpulidae) tubes: new insights into an old problem. *Program of the 12nd International Polychaete Conference, National Museum Wales*. 1-5 August, Cardiff.

Iroso I (1915) Ricerche sui nefridi di *Hydroides pectinata*. *Pubblicazioni della Stazione Zoologica di Napoli* **27**, 102–107.

Iroso I (1921) Revisione dei Serpulidi e Sabellidi del Golfo di Napoli. *Pubblicazioni della Stazione Zoologica di Napoli* **3**, 47–91.

Ishaq S, Mustaquim J (1996) Polychaetous annelids (order Sabellida) from the Karachi Coast, Pakistan. *Pakistan Journal of Marine Sciences* **5**, 161–197.

Jamieson BGM, Rouse GW (1989) The spermatozoa of the Polychaeta (Annelida): an ultrastructural review. *Biological Reviews* **64**, 93–157.

Jennings RM, Halanych K (2005) Mitochondrial genomes of *Clymenella torquata* (Maldanidae) and *Riftia pachyptila* (Siboglinidae): evidence for conserved gene order in Annelida. *Molecular Biology and Evolution* **22**, 210–222.

Jin T, Qian P-Y (2005) Amino acid exposure modulates the bioactivity of biofilms for larval settlement of *Hydroides elegans* by altering bacterial community components. *Marine Ecology – Progress Series* **297**, 169–179.

Jinbo U, Kato T, Ito M (2011) Current progress in DNA barcoding and future implications for entomology. *Entomological Science* **14**, 107–124.

Johnson PG, Vittor BA (1982) Segmented worms. In *A Field Guide to Coral Reefs of the Caribbean and Florida*. (Ed EG Kaplan) pp. 134–140. *The Peterson Field Guide Series*. Houghton Mifflin Company, Boston.

Jones ML (1962) On some polychaetous annelids from Jamaica, the West Indies. *Bulletin of the American Museum of Natural History* **124**, 169–212.

Joydas TV, Qurban MA, Al-Suwailem A, Krishnakumar PK, Nazeer Z, *et al.* (2012) Macrobenthic community structure in the northern Saudi waters of the Gulf, 14 years after the 1991 oil spill. *Marine Pollution Bulletin* **64**, 325–335.

Keough MJ, Raimondi PT (1995) Response of settling invertebrate larvae to bioorganic films: effects of different types of films. *Journal of Experimental Marine Biology and Ecology* **185**, 235–253.

Keppel E, Keith I, Ruiz GM, Carlton JT (2019) New records of native and non-indigenous polychaetes (Annelida: Polychaeta) in the Galapagos Islands. *Aquatic Invasions* **14**, 59–84.

Kinberg JGH (1867) Annulata nova. [Continuatio]. Öfversigt af Kunglige Vetenskabsakademiens Stockholm, Föhrhandlingar **23**, 337–357.

Kitsos M-S, Christodoulou M, Arvanitidis C, Mavidis M, Kirmitzoglou I, *et al.* (2005) Composition of the organismic assemblage associated with *Caretta caretta*. *Journal of Marine Biological Association of the United Kingdom* **85**, 257–261.

Knight-Jones P, Knight-Jones EW (1991) Ecology and distribution of Serpuloidea (Polychaeta) round South America. In *Proceedings of the 2nd International Polychaete Conference*. 18–23 August 1986, Copenhagen (Eds ME Petersen, JB Kirkegaard). *Ophelia Supplement* **5**, 579–586.

Knight-Jones P, Knight-Jones EW, Ergen Z (1991) Sabelliform polychaetes, mostly from Turkey's Aegean coast. *Journal of Natural History* **25**, 837–858. doi:10.1080/00222939100770561

Kubal P, Sukumaran S, Vijapure T (2012) Record of bi-operculate fouling serpulid polychaete, *Hydroides operculatus* (Treadwell, 1929) from Diu, West coast of India. *Indian Journal of Geo-Marine Sciences* **41**, 381–384.

Kudenov JD (1975) Sedentary polychaetes from the Gulf of California, Mexico. *Journal of Natural History*. **9**, 205–231.

Kumaraswamy Achari GP (1969) Catalogue of polychaetes in the reference collections of the Central Marine Fisheries Research Institute. *Bulletin of the Central Marine Fisheries Research Institute* **7**, 31–40.

Kupriyanova EK (2003) Evolution of life history in Serpulimorph polychaetes: a phylogenetic analysis. *Hydrobiologia* **496**, 105–114.

Kupriyanova EK, Nishi E, ten Hove HA, Rzhavsky AV (2001) Life-history patterns in Serpulimorph polychaetes: ecological and evolutionary perspectives. *Oceanography and Marine Biology: an Annual Review* **39**, 1–101.

Kupriyanova EK, McDonald TA, Rouse, GW (2006) Phylogenetic relationships within Serpulidae (Annelida: Polychaeta) inferred from molecular and morphological data. *Zoologica Scripta* **35**, 421–439.

Kupriyanova EK, Bastida-Zavala R, Halt MN, Lee MSY, Rouse GW (2008) Phylogeny of the *Serpula–Crucigera–Hydroides* clade (Serpulidae: Annelida) using molecular and morphological data: implications for operculum evolution. *Invertebrate Systematics* **22**, 425–437.

Kupriyanova EK, Sun Y, ten Hove HA, Wong E, Rouse GW (2015) Serpulidae (Annelida) of Lizard Island, Queensland, Australia. *Zootaxa Special Issue* **4019**, 275–353.

La Greca M (1950) Sullo sviluppo dello pseudopercolo di *Hydroides norvegica* Gunn. Anche in presenza dell'opercolo. *Annuario dell' Istituto e Museo di Zoologico dell' Università di Napoli* **2**, 1–14.

Lakshmana Rao MV (1969) Fouling serpulids from some Indian harbours. *Journal of the Timber Development Association of India* **15**, 1–20.

Lam C, Harder T, Qian P-Y (2003) Induction of larval settlement in the polychaete *Hydroides elegans* by surface-associated settlement cues of marine benthic diatoms. *Marine Ecology – Progress Series* **263**, 83–92.

Lam C, Harder T, Qian P-Y (2005) Induction of larval settlement in the polychaete *Hydroides elegans* by extracellular polymers of benthic diatoms. *Marine Ecology – Progress Series* **286**, 145–154.

Lamarck J-B M de (1816) *Histoire Naturelle des Animaux sans Vertèbres*. Tome second. Verdière, Paris.

Lambe LM (1893 [1894]) Sponges from the Pacific coast of Canada. *Transactions of the Royal Society of Canada* **11** (4), 25–43.

Lane AC, Mukherjee J, Chan VBS, Thiyagarajan V (2013) Decreased pH does not alter metamorphosis but compromises juvenile calcification of the tube worm *Hydroides elegans*. *Marine Biology* **160**, 1983–1993.

Lane A, Campanati C, Dupont S, Thiyagarajan V (2015) Transgenerational responses to low pH depend on parental gender in a calcifying tubeworm. *Science Reports* **5**, 10847.

Langeneck J, Lezzi M, del Pasqua M, Musco L, Gambi MC, *et al.* (2020) Non-indigenous polychaetes along the coasts of Italy: a critical review. *Mediterranean Marine Science* **21**, 238–275.

Lau MC, Chan KM, Leung KMY, Luan TG, Yang MS, *et al.* (2007) Acute and chronic toxicities of tributyltin to various life stages of the marine polychaete *Hydroides elegans*. *Chemosphere* **69**, 135–144.

Lau SCK, Qian PY (1997) Phlorotannins and related compounds as larval settlement inhibitors of the tube-building polychaete *Hydroides elegans*. *Marine Ecology – Progress Series* **159**, 219–227.

Lau SCK, Mak KKW, Chen F, Qian P-Y (2002) Bioactivity of bacterial strains isolated from marine biofilms in Hong Kong waters for the induction of larval settlement in the marine polychaete *Hydroides elegans*. *Marine Ecology – Progress Series* **1226**, 301–310.

Lau SCK, Harder T, Qian PY (2003) Induction of larval settlement in the serpulid polychaete *Hydroides elegans* (Haswell): role of bacterial extracellular polymers. *Biofouling* **19**, 197–204.

Laubier L (1966) Sur quelques Annélides Polychètes de la région de Beyrouth. *American University of Beirut Miscellanean Papers on Natural Sciences* **5**, 9–22.

Laverde-Castillo JJA (1988) Notas sobre algunos serpulidos (Annelida, Polychaeta) de Bahía Malaga, Pacífico colombiano. *Anales del Instituto de Investigaciones Marinas de Punta Betín* **18,** 83–93.

Lee OO, Qian P-Y (2003) Chemical control of bacterial epibiosis and larval settlement of *Hydroides elegans* in the Red Sponge *Mycale adherens*. *Biofouling* **19**, 171–180.

Lehrke J, ten Hove HA, Macdonald TA, Bartolomaeus T, Bleidorn C (2007) Phylogenetic relationships of Serpulidae (Annelida, Polychaeta) based on 18S rDNA sequence data and implications for opercular evolution. *Organisms. Diversity and Evolution* **7**, 195–206.

Lema KA, Constancias F, Rice SA, Hadfield MG (2019) High bacterial diversity in near-shore and oceanic biofilms and their influence on larval settlement by *Hydroides elegans* (Polychaeta). *Environmental Microbiology* **21**, 3472–3488.

Leone DE (1970) The maturation of *Hydroides dianthus*. *The Biological Bulletin* **138**, 306–315.

Leung JYS, Cheung NKM (2017) Feeding behaviour of a serpulid polychaete: turning a nuisance species into a natural resource to counter algal blooms? *Marine Pollution Bulletin* **115**, 376–382.

Leung JYS, Cheung NKM, Qiu JW, Ang PO, Chiu JMY, *et al.* (2013) Effect of parental hypoxic exposure on embryonic development of the offspring of two serpulid polychaetes: Implication for transgenerational epigenetic effect. *Marine Pollution Bulletin* **74**, 149–155.

Lewis JA, Smith BS (1991) *Hydroides* settlement in Sydney Harbour (Australia) and its control in sea-water cooling systems. In *Biodeterioration and Biodegradation 8* (Ed HW Rossmoore). pp. 464–466. Elsevier Applied Science, London.

Lewis JA, Watson C, ten Hove HA (2006) Establishment of the Caribbean serpulid tubeworm *Hydroides sanctaecrucis* Krøyer [in] Mörch, 1863, in northern Australia. *Biological Invasions* **8**, 665–671.

Li C, Chan VBS, He C, Meng Y, Yao H, *et al.* (2014) Weakening mechanisms of the serpulid tube in a high-CO_2 world. *Environmental Science and Technology* **48**, 14158–14167.

Li H-L, Song L-S, Qian P-Y (2008) Cyclic AMP concentration and protein kinase A (*PKA*) gene expression at different developmental stages of the polychaete *Hydroides elegans*. *Journal of Experimental Zoology Part B: Molecular and Developmental Evolution* **310B**, 417–427.

Link H, Nishi E, Tanaka K, Bastida-Zavala R, Kupriyanova EK, *et al.* (2009) *Hydroides dianthus* (Polychaeta: Serpulidae), and alien species introduced into Tokyo Bay, Japan. *Marine Biodiversity Records* **2**, 1–5.

Linnaeus C (1758) Systema Naturae 10 ed. Vol. 1. L. Salvius, Holmiae 1758.

Linnaeus C (1767) Systema Naturae per regna tria naturae, secundum classes, ordines, genera, species cum characteribus, differentiis, synonymis, locis: 12 ed. Vol. 1, Part 2. L. Salvius, Holmiae 1767, 533–1327.

Liu JY (2008) *Checklist of Marine Biota of China Seas*. China Science Press, Beijing. (In Chinese, with English preface).

Lommerzheim A (1981) Paläozäne Serpulidae und Spirorbidae (Polychaeta) von den Emperor Seamounts, NW-Pazifik. *Zitteliana* **7**, 31–54.

Long ER (1974) Marine fouling studies off Oahu, Hawaii. *Veliger* **17**, 23–36.

Lowenstam HA (1954) Status of invertebrate paleontology, 1953. XI. Systematic, paleoecologic and evolutionary aspects of skeletal building materials. *Bulletin of the Museum of Comparative Zoology Harvard* **112**, 287–317.

Lowenstam HA, Weimer S (1989) *On Biomineralization*. Oxford University Press, New York.

Ludwig HW (1957a) Morphologische und physiologische Untersuchungen zur Regeneration der Opercula der Serpuliden. *Verhandlungen der Deutschen Zoologischen Gesellschaft* **1956**, 272–277.

Ludwig HW (1957b) Weiteres zur Regeneration der Opercula von *Hydroides norvegica* (Gunn.). *Wilhelm Roux' Archiv für Entwicklungsmechanik* **149**, 602–612.

Ludwig W (1941) Untersuchungen zum Rechts-Links-Problem an *Hydroides norvegica*. *Die Naturwissenschaften* **29**, 423–424.

Ludwig W, Ludwig HW (1954) Untersuchungen zur kompensatorischen Regeneration an *Hydroides norvegica*. *Wilhelm Roux' Archiv für Entwicklungsmechanik* **147**, 259–287.

Ma Y, Liu P, Yu S, Li D, Cao S (2009) Inhibition of common fouling organisms in mariculture by epiphytic bacteria from the surfaces of seaweeds and invertebrates. *Acta Ecologica Sinica* **29**, 222–226.

Mak PMS (1982) The coral associated polychaetes of Hong Kong, with special reference to the serpulids. In *The marine flora and fauna of Hong Kong and Southern China* (eds BS Morton, CK Tseng) pp. 595–617. Proceedings of the First International Marine Biology Workshop 1980. Hong Kong University Press, Hong Kong.

Mak PMS, Huang ZG (1982) The salinity tolerance of the serpulid polychaete, *Hydroides elegans* (Haswell, 1883), and its possible applications in bio-antifouling. In *The marine flora and fauna of Hong Kong and Southern China* (eds BS Morton, CK Tseng) pp. 817–823. Proceedings of the First International Marine Biology Workshop 1980. Hong Kong University Press, Hong Kong.

Marenzeller E von (1884) Südjapanische Anneliden. II. Ampharetea, Terebellacea, Sabellacea, Serpulacea. *Denkschriften der Akademie der Wissenschaften, Wien* **49**, 197–224. [Note: often as Marenzeller (1885), in journal, but it was published as separate reprint in 1884].

Marion AF (1875a) Sur les Annélides de Marseille. *Revue des Sciences Naturelles, Montpellier* **4**, 301–312.

Marion AF (1875b) Draguages profonds au large de Marseille (Juillet-Octobre 1875). (Note préliminaire). *Revue des Sciences Naturelles de Montpellier* **4**, 469–477.

Matsuo R, Ko Y (1981) Preliminary notes on the development, growth and maturation of laboratory reared three species of *Hydroides* (Annelida: Polychaeta). *Bulletin of the Faculty of Fisheries, Nagasaki University* **51**, 23–28. (In Japanese).

Matsuo R, Yoshioshi K (1983) A scanning electron microscopical observation on the spermatozoa of three species of *Hydroides* (Polychaeta, Serpulidae). *Marine Fouling* **4**, 23–25.

McCloskey LR (1970) The dynamics of the community associated with a marine scleractinian coral. *Internationale Revue der gesamten Hydrobiologie und Hydrographie* **55**, 13–81.

McEdward LR, Qian P-Y (2001) Effects of the duration and timing of starvation during larval life on the metamorphosis and initial juvenile size of the polychaete *Hydroides elegans* (Haswell). *Journal of Experimental Marine Biology and Ecology* **261**, 185–197.

McIntosh WC (1923) *A Monograph of the British Marine Annelids. 4, 2. Polychaeta – Sabellidae to Serpulidae.* Ray Society, London.

McIntosh WC (1925) A second contribution to the marine polychaetes of South Africa. *Union South African Fisheries Marine Biological Survey, Report* **4**, 1–93.

Meng Y, Li C, Li H, Shih K, He C, *et al.* (2019) Recoverable impacts of ocean acidification on the tubeworm, *Hydroides elegans*: implication for biofouling in future coastal oceans. *Biofouling* **35**, 945–957.

Mesnil F, Fauvel P (1939) Polychètes sédentaires de l'Expédition du "Siboga". Maldanidae, Cirratulidae, Capitellidae, Sabellidae et Serpulidae. *Siboga Expedition* **24** (2), 1–42.

Miles CM, Wayne ML (2009) Life history trade-offs and response to selection on egg size in the polychaete worm *Hydroides elegans. Genetica* **135**, 289–298.

Miles CM, Hadfield MG, Wayne ML (2007) Heritability for egg size in the serpulid polychaete *Hydroides elegans. Marine Ecology – Progress Series* **340**, 155–162.

Miller JA, Gillman R, Carlton JT, Clarke Murray C, Nelson JC, *et al.* (2018) Trait-based characterization of species transported on Japanese tsunami marine debris: effect of prior invasion history on trait distribution. *Marine Pollution Bulletin* **132**, 90–101.

Mishra P, Kumar A, Nagireddy A, Mani DN, Shukla AK, *et al.* (2016) DNA barcoding: an efficient tool to overcome authentication challenges in the herbal market. *Plant Biotechnology Journal* **14**, 8–21.

Mitchell R, Kirchman D (1984) The microbial ecology of marine surfaces. In *Marine biodeterioration: an Interdisciplinary Study.* (Eds JD Costlow, RC Tipper) pp. 49–56. Naval Institute Press, Maryland.

Miura T, Kajihara T (1981) The development of a serpulid worm, *Hydroides ezoensis* (Annelida, Polychaeta). *Proceedings of the Japanese Society of Systematic Zoology* **20**, 7–12.

Miura T, Kajihara T (1984) An ecological study of the life histories of two Japanese serpulid worms, *Hydroides ezoensis* and *Pomatoleios kraussii.* In *Proceedings of the First International Polychaete Conference.* 4–9 July 1983, Sydney. (Ed. PA Hutchings) pp. 338–354. Linnaean Society of New South Wales, Sydney.

Moen TL (2006) A translation of Bishop Gunnerus' description of the species *Hydroides norvegicus* with comments on his *Serpula triqvetra. Scientia Marina* **70** (Supplement 3), 115–123.

Mohammad M-BM (1975) Competitive relationship between *Balanus amphitrite amphitrite* and *Pomatoleios kraussii* with special reference to their larval settlement. *Hydrobiologia* **46**, 1–15.

Mohammad M-BM (1981) Malformations in some polychaete annelids from Kuwait, Arabian Gulf. *Hydrobiologia* **78**, 129–131.

Mona MH (1992) Gametogenic cycles of the tube worm *Hydroides dirampha* (Polychaeta Serpulidae) from Lake Timsah (Suez Canal, Egypt). *Delta Journal of Science* **16**, 244–261.

Mona MH, Eissa SHH, Abdel-Gawad AM, Barbary MS (1994) Ultrastructural investigation of spermatogenesis in the tubeworm *Hydroides dirampha* (Polychaeta, Serpulidae). *Journal of the Egyptian German Society of Zoology* **13I**, 115–128.

Monro CCA (1930) Polychaete worms. *'Discovery' Reports* **2**, 1–222.

Monro CCA (1933) The Polychaeta Sedentaria collected by Dr. C. Crossland at Colón, in the Panama Region, and the Galapagos Islands during the Expedition of the S.Y. 'St. George'. *Proceedings of the Zoological Society of London* **103**, 1039–1092.

Monro CCA (1937) The John Murray Expedition 1933–34. Polychaeta. *Scientific Reports, Zoology* **4**, 243–321.

Monro CCA (1939) On a collection of Polychaeta from near the mouth of the River Congo. *Revue de Zoologie et de Botanique Africaines, Bruxelles* **32**, 213–225.

Monroy A (1954) Observations on the fertilization reaction in the eggs of some polychaetes and an ascidian, with special reference to the cortical changes. *Pubblicazioni della Stazione Zoologica di Napoli* **25**, 188–197.

Moran PJ (1984) Variability in the opercular structures of the serpulid polychaete *Hydroides elegans* (Haswell). *Proceedings of the Linnean Society of NSW* **107**(4), 487–492.

Mörch OAL (1863) Revisio critica Serpulidarum. *Et Bidrag til Rørormenes Naturhistorie. Naturhistorisk Tidsskrift, København, Series 3* **1**, 347–470.

Morton B, Britton JC (2000) Origins of the Azorean intertidal biota: the significance of introduced species, survivors of chance events. *Arquipelago (Life and Marine Science). Supplement* **2** (Part A), 29–51.

Morton B, Salvador A (2009) The biology of the zoning subtidal polychaete *Ditrupa arietina* (Serpulidae) in the Açores, Portugal, with a description of the life history of its tube. *Açoreana Suplemento* **6**, 145–156.

Mullineaux LS, Garland ED (1993) Larval recruitment in response to manipulated field flows. *Marine Biology* **116**, 667–683.

Murray A, Hutchings P, Pillai TG (2010) Note on *Hydroides malleolaspinus* from the Kimberleys of Western Australia (Polychaeta: Serpulidae). *Records of the Australian Museum* **62**, 393–394.

Nedved BT, Hadfield MG (2008) *Hydroides elegans* (Annelida: Polychaeta): a model for biofouling research. In *Marine and Industrial Biofouling* (Eds H-S Flemming, P Sriyutha Murthy, R Venkatesan, K Cooksey) pp. 203–218. Springer Series on Biofilms, Volume 4, Los Angeles.

Neff JM (1969) Mineral regeneration by serpulid polychaete worms. *Biological Bulletin* **136**, 76–90.

Neff JM (1971) Ultrastructural studies of the secretion of calcium carbonate by the serpulid polychaete worm *Pomatoceros caeruleus. Zeitschrift fur Zellforschung und Mikroskopische Anatomie* **120**, 160–186.

Nishi E (1992) Sperm morphology of serpulid polychaetes, *Pomatoleios kraussii* (Baird), *Spirobranchus giganteus corniculatus* Pallas, *Hydroides elegans* Haswell and *Salmacina dysteri* (Huxley). *Galaxea* **11**, 9–14.

Nishi E (1993) On the internal structure of calcified tube walls in Serpulidae and Spirorbidae (Annelida, Polychaeta). *Marine Fouling* **10**, 17–20.

Nonato EF, Luna JAC (1970) Anélideos poliquetas do nordeste do Brasil. I. Poliquetas bentônicos da costa de Alagoas e Sergipe. *Boletim do Instituto Oceanográfico* **19**, 57–130.

Nordback K (1956) On the oogenesis and fertilization of the serpulid *Hydroides norvegica* (Gunnerus). *Nytt Magasin for Zoologii* **4**, 121–123.

Nott JA, Parkes KR (1975) Calcium accumulation and secretion in the serpulid polychaete *Spirorbis spirorbis* L. at settlement. *Journal of Marine Biological Association of the United Kingdom* **55**, 911–921.

Nygren A, Pleijel F (2011) From one to ten in a single stroke – resolving the European *Eumida sanguinea* (Phyllodocidae, Annelida) species complex. *Molecular Phylogenetics and Evolution* **58**, 132–141.

Okada YK (1933) Remarks on the reversible asymmetry in the opercula of the polychaete *Hydroides. Journal of Marine Biological Association of the United Kingdom* **18**, 655–670.

Okamoto K, Watanabe A, Watanabe N, Sakata K (1995) Induction of larval metamorphosis in serpulid polychaetes by L-DOPA and catecholamines. *Fisheries Science (Tokyo)* **61**, 69–74.

Okamoto K, Watanabe A, Sakata K, Watanabe N (1998) Chemical signals involved in larval metamorphosis in *Hydroides ezoensis* (Serpulidae; Polychaeta). Part I: Induction of larval metamorphosis by extract of adult tube clumps. *Journal of Marine Biotechnology* **6**, 7–10.

Oken L (1832) Isis von Oken, Heft I-XII (mit 28 Tafeln). Leipzig beim Brockhaus.

Okuda S (1934) Some tubicolous annelids from Hokkaido. *Journal of the Faculty of Science, Hokkaido University, Series 6, Zoology* **3**, 233–246.

Okuda S (1937) Annelida polychaeta in Onagawa Bay and its vicinity. I. Polychaeta Sedentaria. *Science Reports of the Tohoku Imperial University. Fourth Series Biology* **12**, 45–69.

Otani M, Yamanishi R (2010) Distribution of the alien species *Hydroides dianthus* (Verrill, 1873) (Polychaeta: Serpulidae) in Osaka Bay, Japan, with comments on the factors limiting its invasion. *Plankton and Benthos Research* **5**, 62–68.

Palero F, Torrado H, Perry O, Kupriyanova EK, Ulman A, *et al.* (2020) Following the Phoenician example: Western Mediterranean colonization by *Spirobranchus* cf. *tetraceros* (Annelida: Serpulidae). *Scientia Marina* **84**, 83–92.

Palmer TA, Breaux N, Lebreton B, Guillou G, Beseres Pollack J (2022) Importance of serpulid reef to the functioning of a hypersaline estuary. *Estuaries and Coasts* **45**, 603–618.

Paul MD (1937) Sexual maturity of some organisms in the Madras Harbour. *Current Science Bangalore* **5**, 478–479.

Paul MD (1942) Studies on the growth and breeding of certain sedentary organisms in the Madras Harbour. *Proceedings of the Indian Academy of Sciences* **15**, 11–42.

Paul M, Abraham MJ, Menon NR (1998) Settlement pattern of the larvae of *Hydroides elegans* (Polychaeta) in the presence of petroleum hydrocarbon compounds. *Indian Journal of Marine Science* **27**, 489–491.

Pawlik JR (1992) Chemical ecology of the settlement of benthic marine invertebrates. *Oceanography and Marine Biology: an Annual Review* **30**, 273–335.

Pazoki S, Rahimian H, Struck TH, Katouzian AR, Kupriyanova EK (2020) A new species of the *Spirobranchus kraussii* complex (Annelida, Serpulidae) from the Persian Gulf and Gulf of Oman. *Zootaxa* **4748**, 401–430.

Pazoki S, Rahimian H, Struck TH (2021) Genetic diversity and population structure of three *Hydroides* species (Sedentaria, Serpulidae) in the Persian Gulf and Gulf of Oman, with the possible indication of heteroplasmy. *Systematics and Biodiversity* **19**, 993–1011.

Pechenik JA, Qian PY (1998) Onset and maintenance of metamorphic competence in the marine polychaete *Hydroides elegans* Haswell in response to three chemical cues. *Journal of Experimental Marine Biology and Ecology* **226**, 51–74.

Pechenik JA, Pearse JS, Qian P-Y (2007) Effects of salinity on spawning and early development of the tube-building polychaete *Hydroides elegans* in Hong Kong: not just the sperm's fault? *The Biological Bulletin* **212**, 151–160.

Pequegnat WE, Pequegnat LH (1968) 'Ecological aspects of marine fouling in the northeastern Gulf of Mexico'. Texas A&M University, Department of Oceanography, College Station, Texas.

Perkins TH, Savage T (1975) A bibliography and checklist of polychaetous annelids of Florida, the Gulf of Mexico, and the Caribbean region. *Florida Marine Research Publications* **14**, 1–62.

Pettengill JB, Wendt DE, Schug MD, Hadfield MG (2007) Biofouling likely serves as a major mode of dispersal for the polychaete tubeworm *Hydroides elegans* as inferred from microsatellite loci. *Biofouling* **23**, 161–169.

Philippi A (1844) Einige Bemerkungen über die Gattung *Serpula*, nebst Aufzählung der von mir im Mittelmeer mit dem Thier beobachteten Arten. *Archiv für Naturgeschichte Berlin* **10**, 186–198.

Phuttapreecha R, Kajonwattanakul S, Songkai P, Choamanee C (2018) Survey of fouling organisms at Songkhla Port in Thailand. *Asean Journal on Science & Technology for Development* **35**, 147–152.

Pillai TG (1960) Some marine and brackish water serpulid polychaeta from Ceylon, including new genera and species. *Ceylon Journal of Science (Biological Sciences)* **3**, 1–40.

Pillai TG (1961) Annelida Polychaeta of Tambalagam Lake, Ceylon. *Ceylon Journal of Science (Biological Sciences)* **4**, 1–40.

Pillai TG (1965) Annelida Polychaeta from the Philippines and Indonesia. *Ceylon Journal of Science (Biological Sciences)* **5**, 110–177.

Pillai TG (1971) Studies on a collection of marine and brackish-water polychaete annelids of the family Serpulidae from Ceylon. *Ceylon Journal of Science (Biological Sciences)* **9**, 88–130.

Pillai TG (1972) A review and revision of the systematics of the genera *Hydroides* and *Eupomatus* together with an account of their phylogeny and zoogeography. *Ceylon Journal of Science (Biological Sciences)* **10**, 7–31.

Pillai TG (2009) Descriptions of new serpulid polychaetes from the Kimberleys of Australia and discussion of Australian and Indo-West Pacific species of *Spirobranchus* and superficially similar taxa. *Records of the Australian Museum* **61**, 93–199.

Pixell HLM (1913) Polychaeta of the Indian Ocean, together with some species from the Cape Verde Islands. The Serpulidae, with a classification of the genera *Hydroides* and *Eupomatus*. *The Transactions of the Linnean Society of London (2, Zoology)* **16**, 69–92.

Pleijel F, Rouse G, Nygren A (2009) Five colour morphs and three new species of *Gyptis* (Hesionidae, Annelida) under a jetty in Edithburgh, South Australia. *Zoologica Scripta* **38**, 89–99.

Poore GCB, Rainer SF, Spies RB, Ward E (1975) The zoobenthos program in Port Phillip Bay, 1969-'73. *Fisheries and Wildlife Paper, Victoria* **7**, 1–78.

Pratt HS (1916) *A Manual of the Common Invertebrate Animals, Excluding Insects*. AC McClurg and Co., Chicago.

Puccia E, Durante M (1973) The compensatory regeneration of the operculum of *Hydroides norvegica*: identification of the inhibiting substance. *Journal of Experimental Zoology* **184**, 1–6.

Qian PY (1999) Larval settlement of polychaetes. *Hydrobiologia* **402**, 239–253.

Qian PY, Pechenik JA (1998) Effects of larval starvation and delayed metamorphosis on juvenile survival and growth of the tube-dwelling polychaete *Hydroides elegans*. *Journal of Experimental Marine Biology and Ecology* **227**, 169–185.

Qian PY, Xu SY (2012) Antifouling activity of marine natural products. In *Handbook of Marine Natural Products* (Eds E Fattorusso, O Taglialatela-Scafati, WH Gerwick). pp. 1–73. Springer, Dordrecht.

Qian PY, Lau SCK, Dahm HU, Dobretsov S, Harder T (2007) Marine biofilms as mediators of colonization by marine macroorganisms: implications for antifouling and aquaculture. *Marine Biotechnology* **9**, 399–410.

Qiu JW, Qian PY (1997) Combined effects of salinity, temperature and food on early development of the polychaete *Hydroides elegans*. *Marine Ecology – Progress Series* **152**, 79–88.

Qiu JW, Qian PY (1998) Combined effects of salinity and temperature on juvenile survival, growth and maturation in the polychaete *Hydroides elegans*. *Marine Ecology – Progress Series* **168**, 127–134.

Qiu JW, Tang X, Zheng C, Li Y, Huang Y (2007) Copper complexation by fulvic acid affects copper toxicity to the larvae of the polychaete *Hydroides elegans*. *Marine Environmental Research* **64**, 563–573.

Quatrefages A de (1866) [1865] *Histoire Naturelle des Annelés Marins e' d'eau Douce.* Annélides et Géphyriens. Librairie Encyclopédique de Rôret, Paris.

Radwańska U (1994) Tube-dwelling polychaetes from the Korytnica Basin (Middle Miocene; Holy Cross Mountains, Central Poland). *Acta Geologica Polonica* **44**, 35–83.

Rafinesque CS (1815) *Analyse de la Nature ou Tableau de l'Univers et des Corps Organisés*. Aux dépens de l'Auteur, Palerme.

Ranzoli F (1962) Il falso gonocorismo nel ciclo biologico del polichete *Hydroides norvegica* (Gunn). *Archivio Zoologico, Napoli* **47**, 1–19.

Read GB, Fauchald K (2020) World Register of Marine Species. Polychaeta (2021). *Hydroides* Gunnerus, 1768, <http://www.marinespecies.org>.

Read GB, ten Hove HA, Sun Y, Kupriyanova EK (2017) *Hydroides* Gunnerus, 1768 (Annelida, Serpulidae) is feminine: a nomenclatural checklist of updated names. *Zookeys* **642**, 1–52.

Rech S, Salmina S, Pichs YJB, García-Vazquez E (2018) Dispersal of alien invasive species on anthropogenic litter from European mariculture areas. *Marine Pollution Bulletin* **131**, 10–16.

Reish DJ (1961) The relationship of temperature and dissolved oxygen to the seasonal settlement of the polychaetous annelid *Hydroides norvegica* (Gunnerus). *Bulletin of the Southern California Academy of Sciences* **60**, 1–11.

Reish DJ (1968) Polychaetous annelids of the Marshall Islands. *Pacific Science* **22**, 208–231.

Reish DJ, Pernet B (2009) Chapter 4. Annelid life cycle cultures. In *Annelids in Modern Biology.* (Ed DH Shain) pp. 47–62. John Wiley & Sons, New York.

Rioja E (1919) Una curiosa anomalía del '*Hydroides norvegica*' Gunn. y algunas consideraciones acerca de la filogenia de los serpúlidos. *Boletín de la Sociedad Española de Historia Natural, Madrid* **19**, 445–449.

Rioja E (1934) Consideraciones acerca de la sistematica de los generos *Serpula, Crucigera* e *Hydroides* (Annelida, Serpulidae). *Boletín de la Sociedad Española de Historia Natural, Madrid* **34**, 205–209.

Rioja E (1941a) Estudios anelidológicos. II. Observaciones acerca de varias especies del genero *Hydroides* Gunnerus (*sensu* Fauvel) de las costas Mexicanas del Pacifico. *Anales del Instituto de Biologia, Mexico* **12**, 161–175.

Rioja E (1941b) Estudios Anelidologicos. III. Datos para el conocimiento de la fauna de poliquetos de las costas del pacifico de Mexico. *Anales del Instituto de Biologia, Mexico* **12**, 669–746.

Rioja E (1942) Estudios anelidológicos. IV. Observaciónes sobre especies de Serpulidos de las costas del Pacífico de México, con descripción de una especie nueva del genero *Hydroides. Anales del Instituto de Biologia, Mexico* **13**, 125–135.

Rioja E (1944) Estudios anelidológicos. XII. Observaciónes acerca del operculo de *Hydroides crucigera* Mörch y descripcíon de un caso de duplicidad de este organo. *Anales del Instituto de Biologia, Mexico* **15**, 409–414.

Rioja E (1946) Estudios anelidológicos. XIV. Observaciónes sobre algunos poliquetos de las costas del Golfo de México. *Anales del Instituto de Biologia, Mexico* **17**, 193–204.

Rioja E (1947) Estudios anelidológicos. XVI. Evolución y significado sistemática del opérculo de los Serpulidae. *Anales del Instituto de Biologia, Mexico* **18**, 189–196.

Rioja E (1958) Estudios anelidologicos. XXII. Datos para el conocimiento de la fauna de anelidos poliquetos de las costas orientales de Mexico. *Anales del Instituto de Biologia, Mexico* **29** (1/2), 219–301.

Rioja E (1959) Estudios anelidológicos. XXI. Observaciónes acerca de algunas especies de Serpulidos de los generos *Hydroides* y *Eupomatus* de las costas del Golfo de México. *Anales del Instituto de Biologia, Mexico* **28**, 247–266.

Robertson R, Mau-Lastovicka T (1979) The ectoparasitism of *Boonea* and *Fargoa* (Gastropoda: Pyramidellidae). *The Biological Bulletin* **157**, 320–333.

Rodrigues AR, Skinner LF, Brasil ACdS (2020) Do morphological similarities and human-induced dispersal explain the non-native occurrence of Serpulidae (Annelida) in Southwest Atlantic? Taxonomic detailing is the key. *Papéis Avulsos de Zoologia* **60**, 1–15.

Rovereto G (1904) Studi monografici sugli Anellidi fossili, 1, Terziario. *Palaeontographia Italica. Memorie di Paleontologia* **10**, 1–73.

Rullier F (1972) Annélides Polychètes de Nouvelle Calédonie recueillies par Y. Plessis et B. Salvat. *Expédition Française sur les Récifs Coralliens de la Nouvelle-Calédonie* **6**, 1–169.

Rullier F, Amoureux L (1979) Campagne de la Calypso au large des côtes Atlantiques de l'Amérique du Sud I. 33. Annélides Polychètes. *Annales de l'Institut Océanographique* **55**, 145–206.

Rzhavsky AV, Kupriyanova EK, Sikorski AV, Dahle S (2014) *Calcareous Tubeworms (Polychaeta, Serpulidae) of the Arctic Ocean.* KMK Scientific Press, Moscow.

Saint-Joseph A de (1895) Les annélides polychètes des Côtes de Dinard. IV. Appendice. *Annales des Sciences Naturelles, Paris* **20**, 185–272.

Sandonnini J, Ruso YD, Melendreras EC, Casalduero FG (2021) Massive aggregations of Serpulidae associated with eutrophication of the Mar Menor, Southeast Iberian Peninsula. *Frontiers in Marine Science* **7**, 531726

Satheesh S, Wesley SG (2006) Polychaetes associated with the brown alga *Sargassum* sp. in Kudankulam waters, east coast of India. *Ecology, Environment and Conservation Paper* **12**, 323–326.

Scheltema RS (1974) Biological interactions determining larval settlement of marine invertebrates. *Thalassia Jugoslavica* **10**, 263–269.

Scheltema RS, Williams IP, Shaw MA, Loudon C (1981) Gregarious settlement by the larvae of *Hydroides dianthus* (Polychaeta: Serpulidae). *Marine Ecology – Progress Series* **5**, 69–74.

Schmarda LK (1861) *Neue wirbellose Thiere beobachtet und gesammelt auf einer Reise um die Erde 1853 bis 1857.* I. *Turbellarien, Rotatorien und Anneliden*, Wilhelm Engelman, Leipzig.

Schmidt O (1870) *Grundzüge einer Spongien-Fauna des atlantischen Gebietes.* Wilhelm Engelmann, Leipzig.

Schmidt W (1950) Neue *Serpula*-Arten aus dem Material des Naturhistorischen Museums in Wien. *Annalen des Naturhistorischen Museums in Wien* **57**, 159–162.

Schmidt WJ (1951) Neue Serpulidae aus dem tertiären Wiener-Becken. *Annalen des Naturhistorischen Museums in Wien* **58**, 77–84.

Schmidt WJ (1955) Die Tertiären Würmer Österreichs. Österreichische Akademie der Wissenschaften. Mathematisch-Naturwissenschaftliche Klasse. Denkschriften **109**, 1–121.

Schochet J (1973a) Opercular regulation in the polychaete *Hydroides dianthus* (Verrill, 1873) I. Opercular ontogeny, distribution and flux. *The Biological Bulletin* **144**, 400–420.

Schochet J (1973b) Opercular regulation in the polychaete *Hydroides dianthus* (Verrill, 1873) 2. Control of opercular regulation. *Journal of Experimental Zoology* **184**, 259–280.

Schwan IDS, Brasil ACDS, Neves D, Dias GM (2016) The invasive worm *Hydroides elegans* (Polychaeta: Serpulidae) in southeastern Brazil and its potential to dominate hard substrata. *Marine Biology Research* **12**, 96–103.

Seaver EC, Thamm K, Hill SD (2005) Growth patterns during segmentation in the two polychaete annelids, *Capitella* sp. I and *Hydroides elegans*: comparisons at distinct life history stages. *Evolution and Development* **7**, 312–326.

Seixas VC, Russo CAM, Paiva PC (2017) Mitochondrial genome of the Christmas tree worm *Spirobranchus giganteus* (Annelida: Serpulidae) reveals complete different aspects within mitochondrial Annelida evolution. *Gene* **605**, 43–53.

Sentz É (1962) Étude morphologique des Serpulidae *Hydroides norvegica* (Gunnerus) et *Serpula concharum* Langerhans et de leur formes intermédiaires. *Vie et Milieu* **13**, 441–452

Sentz-Braconnot É (1964) Sur le développement des Serpulidae *Hydroides norvegica* (Gunnerus) et *Serpula concharum* Langerhans. *Cahiers de Biologie Marine* **5**, 385–389.

Sentz-Braconnot É (1968) Données écologiques et biologiques sur la fixation des Serpulidae dans la rade de Villefranche-sur-Mer (Alpes Maritimes). *Vie et Milieu (B, Océanographie)* **1**, 109–132.

Shikuma NJ, Hadfield MG (2006) Temporal variation of an initial marine biofilm community and its effects on larval settlement and metamorphosis of the tubeworm *Hydroides elegans*. *Biofilms* **2**, 23–238.

Shikuma NJ, Antoshechkin I, Medeiros JM, Pilhofer M, Newman DK (2016) Stepwise metamorphosis of the tubeworm *Hydroides elegans* is mediated by a bacterial inducer and MAPK signalling. *Proceedings of the National Academy of Sciences of the United States of America* **113**, 10097–10102.

Shin PKS, Leung JYS, Qiu JW, Ang PO, Chiu JMY, *et al.* (2013) Hypoxia induces abnormal larval development and affects biofilm–larval interaction in the serpulid polychaete *Hydroides elegans*. *Marine Pollution Bulletin* **76**, 291–297.

Sigovini M, Keppel E, Tagliapietra D (2016) Open nomenclature in the biodiversity era. *Methods in Ecology and Evolution* **7**, 1217–1225.

Simkiss K, Wilbur K (1989) *Biomineralization: Cell Biology and Mineral Deposition*. Academic Press, New York.

Simon C, van Niekerk H, Burghardt I, ten Hove H, Kupriyanova EK (2019) Not out of Africa: *Spirobranchus kraussii* (Baird, 1865) is not a global fouling and invasive serpulid of Indo-Pacific origin. *Aquatic Invasions* **14**, 221–249.

Smith AM, Riedi MA, Winter DJ (2013) Temperate reefs in a changing ocean: skeletal carbonate mineralogy of serpulids. *Marine Biology* **160**, 2281–2294.

Smith SV, Haderlie EC (1969) Growth and longevity of some calcareous fouling organisms, Monterey Bay, California. *Pacific Science* **23**, 447–451.

Solis-Weiss V, Bertrand Y, Helleouet M-N, Pleijel F (2004) Types of polychaetous annelids at the *Museum National d'Histoire Naturelle, Paris*. *Zoosystema* **26**, 377–384.

Spalding MD, Fox HE, Allen GR, Davidson N, Ferdaña ZA, *et al.* (2007) Marine ecoregions of the world: a bioregionalization of coastal and shelf areas. *BioScience*, **57** (7), 573–583.

Srinivasagam RT (1966) Effect of biological conditioning of sea water on development of a sedentary polychaete. *Nature* **212**, 742–743.

Stock JH (1995) Two new copepods parasitic on Caribbean polychaetes. *Studies on the Natural History of the Caribbean Region* **72**, 1–11.

Stossich M (1882) Prospetto della Fauna del mare Adriatico. Parte IV. Vermes. *Bollettino della Società Adriatica di Scienze Naturali in Trieste* **7**, 97–171.

Straughan D (1967a) Marine Serpulidae (Annelida: Polychaeta) of eastern Queensland and New South Wales. *Australian Journal of Zoology* **15**, 201–261.

Straughan D (1967b) Some Serpulidae (Annelida: Polychaeta) from Heron Island, Queensland. *University of Queensland Papers* **1**, 27–45.

Straughan D (1967c) A small collection of serpulid worms (Annelida: Polychaeta) from Darwin, Northern Territory, Australia. *Australian Zoologist* **14**, 222–225.

Straughan D (1969) Serpulidae (Annelida, Polychaeta) from Oahu, Hawaii. *Bulletin of Southern California Academy of Sciences* **68**, 229–240.

Streftaris N, Zenetos A (2006) Alien marine species in the Mediterranean – the 100 'worst invasives' and their impact. *Mediterranean Marine Science* **7**, 87–118.

Sun R, Yang D (2000) Study on *Hydroides* (Polychaeta: Serpulidae) from waters off China. 1. *Studia Marina Sinica* **42**, 116–135.

Sun R, Yang D (2014) Annelida Polychaeta III Sabellida. *Fauna Sinica Invertebrata* **54**, 1–493.

Sun Y (2011) Biodiversity and phylogeny of coral-associated polychaetes. MSc thesis. Hong Kong Baptist University, Hong Kong.

Sun Y (2017) Taxonomy, barcoding and phylogeny of *Hydroides* (Serpulidae, Annelida), the largest genus of notorious fouling and invading calcareous tubeworms. PhD thesis. Macquarie University, Australia.

Sun Y, ten Hove HA, Qiu J-W (2012a) Serpulidae (Annelida: Polychaeta) from Hong Kong. *Zootaxa* **3424**, 1–42.

Sun Y, Kupriyanova EK, Qiu JW (2012b) COI barcoding of *Hydroides*: a road from impossible to difficult. *Invertebrate Systematics* **26**, 539–547.

Sun Y, Wong E, ten Hove HA, Hutchings PA, Williamson JE, *et al.* (2015) Revision of the genus *Hydroides* (Annelida: Serpulidae) from Australia. *Zootaxa* **4009**, 1–99.

Sun Y, Wong E, Tovar-Hernández MA, Williamson JE, Kupriyanova EK (2016) Is *Hydroides brachyacantha* (Serpulidae: Annelida) a widespread species? *Invertebrate Systematics* **30**, 41–59.

Sun Y, Wong E, Keppel E, Williamson J, Kupriyanova EK (2017a) A global invader or a complex of regionally distributed species? Clarifying the status of an invasive calcareous tubeworm *Hydroides dianthus* (Verrill, 1873) using barcoding. *Marine Biology* **164**, 28-1 – 28-12.

Sun Y, Al-Kandari M, Kubal P, Walmiki N, Kupriyanova EK (2017b) Cutting a Gordian knot of tubeworms with DNA data: the story of the *Hydroides operculata*-complex (Annelida, Serpulidae). *Zootaxa* **4323**, 39–48.

Sun Y, Wong E, Williamson JE, Hutchings PA, Ahyong ST, *et al.* (2018) Barcoding and multi-gene based phylogeny and biogeography of the globally distributed calcareous tubeworm genus *Hydroides* Gunnerus, 1768 (Annelida, Serpulidae). *Molecular Phylogenetics and Evolution* **127**, 732–745.

Sun Y, Daffe G, Zhang Y, Qiu JW, Kupriyanova EK (2021) Another blow to the conserved gene order in Annelida: mitogenomics reveals extensive gene rearrangement in the tubeworm genus *Hydroides* (Serpulidae). *Molecular Phylogenetics and Evolution* **160**, 107124.

Tanur A, Gunari N, Sullan RM, Kavanagh C, Walker G (2010) Insights into the composition, morphology, and formation of the calcareous shell of the serpulid *Hydroides dianthus*. *Journal of Structural Biology* **169**, 145–160.

ten Hove HA (1970) Additional notes on the polychaetes of the Iranian Gulf. *Steenstrupia* **1**, 55–60.

ten Hove HA (1973) Serpulinae (Polychaeta) from the Caribbean: II. The genus *Sclerostyla*. *Studies on the Fauna of Curaçao and other Caribbean Islands* **43,** 1–21.

ten Hove HA (1974) Notes on *Hydroides elegans* (Haswell, 1883) and *Mercierella enigmatica* Fauvel, 1923, alien serpulid polychaetes introduced into the Netherlands. *Bulletin Zoölogisch Museum, Universiteit van Amsterdam* **4**, 45–51.

ten Hove HA (1984) Towards a phylogeny in serpulids (Annelida; Polychaeta). In *Proceedings of the First International Polychaete Conference*. 4–9 July 1983, Sydney. (Ed. PA Hutchings) pp. 181–196. Linnaean Society of New South Wales, Sydney.

ten Hove HA (1989) Serpulinae (Polychaeta) from the Caribbean: IV – *Pseudovermilia madracicola* sp. n., a symbiont of corals. *Studies in honour of Dr. Pieter Wagenaar Hummelinck. Foundation for Scientific Research in Surinam and the Netherlands* **123**, 135–144.

ten Hove HA (1990) Description of *Hydroides bulbosus* sp. nov. (Polychaeta, Serpulidae), from the Iranian Gulf, with a terminology for opercula of *Hydroides*. *Beaufortia* **41,** 115–120.

ten Hove HA (1994) Serpulidae (Annelida: Polychaeta) from the Seychelles and Amirante Islands. In *Oceanic Reefs of the Seychelles. Report on a cruise of RV Tyro to the Seychelles in 1992 and 1993.* (Ed. J van der Land). *Netherlands Indian Ocean Programme Cruise Reports* 2, 107–116.

ten Hove HA, Ben-Eliahu MN (2005) On the identity of *Hydroides priscus* Pillai 1971 – taxonomic confusion due to ontogeny in some serpulid genera (Annelida: Polychaeta: Serpulidae). *Senckenbergiana Biologica* **85**, 127–145.

ten Hove HA, Jansen-Jacobs MJ (1984) A revision of the genus *Crucigera* (Polychaeta, Serpulidae); a proposed methodical approach to serpulids, with special reference to variation in *Serpula* and *Hydroides*. In *Proceedings of the First International Polychaete Conference*. 4–9 July 1983, Sydney. (Ed. PA Hutchings) pp. 143–180. Linnaean Society of New South Wales, Sydney.

ten Hove HA, Kupriyanova EK (2009) Taxonomy of Serpulidae (Annelida, Polychaeta): the state of affairs. *Zootaxa* **2036**, 1–126.

ten Hove HA, Nishi E (1996) A redescription of the Indo-West Pacific *Spirobranchus corrugatus* Straughan, 1967 (Serpulidae, Polychaeta), and an alternative hypothesis on the nature of a group of Middle Miocene microfossils from Poland. *Beaufortia* **46**, 83–96.

ten Hove HA, San Martín G (1995) Serpulidae (Polychaeta) procedentes de la I Expedición Cubano-Española a la Isla de la Juventud y Archipiélago de los Canarreos (Cuba). *Studies Natural History of the Caribbean Region* **72**, 13–24.

ten Hove HA, Wolf PS (1984) Family Serpulidae Johnston, 1865. In *Taxonomic Guide to the Polychaetes of the Northern Gulf of Mexico* (Eds JM Uebelacker, PG Johnson) pp. 55–1 to 55–34. Barry A. Vittor & Associates, Mobile, Alabama.

Thilagam H, Gopalakrishnan S, Vijayavel K, Raja PV (2008) Effluent toxicity test using developmental stages of the marine polychaete *Hydroides elegans*. *Archives of Environmental Contamination and Toxicology* **54**, 674–683.

Thorp CH, Pyne S, West SA (1987) *Hydroides ezoensis* Okuda, a fouling serpulid new to British coastal waters. *Journal of Natural History* **21**, 863–877.

Toonen RJ, Pawlik JR (1994) Foundations of gregariousness. *Nature* **370**, 511–512.

Toonen RJ, Pawlik JR (1996) Settlement of the gregarious tube worm *Hydroides dianthus* (Polychaeta: Serpulidae): cues for gregarious settlement. *Marine Biology* **126**, 725–733.

Toonen RJ, Pawlik JR (2001a) Settlement of the gregarious tube worm *Hydroides dianthus* (Polychaeta: Serpulidae). I. Gregarious and nongregarious settlement. Marine Ecology – *Progress Series* **244**, 103–114.

Toonen RJ, Pawlik JR (2001b) Settlement of the gregarious tube worm *Hydroides dianthus* (Polychaeta: Serpulidae). II. Testing the desperate larva hypothesis. Marine Ecology – *Progress Series* **224**, 115–131.

Toonen RJ, Bowen BW, Iacchei M, Briggs JC (2016) Biogeography, Marine. In *Encyclopedia of Evolutionary Biology* (Ed. RM Kliman) pp. 166–178. Academic Press, Oxford.

Tovar-Hernández MA, Méndez N, Villalobos-Guerrero TF (2009) Fouling polychaete worms from the Southern Gulf of California: Sabellidae and Serpulidae. *Systematics and Biodiversity* **7**, 319–336.

Tovar-Hernández MA, Villalobos TF, Kupriyanova EK, Sun Y (2016) A new fouling *Hydroides* (Serpulidae, Sabellida, Annelida) from southern Gulf of California. *Journal of Marine Biological Association of the United Kingdom* **96**, 693–705.

Treadwell AL (1902) The polychaetous annelids of Porto Rico. *Bulletin of the United States Fish Commission* **20**, 181–210.

Treadwell AL (1929) New species of polychaetous annelids in the collections of the American Museum of Natural History from Porto Rico, Florida, Lower California, and British Somaliland. *American Museum Novitates* **392**, 1–13.

Treadwell AL (1931) Three new species of polychaetous annelids in the collections of the United States National Museum. *Proceedings of the United States National Museum* **80**, 1–6.

Turner HJ, Hanks JE (1960) Experimental stimulation of gametogenesis in *Hydroides dianthus* and *Pecten irradians* during the winter. *Biological Bulletin* **119**, 145–152.

Uchida H (1978) Serpulid tube worms (Polychaeta, Sedentaria) from Japan with the systematic review of the group. *Bulletin of the Marine Park Research Stations* **2**, 1–98.

Ulman A, Ferrario J, Forcada A, Arvanitidis C, Occhipinti Ambrogi A, Marchini A (2019) A hitchhiker's guide to Mediterranean marina travel for alien species. *Journal of Environmental Management* **241**, 328–339.

Unabia CRC, Hadfield MG (1999) Role of bacteria in larval settlement and metamorphosis of the polychaete *Hydroides elegans*. *Marine Biology* **133**, 55–64.

Uschakov PV [1955] (1965) *Polychaeta of the Far Eastern Seas of the U.S.S.R.* Israel Program for Scientific Translations, Jerusalem.

Uschakov PV (1970) Observations sur la répartitions de la fauna benthique du littoral guinéen. *Cahiers de Biologie Marine* **11**, 435–457.

Uschakov PV, Wu BL (1963) A preliminary zoogeographical studies (sic) on Polychaeta of the Yellow Sea. *Oceanologia Limnologia Sinica* **5**, 154–164. [in Chinese and Russian].

Vasconcelos P, Cúrdia J, Castro M, Gaspar MB (2007) The shell of *Hexaplex (Trunculariopsis) trunculus* (Gastropoda: Muricidae) as a mobile hard substratum for epibiotic polychaetes (Annelida: Polychaeta) in the Ria Formosa (Algarve coast – southern Portugal). *Hydrobiologia* **575**, 161–172.

Verrill AE (1873) 'VIII. Report upon the invertebrate animals of Vineyard Sound and the adjacent waters, with an account of the physical characters of the region'. United States Government Printing Office, Washington.

Vijayan N, Hadfield MG (2020) Bacteria known to induce settlement of larvae of *Hydroides elegans* are rare in natural inductive biofilm. Aquatic Microbial Ecology **84**, 31–42.

Vijayan N, Lema KA, Nedved BT, Hadfield MG (2019) Microbiomes of the polychaete *Hydroides elegans* (Polychaeta: Serpulidae) across its life-history stages. *Marine Biology* **166**, 19.

Vijayaragavan S, Vivek Raja P (2018) The toxic effects of phorate on early embryonic stages of sedentary polychaete *Hydroides elegans* (Haswell, 1883). *The Journal of Basic and Applied Zoology* **79**, 45.

Vine PJ, Bailey-Brock JH (1984) Taxonomy and ecology of coral reef tube worms (Serpulidae, Spirorbidae) in the Sudanese Red Sea. *Zoological Journal of the Linnean Society* **80**, 135 156.

Vinn O (2011) The role of an internal organic tube lining in the biomineralization of serpulid tubes. *Carnets de Géologie* CG2011_L01.

Vinn O, Hove HA ten, Mutvei H, Kirsimäe K (2008) Ultrastructure and mineral composition of serpulid tubes (Polychaeta, Annelida). *Zoological Journal of the Linnean Society, London* **154**, 633–650.

Vinn O, Kirsimiäe K, ten Hove HA (2009) Tube ultrastructure of *Pomatoceros americanus* (Polychaeta, Annelida): implications for the tube formation of serpulids. *Estonian Journal of Earth Sciences* **58**, 148–152.

Vittor BA (1975) Preliminary estimates of coral reef infaunal populations at Hydro-Lab, Grand Bahama Island. *Hydro-Lab Journal* **3**, 76–79.

Vittor BA, Johnson PC (1977) Polychaete abundance, diversity and trophic role in coral reef communities at Grand Bahama Island and the Florida Middle Grounds. In *Proceedings of the Third International Coral Reef Symposium*. May. (Ed. DL Taylor) pp. 163–168. Rosenstiel School of Marine and Atmospheric Science of University of Miami, Miami.

von Lendenfeld R (1884) The scyphomedusae of the southern hemisphere. Part III. *Proceedings of the Linnean Society of New South Wales* **9**, 259–306.

Vuillemin S (1965) Contribution à l'étude écologique du lac de Tunis. Biologie de *Mercierella enigmatica* Fauvel. PhD Thesis. l'Université de Paris, France.

Wagner R (1832) Weichthiere und Würmer [summary of and comment upon delle Chiaje (1822–28)]. In: Oken L (1832) Isis von Oken, Heft I–XII (mit 28 Tafeln). Leipzig beim Brockhaus [column 541–559, 631–655, 658, pl. X figs 1–12].

Walters LJ, Hadfield MG, Smith CM (1996) Waterborne chemical compounds in tropical macroalgae: positive and negative cues for larval settlement. *Marine Biology* **126**, 383–393.

Walters LJ, Hadfield MG, Del Carmen KA (1997) The importance of larval choice and hydrodynamics in creating aggregations of *Hydroides elegans* (Polychaeta: Serpulidae). *Invertebrate Biology* **116**, 102–114.

Wang H, Qian P-Y (2010) Involvement of a novel p38 mitogen-activated protein kinase in larval metamorphosis of the polychaete *Hydroides elegans* (Haswell). *Journal of Experimental Zoology Part B Molecular and Developmental Evolution* **314B**, 390–402.

Wang J, Huang Z (1993) Fouling polychaetes of Hong Kong and adjacent waters. *Asian Marine Biology* **10**, 1–12.

Watanabe N, Watanabe S, Ide J, Watanabe Y, Sakata K, *et al.* (1998) Chemical signals involved in larval metamorphosis in *Hydroides ezoensis* (Serpulidae; Polychaeta). Part II: Isolation and identification of a new monoacyl glycerol from adult tube clumps as a metamorphosis-inducing substance. *Journal of Marine Biotechnology* **6**, 11–15.

Wehe T, Fiege D (2002) Annotated checklist of the polychaete species of the seas surrounding the Arabian Peninsula: Red Sea, Gulf of Aden, Arabian Sea, Gulf of Oman, Arabian Gulf. *Fauna of Arabia* **19**, 7–238.

Weigert A, Golombek A, Gerth M, Schwarz F, Struck TH, *et al.* (2016) Evolution of mitochondrial gene order in Annelida. *Molecular Phylogenetics and Evolution* **94**, 196–206.

Wells HW, Wells MJ (1961) Three species of *Odostomia* from North Carolina, with description of new species. *The Nautilus* **74**, 149–157.

Wells HW, Wells MJ (1969) New host and distribution records of *Odostomia dianthophila*. *The Nautilus* **82**, 109–110.

Wells HW, Wells MJ, Gray IE (1964) The calico scallop community in North Carolina. *Bulletin of Marine Science of the Gulf and Caribbean* **14**, 561–593.

Wesenberg-Lund E (1949) Polychaetes of the Iranian Gulf. *Danish Scientific Investigations in Iran* **4**, 247–400.

Whitin N, Azariah J (1982) Can the development of *Hydroides elegans* be affected if the eggs and sperms are treated separately with the extract of *Ptychodera flava* (Hemichordate)? *Atlantica* **5**, 129.

Wiktor J (1980) Type specimens of Annelida Polychaeta in the Museum of Natural History of the Wroclaw University. *Annales Zoologici* **35**, 267–283.

Willey A (1905) 'Report on the Polychaeta collected by Professor Herdman, at Ceylon, in 1902'. In: 'Report to the government of Ceylon on the pearl oyster fisheries of the Gulf of Manaar' (Ed. WA Herdman) pp. 243–324. London, Royal Society, 4, Suppl. Rep. 30.

Winkley HW (1909) New England Pyramidellidae, with description of a new species. *The Nautilus* **23** (3), 39–40.

Wisely B (1958) The development and settling of a serpulid worm, *Hydroides norvegica* Gunnerus (Polychaeta).

Australian Journal of Marine and Freshwater Research **9**, 351–361.

Wisely B (1959) Factors influencing the settling of the principal marine fouling organisms in Sydney Harbour. *Australian Journal of Marine and Freshwater Research* **1**, 30–44.

Wolff WJ (2005) Non-indigenous marine and estuarine species in The Netherlands. *Zoologische Mededelingen Leiden* **79**, 1–116.

Wong E, Kupriyanova EK, Hutchings P, Capa M, Radashevsky VI, *et al.* (2014) A graphically illustrated glossary of polychaete terminology: invasive species of Sabellidae, Serpulidae and Spionidae. *Memoirs of Museum Victoria* **71**, 327–342.

Wood EJF, Allen FE (1958) *Common Marine Fouling Organisms of Australian Waters*. Department of the Navy, Melbourne.

Wu BL, Chen M (1981) Two new species of *Hydroides* (Polychaeta: Serpulidae) from South China Sea. *Oceanologia et Limnologia Sinica* **12**, 354–357.

Xie Z-C, Wong NC, Qian P-Y, Qiu J-W (2005) Responses of polychaete *Hydroides elegans* life stages to copper stress. *Marine Ecology – Progress Series* **285**, 89–96.

Xu Y, Li H, Xiancui Li, Xiao X, Qian P-Y (2009) Inhibitory effects of a branched-chain fatty acid on larval settlement of the polychaete *Hydroides elegans*. *Marine Biotechnology* **11**, 495–504.

Yan T, Zhang H, Li YQ, Cao WH, Hu B (2014) An overview of fouling sedentary polychaetes. *Acta Ecologica Sinica* **34**, 6049–6057.

Zachs IG (1933) [Polychaeten vom Nordjapanischen Meer. (Bucht 'Peter der Grosse').] *Issledovaniya Morei SSSR* **19**, 125–137.

Zeleny C (1902) A case of compensatory regulation in the regeneration of *Hydroides dianthus*. *Wilhelm Roux' Archiv für Entwicklungsmechanik der Organismen* **13**, 597–609.

Zeleny C (1905a) The rearing of serpulid larvae with notes of the behavior of young animals. *Biological Bulletin* **8**, 308–312.

Zeleny C (1905b) Compensatory regulation. *Journal of Experimental Zoology* **2**, 1–102.

Zeleny C (1911) Experiments on the control of asymmetry in the development of the serpulid, *Hydroides dianthus*. *Journal of Morphology* **22**, 927–944.

Zeleny C (1927) The change in the developmental value of a rudiment with age in the serpulid worm, *Hydroides pectinata*. *Anatomical Record, Philadelphia* **37**, 162.

Zibrowius H (1968) Étude morphologique, systématique et écologique des Serpulidae (Annelida Polychaeta) de la région de Marseille. *Recueil des Travaux de la Station Marine d'Endoume* **43**, 81–252.

Zibrowius H (1969) *Hydroides gairacensis* Augener, 1934, a little-known serpulid polychaete from Central and South America. *Bulletin of Marine Science* **19**(2), 366–376.

Zibrowius H (1971a) Les espèces Méditerranéennes du genre *Hydroides* (Polychaeta Serpulidae). Remarques sur le pretendu polymorphisme de *Hydroides uncinata*. *Tethys* **2**, 691–746.

Zibrowius H (1971b) Contribution à l'étude des Serpulidae (Polychaeta sedentaria) du Brésil. *Boletim do Instituto Oceanográfico da Universidade de São Paulo* **19**, 1–32.

Zibrowius H (1972a) Deux espèces nouvelles du genre *Hydroides* (Polychaeta, Serpulidae) de la Mer Jaune et des Iles Banda. *Bulletin de la Société Zoologique de France* **97**, 89–93.

Zibrowius H (1972b) *Hydroides norvegica* Gunnerus, *Hydroides azorica* n. sp. et *Hydroides capensis* n. sp. (Polychaeta Serpulidae) espèces vicariantes dans l'Atlantique. *Bulletin du Muséum d'Histoire Naturelle, Paris, Series 3* **33**, 433–446.

Zibrowius H (1972c) *Hydroides plateni* (Kinberg, 1867) (Polychaeta Serpulidae), espèce des côtes atlantiques d'Amérique du Sud, Redescription, remarques sur la répartition et l'écologie. *Bulletin de la Société Zoologique de France* **96**, 153–160.

Zibrowius H (1972d) Mise au point sur les espèces Méditerranéennes de Serpulidae (Annelida Polychaeta) décrites par Stefano delle Chiaje (1822–1829, 1841–1844) et Orionzo Costa (1861). *Téthys* **4**, 113–126.

Zibrowius H (1973a) Serpulidae (Annelida Polychaeta) des côtes ouest de l'Afrique et des Archipels Voisins. *Annales Musee Royal de l'Afrique Centrale, Series. 8, Sciences Zoologique* **207**, 1–93.

Zibrowius H (1973b) Remarques sur trois espèces de Serpulidae acclimatées en Méditerranée: *Hydroides dianthus* (Verrill, 1873), *Hydroides dirampha* Mörch, 1863, et *Hydroides elegans* (Haswell, 1883). *Commission Internationale pour l'Exploration Scientifique de la Mer Méditerranée [CIESM], Rapports et Procès-Verbaux* **21**, 683–686.

Zibrowius H (1977) Review of Serpulidae (Polychaeta) from depths exceeding 2000 meters. In *(1977) Essays on polychaetous annelids in memory of dr. Olga Hartman* (Eds DJ Reish, K Fauchald) pp. 289–306. Allan Hancock Foundation Press, Los Angeles.

Zibrowius H (1978) Introduction du polychète Serpulidae japonaise *Hydroides ezoensis* sur la côte atlantique française et remarques sur la repartition d'autres espèces de Serpulidae. *Téthys* **8**, 141–150.

Zibrowius H (1979) Serpulidae (Annelida Polychaeta) de l'Océan Indien arrivés sur des coques de bateaux à Toulon (France, Mediterranée). *Rapport de la Commission de l'Exploration de la Mer Méditerranée, Monaco* **25/26**, 133–134.

Zibrowius H (1992) Ongoing modification of the Mediterranean marine fauna and flora by the establishment of exotic species. *Mésogée* **51**, 83–107.

Zibrowius H (1994) Introduced invertebrates: examples of success and nuisance in the European Atlantic and in the Mediterranean. In *Introduced Species in European Coastal Waters*. (Eds CF Boudouresque, F Briand, C Nolan) pp. 44–49. Report on an international workshop organized jointly by the Environment Programme of DGXII of the European Commission and the Commission Internationale pour l'Exploration Scientifique de la Mer Méditerranée (CIESM), 2–3 March, Monaco. European Commission, Ecosystem Research Report 8.

Zibrowius H (2002) Assessing scale and impact of ship-transported alien fauna in the Mediterranean? In *CIESM Workshop Monograph no. 20, Istanbul, 6–9 November 2002. Alien Marine Organisms Introduced by Ships in the Mediterranean and Black Sea*. (Ed. F Briand) pp. 63–68. CIESM, Monaco.

Zibrowius H, Bitar G (1981) Serpulidae (Annelida Polychaeta) indo-pacifiques établis dans la région de Beyrouth, Liban. *Rapport Commission Internationale pour l'Exploration Scientifique de la Mer Méditerranée, Monaco* **27**, 159–160.

Zibrowius H, Thorp CH (1989) A review of the alien serpulid and spirorbid polychaetes in the British Isles. *Cahiers de Biologie Marine* **30**, 271–285.

Zuraw EA, Leone DE (1968) 'Laboratory culture of the tubeworm *Hydroides (Eupomatus) dianthus* Verril 1873'. General Dynamics Electric Boat Division, Connecticut.

Index